AF572734

Water Transport in Biological Membranes

Volume II
From Cells to Multicellular Barrier Systems

Editor
Gheorghe Benga, M.D., Ph.D.
Chairman
Department of Cell Biology
Faculty of Medicine
Medical and Pharmaceutical Institute
Cluj-Napoca, Romania

CRC Press, Inc.
Boca Raton, Florida

LIBRARY OF CONGRESS
Library of Congress Cataloging-in-Publication Data

Water transport in biological membranes/editor, Gheorghe Benga.
p. cm.
Includes bibliographies and indexes.
Contents: v. 1. From model membranes to isolated cells — v.
2. From cells to multicellular barrier systems.
ISBN 0-8493-6082-X (v. 1). ISBN 0-8493-6083-8 (v. 2)
1. Water—Physiological transport. 2. Cell membranes. I. Benga,
Gheorghe.
QH509.W38 1989 88-16654
574.87'5—dc19 CIP

This book represents information obtained from authentic and highly regarded sources. Reprinted material is quoted with permission, and sources are indicated. A wide variety of references are listed. Every reasonable effort has been made to give reliable data and information, but the author and the publisher cannot assume responsibility for the validity of all materials or for the consequences of their use.

Direct all inquiries to CRC Press, Inc., 2000 Corporate Blvd., N.W., Boca Raton, Florida, 33431.

International Standard Book Number 0-8493-6082-X (v.1)
International Standard Book Number 0-8493-6083-8 (v.2)

Library of Congress Card Number 88-16654
Printed in the United States

PREFACE

There is general agreement that investigating the permeability characteristics of biological membranes not only has intrinsic value but could also contribute to our understanding of the function and structure of membranes. This is documented by many studies starting with the early work of Overton on the movement of water and nonelectrolytes across cell membranes. Indeed, one of the main functions of the plasma membranes of living cells is to control the transport processes into and out of the cell and thus to regulate the cell volume and the composition of the intracellular fluid. In fact, continued existence of the cell is critically dependent on a functional cell membrane. Aside from being of considerable theoretical importance, the process of water transport across biological membranes is of practical importance in a broad range of processes, from protection of cells in cryogenic preservation to the effects of certain hormones in some tissues.

In recent years there have been significant advances in studies of several natural membranes. An important role in these advances has been the application of physical techniques such as nuclear magnetic resonance, X-ray diffraction, reflectance or fluorescence measurements, and light-scattering or neutron scattering techniques.

The main purpose of this book is to provide in-depth presentations of *physical techniques for measuring water transport and their applications* to a variety of biological membranes, from model membrane systems to cell membranes, and then from isolated cells to multicellular barrier systems, such as epithelia or even whole organisms. This survey of water transport in such a broad range of membrane systems will hopefully contribute to understanding of the *structure-function relationships and molecular mechanisms of water permeation*. Moreoever, the description of various techniques, together with a review of literature will enable the readers to assess whether a technique would be useful in helping to solve his or her particular problem of research and will also expand their competence in these techniques. The book consists of two volumes.

Volume I mainly covers aspects of the lipid bilayer as a major barrier for water in several model systems or in cells. Some chapters are devoted to interaction of water with the lipid bilayer, others to water transport in model membrane systems (planar bilayers, liposomes), while some other chapters describe the effects of water and sugars on stability of phospholipid bilayers, or the role of local membrane dehydration in fusion. The last chapters present the plasma membrane ultrastructural changes produced by freezing and the very interesting phenomenon of anhydrobiosis: the ability of some unicellular organisms to lose essentially all of their intracellular water in a fully reversible fashion.

Volume II covers water transport in a broad range of systems, from subcellular systems (subcellular organelles, thylakoid membranes, chromaffin granules) to isolated cells and multicellular barrier systems. Since the red blood cells have been a favorite object for permeability studies several chapters describe the particular features of water transport across erythrocyte membranes. Other cellular systems such as the lung cells, Novikoff hepatoma cells, or virally infected cells are also discussed. Several chapters in this volume cover water transport in epithelia. One of these presents theoretical aspects of solute-solvent coupling in epithelia. Other chapters describe the water permeability of the antidiuretic hormone-sensitive epithelia, the corneal-limiting layers, the gill epithelium, or the invertebrate water permeability with whole organisms. The last chapters are devoted to peculiarities of water transport in arthropods and insects.

The volumes provide in-depth presentations of well-defined topics related to water transport across biological membranes. The discussion of key areas by specialists of international repute is based both on their own expertise as well as on critical review of the literature. The volumes, providing contributions for reference purposes at the professional level, are broadly aimed at biologists, biophysicists, biochemists, physicists, etc., active investigators, working on transport processes in various biological membranes. It is also hoped that such a book could be of help to teachers and students, mainly at the postgraduate level.

Gheorghe Benga

CONTRIBUTORS

Volume II

Gheorghe Benga, M.D., Ph.D.
Chairman
Department of Cell Biology
Medical & Pharmaceutical Institute
Cluj-Napoca, Romania

Stephen R. L. Bolt, Ph.D.
Research Fellow
Department of Oceanography
Southampton University
Southampton, England

Jacques Bourguet, M.D.
Head, Biomembranes
Department of Biology
Centre d'Etudes Nucléaires de Saclay
Gif Sur Yvette, France

Jesper Brahm, Ph.D.
Associate Professor
Department of General Physiology & Biophysics
The Panum Institute
University of Copenhagen
Copenhagen, Denmark

J. Chevalier, D.Sc.
Head
Biomembrane Group
National Institute of Health
Institut National de la Santé et de la Recherche Medicale
Paris, France

Dieter Coenen-Stass, Dr. Habil.
Zoologisches Institut
Universitat Karlsruhe
Karlsruhe, West Germany

Bernhard Deuticke, Ph.D.
Professor
Department of Medicine
Abteilung Physiologie
RWTH Aachen
Aachen, West Germany

Jorge Fischbarg, M.D., Ph.D.
Professor
Departments of Physiology and Ophthalmology
College of Physicians and Surgeons
Columbia University
New York, New York

Rita Anne Garrick, Ph.D.
Associate Professor
Division of Science & Mathematics
Fordham University
College at Lincoln Center
New York, New York
and
Department of Medicine
UMDNJ-New Jersey Medical School
Newark, New Jersey

Beate Klösgen, Ph.D.
Research Fellow
Department of Chemistry
Institut für Physikalische Chemie
RWTH Aachen
Aachen, West Germany

Josef Küppers, Dr.
Lehrstuhl für Neurophysiologie
Zoologisches Institut der Universitat Münster
Münster, West Germany

Robert I. Macey, Ph.D.
Professor & Chairman
Department of Physiology & Anatomy
University of California
Berkeley, California

Kingsley John Micklem, Ph.D.
Research Fellow
Nuffield Department of Pathology
University of Oxford
Oxford, England

Mario Parisi, M.D.
Professor
Department of Physiology
Faculty of Medicine
University of Buenos Aires
Buenos Aires, Argentina

Charles A. Pasternak, D.Phil.
Professor and Chairman
Department of Biochemistry
St. George's Hospital Medical School
University of London School of Medicine
London, England

Thomas G. Polefka, Ph.D.
Senior Research Biochemist
Department of Periodontal Research
Colgate-Palmolive
Piscataway, New Jersey

P. Ripoche, D.S.
Department of Biology
Centre d'Etudes Nucléaires de Soclay
Gif Sur Yvette, France

Hansjürgen Schoenert, Ph.D.
Professor
Department of Chemistry
Institut für Physikalische Chemie
RWTH Aachen
Aachen, West Germany

Robert R. Sharp, Ph.D.
Professor
Department of Chemistry
University of Michigan
Ann Arbor, Michigan

Peter Maving Taylor, Ph.D.
Postdoctoral Research Fellow
Department of Physiology
University of Dundee
Dundee, Scotland

Ulrich Thurm, Dr.
Lehrstuhl für Neurophysiologie
Zoologisches Institut der Universität Münster
Münster, West Germany

Vincent B. Wigglesworth, F.R.S.
Professor
Department of Zoology
University of Cambridge
Cambridge, England

TABLE OF CONTENTS

Volume I

TABLE OF CONTENTS

Volume II

Chapter 1

THE WATER PERMEABILITY OF INTACT SUBCELLULAR ORGANELLES

Robert R. Sharp

TABLE OF CONTENTS

I. INTRODUCTION

Water permeability measurements in microscopic, osmotically enclosed systems date from the 1950s,[1] but accurate measurements first became possible in the mid-1960s with the development of rapid mixing techniques using tritiated water.[2-4] Discussions of these nonequilibrium, rapid mixing approaches are described elsewhere in these volumes. Early in the 1970s, it was realized that nuclear magnetic resonance (NMR) provides an equilibrium approach to the same problems.[5-16] The power of NMR lies in its ability to monitor, in an equilibirum setting, fast chemical exchange reactions which lead to microscopic reversibility in equilibrium systems. Explicit isotopic labeling of the transported species is no longer required. In NMR experiments, the nuclear spins of molecules in different physical environments are labeled magentically by the action of a perturbing radio frequency magnetic field. The perturbation imposes a nonequilibrium distribution of spin populations or a nonequilibrium phase distribution on the Zeeman energy states, which describe the orientation of nuclear spins in a laboratory field. The relaxation of the spin populations back to their thermal equilibrium distribution, normally a first-order process kinetically, is then describable in the form of a first-order rate constant of the relaxation process. Spin-lattice relaxation describes decay of the spin populations from a perturbed configuration back to the Boltzmann distribution that is characteristic of thermal equilibrium. The first-order rate constant of this process is $R_1 = T_1^{-1}$. In addition to perturbing the spin populations, the radiofrequency (r.f.) magnetic field induces phase coherence, which is also a nonequilibrium situation, in the processing spins. The relaxation of coherence back to the random phase situation that is characteristic of thermal equilibrium is termed transverse relaxation, the first order rate constant of which is $R_2 = T_2^{-1}$.

These two relaxation processes are inherently sensitive to chemical exchange phenomena which transfer spins between magnetically distinct environments, a dependence that can be put to use in studies of water transport. As a simple example of this phenomenon, consider a situation in which the internal water space in a suspension of osmotically tight vesicles is doped magnetically in a manner that provides very efficient spin relaxation in the internal phase relative to that in the external buffer. If chemical exchange processes transfer water molecules between the two environments, and if these processes occur on a time scale comparable to, or faster than, that of magnetic relaxation, then the observed NMR resonances will have relaxation properties reflective of both magnetic environments, as well as of the kinetic processes responsible for water transport.

It is worth stressing that the power of NMR in kinetic investigations lies in its ability to label spins in an equilibrium system and to follow their subsequent movement by means of the decay of nonequilibrium nuclear magnetization. The specific labeling method depends on the NMR technique employed. In pulsed NMR, the r.f. field is applied as a very short pulse, which, depending on its duration, can produce either maximal phase coherence (90° pulse) or spin population inversion (180° pulse). Observation of the decay of the nuclear magnetization following one of these two preparatory pulses forms the basis of the T_2 (90° pulse) or T_1 (180° pulse) measurement. In continuous-wave (cw) NMR, the r.f. magnetic field is applied continuously, while a static magnetic field is swept to define the resonance lineshape. CW NMR methods likewise provide measurements of T_1 and T_2, in this case from the resonance linewidth at half height ($\Delta\nu\ ^1/_2 = (\pi\ T_2)^{-1}$ or from the saturation properties of the signal. Detailed descriptions of NMR techniques are given elsewhere[17-19] and are beyond the scope of this article. It will be evident in the following sections that for water transport in simple systems, pulsed techniques provide a much more versatile approach to NMR relaxation studies than do cw NMR.

The simple picture of water permeability effects given above imagines the solvent to be present in two aqueous environments that are internally uniform but differ in their magnetic properties. These two phases are separated by a membrane which permits relatively slow water exchange between the two aqueous pools. This simple physical model is somewhat reminiscent of a spherical cow, with the milk uniformly distributed throughout, and one might wonder to what degree effects of magnetic nonuniformity will influence the relaxation phenomena. For example, water molecules adjacent to a biological membrane experience a physical environment that is different from that in the bulk solvent. In part, this difference reflects magnetic inequivalence of the surface and bulk environments, and is reflected in spatial variations in the chemical shifts and relaxation times. An additional, and very important, magnetic effect occurs at the lipid-water interface, where a discontinuity in bulk magnetic susceptibility produces a gradient in the static magnetic field used to establish the Zeeman energy levels. It is appropriate in the present context to consider whether the movement of solvent molecules between such magnetically distinct environments by self-diffusion, a process that is magnetically similar to transport across a membrane, influences the magnetic relaxation phenomena considered above.

In general, the answer to this question depends on the spatial dimensions of the physical system relative to a characteristic diffusion length in the NMR experiment. When an ensemble of magnetically distinct environments is sampled rapidly by the nuclear spin in the course of its Brownian motion, then the spin can be viewed as occupying a homogeneous, exchange-averaged environment. ''Rapidly'', in this context, refers to a time scale defined by the physical parameters which characterize the magnetic gradients in the system. The NMR time scale will be defined more precisely in the following sections. As a rough guide, the kinetic window in water transport measurements generally falls in a range of 10^{-4} to 10^{-1} sec, depending on the experimental approach employed.

In suspensions of cells and organelles, simple two-site models have invariably been employed.[5-11,16] The water space is separated conceptually into two regions which are assumed to be well mixed internally by solvent self-diffusion. The accuracy of this approximation is readily assessed. From the self-diffusion coefficient of water, $D \approx 2(10^{-5}\ cm^2 \cdot sec^{-1})$, and the self-diffusion distance in a three-dimensional system, $\bar{r}^2 = 6\ Dt$, it is apparent that diffusion over a time course of 100 μsec (a rough lower limit to the NMR kinetic window) will exchange-average regions of approximately 1 μm in diameter. This length is smaller than the linear dimensions of cells (for example, the mean diameter of human erythrocytes is 8.5 μm[20]), but it is substantially larger than that of the organelles discussed here. Thus, the simple two-site model appears to provide a reasonable description of organelles (and in many practical cases of cells as well), at least in concentrated suspensions, where the average distance between particles is small enough to permit effective diffusive mixing of the external phase. The description of complex structures should be considered with care however. Large magnetic susceptibility effects on T_2 has been reported in structurally heterogeneous tissues such as muscle[21,22] and in model systems such as packed glass beads[23] and clays.[24] In vesicular systems, the presence of diffusion barriers or of structural heterogeneity in the sample can impair the validity of simple two-site models. The description of more complex structures and of methods appropriate for handling structural heterogeneity are described at greater length in Section III below.

II. THEORY OF CHEMICAL EXCHANGE EFFECTS IN NMR

A. Modified Bloch Equations

The phenomenological description of chemical exchange effects in NMR spectroscopy originated with McConnell,[25] who introduced phenomenological rate expressions describing the chemical exchange processes into the classical equations of motion of the spin (the Bloch equations). The Bloch equations describe the motion of the spins due to precession and thermal relaxation, in a coordinate frame that is rotating synchronously with the radio frequency H_1 field, an applied field which induces transitions between the Zeeman levels:

$$u = +(\omega_L - \omega)v - u/T_2 \tag{1a}$$

$$v = [-(\omega_L - \omega)u + \gamma H_1 M_z] - v/T_2 \tag{1b}$$

$$M_z = \gamma H_1 v + (M_O - M_z)/T_1 \tag{1c}$$

The first terms on the right-hand side of these equations describe the precessional driving forces on the spins due to the static and r.f. magnetic fields, and the second terms describe thermal relaxation along the magnetic field direction (T_1) or in a plane transverse to the magnetic field (T_2), respectively. ω_L is the Larmor frequency and γ the gyromagnetic ratio of the resonant nucleus.

In continuous-wave NMR experiments, H_1 is applied continuously during observation of the signal at an amplitude large enough to induce transitions between the Zeeman levels at an observable rate, but not so large that it seriously saturates the spin populations. In pulsed NMR experiments, the H_1 field is applied in brief ($\approx$10 μsec) pulses, and the signal is detected as a transient oscillating voltage produced by stimulated emission of radiation by the coherently precessing spin magnetization. In either experiment, the NMR signal is described as a complex magnetization, G, expressed in the rotating coordinate frame of the applied H_1 field. The real and imaginary parts of G, $G = u + \iota v$, are respectively the in-phase and out-of-phase components of the precessing magnetization with respect to H_1. $u(\omega)$ and $v(\omega)$ are NMR lineshape functions that are often, but not always, of Lorentzian form in liquid systems. They describe the dispersion and absorption mode NMR signals respectively.

To describe chemical exchange phenomena, the Bloch equations are modified by writing one set of equations in the form of Equation 1 for the magnetization vector in each chemically distinct spin population (e.g., for each pool of solvent protons across an osmotically tight biological membrane), and including terms of the form G_j/τ_j to describe, phenomenologically, magnetization transfer between the various environments. The two-site Bloch equations, modified in this way, are written

$$u_1 = (\omega_1 - \omega)v_1 - u_1/t_{2,1} - u_1/\tau_1 + u_2/\tau_2$$

$$u_2 = (\omega_2 - \omega)v_2 - u_2/T_{2,2} - u_2/\tau_2 + u_1/\tau_1$$

$$v_1 = [-(\omega_1 - \omega)u_1 + \gamma H_1 M_{z1}] - v_1/T_{2,1} - v_1/\tau_1 + v_2/\tau_2$$

$$v_2 = [-(\omega_2 - \omega)u_2 + \gamma H_1 M_{z2}] - v_2/T_{2,2} - v_2/\tau_2 + v_1/\tau_1$$

$$M_{z1} = \gamma H_1 v_1 + (M_1^o - M_{z1})/T_{1,1} - M_{z1}/\tau_1 + M_{z2}/\tau_2$$

$$M_{z2} = \gamma H_1 v_2 + (M_2^o - M_{z2})/T_{1,2} - M_{z2}/\tau_2 + M_{z1}/\tau_1 \tag{2}$$

The exchange-modified Bloch equations are phenomenological in origin but can be justified quite rigorously from the equation of motion of the density matrix.[26] Solution of these equations provides the observable NMR parameters, namely, the linewidths, chemical shifts, and relaxation times of the exchanging species, in terms of the NMR parameters which describe the system in the absence of exchange, plus the first-order rate constants, $1/\tau_j$, of the exchange process.

Boundary conditions required for different physical situations must be applied in accord with the physical nature of the experiment. In slow-passage continuous-wave experiments, the magnetization is assumed to be a steady state, and the left-hand side of each of Equation

2 can be set to zero. An analytical solution for the absorption mode signal in the general two-site chemical exchange problem has been given by Rogers and Woodbrey:[27]

$$v(\omega) = -\gamma H_1 M_o[U(1 + \tau[P_2R_{2,1} + P_1R_{2,2}]) + VW]/(U^2 + W^2) \quad (3)$$

where
$$U = \tau[R_{2,1}R_{2,2} - (\omega - \omega_o)^2 + (\delta^2/4)] + P_2R_{2,2} + P_1R_{2,1}$$
$$V = \tau[(\omega - \omega_o) - (\delta/2)(P_1 - P_2)]$$
$$W = (\omega - \omega_o)[(1 + \tau R_2^o) + (\tau/2)(R_{2,2} - R_{2,1}) + (P_1 - P_2)/2]$$
$$\omega_o = (\omega_1 + \omega_2)/2, \quad \text{and} \quad \tau^{-1} = \tau_1^{-1} + \tau_2^{-1}$$

Equation 3 describes the lineshape of the absorption mode signal and is Lorentzian in the fast-exchange region.

An important special case of the two-site solution, which is virtually always appropriate for studies of organelle suspensions, occurs when site 2 contains a very small fraction of the total spin population. In this case, the observed resonance is a Lorentzian peak with an intensity due solely to site 1, but with relaxation properties and a chemical shift influenced by site 2. In this physical situation, the transverse relaxation rate is given by

$$R_{2,1} = R_{2,1}^o + P_2 \frac{[R_{2,2}(R_{2,2} + \tau_2^{-1}) + \Delta\omega_2^2]}{[\tau_2(R_{2,2} + \tau_2^{-1})^2 + \tau_2\Delta\omega_2^2]} \quad (4)$$

and the chemical shift, relative to the unperturbed resonance (in the absence of chemical exchange) is

$$\Delta\omega = P_2\delta[(\tau_2R_{2,2} + 1)^2 + \tau_2^2\delta^2]^{-1} \quad (5)$$

In the limit of very rapid exchange, the resonances of sites 1 and 2 are exchange-averaged into a single Lorentzian peak, the half-height line width of which contains a chemical exchange contribution of

$$\Delta\nu_{1/2} = (\pi T_2)^{-1} = \pi^{-1}P_1^2P_2^2\delta^2(\tau_1 + \tau_2)$$
$$\approx \pi^{-1}P_1P_2\delta^2\tau_2 \quad (6)$$

Swift and Connick[28] first applied the fast-exchange equations in an analysis of chemical exchange of $H_2{}^{17}O$ between the bulk solvent and the first coordination sphere of dissolved paramagnetic metal ions.

B. Chemical Exchange Effects on the Carr-Purcell Train

Pulsed techniques offer a more accurate and versatile probe of chemical exchange effects in rapidly exchanging systems than does the NMR linewidth in a high resolution experiment. The T_2 measurement is normally based on the Carr-Purcell sequence, 90-t_1-(180-t_1-echo-t_1-$)_n$, which measures transverse decay directly and suppresses unwanted effects of coherent spin dephasing due to magnetic field inhomogeneity.[29] In the C-P pulse train, the initial 90° pulse tilts the magnetization vector to the XY plane, and the sequence of 180° pulses repetitively refocuses the spins, maintaining coherence in the magnetization through the formation of a train of spin echos. The envelope of echo maxima, measured at jt_{cp} (j = integral, $t_{cp} = 2t_1$), defines the T_2 decay process.

The presence of stochastic chemical exchange events which transfer spins between sites with differing chemical shifts produces random fluctuations in the Larmor precession frequency of the exchanging spin. These fluctuations act to dephase the spin packet in a manner that is irreversible within the spin echo sequence. In this way, chemical exchange alters the apparent T_2 through a process that is physically related to chemical exchange line broadening effects observed in cw NMR experiments. There is, however, a crucial difference between the cw and pulsed NMR experiments. While the spin echo sequence can in no way suppress irreversible dephasing events, the spacing between 180° pulses does control the cumulative phase error produced by single jumps between chemically shifted sites. For example, if the spacing between 180° pulses is 100 msec and the chemical shift difference between sites is 10 Hz, then a single jump can lead to a cumulative phase error of at most 20 π radians prior to the application of the next rephasing 180° pulse. In contrast, if the spacing between 180° pulses is 1 msec and the chemical shift difference is 10 Hz, then the maximum cumulative phase error prior to a refocusing pulse is 0.2 π radians. In this way, the phase error produced by a sequence of random dephasing events can be "pulsed away" by a decrease in pulse spacing in the Carr-Purcell experiment. Kinetic information is obtained from an analysis of the appraent T_2 measured as a function of t_{cp}, the Carr-Purcell pulse spacing.

The use of the Carr-Purcell technique offers a number of advantages, particularly in the fast-exchange region, relative to linewidth studies of chemical exchange phenomena. Spin-echo techniques are not limited by the magnet linewidth and are capable of a broad kinetic window. But perhaps the most substantial advantage for studies of complex biological structures stems from the availability of an additional kinetically sensitive experimental parameter, t_{cp}, in the spin echo methods. Systematic variation of t_{cp} provides kinetic information that is independent of knowledge of the nonkinetic parameters of the system, specifically, of the internal volume of the structure or of the chemical shift gradient across the membrane.

Analyses of chemical exchange effects on the Carr-Purcell pulse train have followed two conceptual approaches. Bloom et al.[30] have calculated the classical statistical phase error due to chemical exchange phenomena and its effect on the Carr-Purcell train. This calculation is a quantitative description of the dephasing process described physically above. Allerhand and Gutowsky[31,32] have based a parallel calculation on the modified Bloch equations (Equation 2) and have obtained general solutions in closed form for the two-site exchange case. Subsequently, Gutowsky et al.[33] developed a very general matrix formulation that describes effects of chemical exchange, spin-spin coupling and quadrupolar relaxation on the Carr-Purcell experiment.

Allerhand and Gutowsky's[32] solution of the general two-site case can be written in closed form:

$$R_2 = R_2^\circ + (2\tau)^{-1} - t_{cp}^{-1} \sinh^{-1} F \tag{7}$$

where

$$F = [D_+ \sinh^2(t_{cp}S_r/2) + D_- \sin^2(t_{cp}S_i/2)]^{1/2} \tag{8}$$

$$2D\pm = \pm 1 + [\tau^{-2} + \delta^2](S_r^2 + S_i^2)^{-1}$$

and S_r and S_i are real and imaginary parts of

$$S = [\tau^{-2} - \delta^2 + 2\iota(P_1 - P_2)(\delta/\tau)]^{1/2}$$

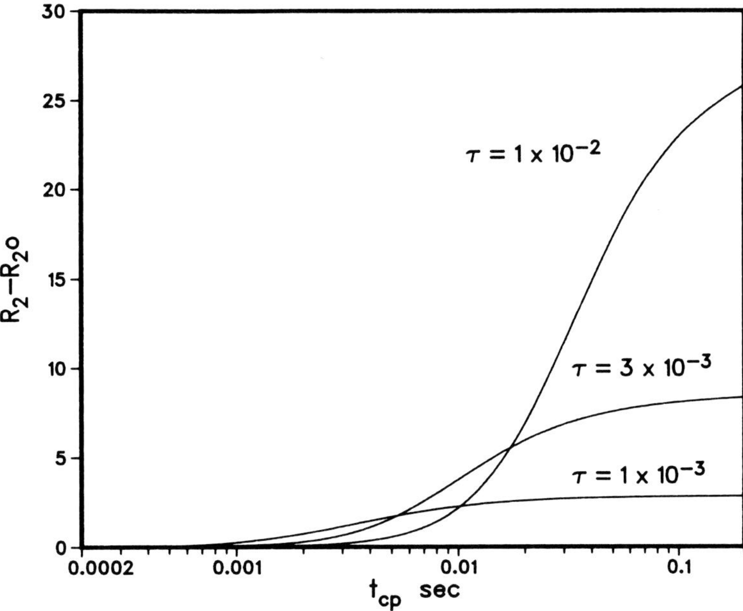

FIGURE 1. Chemical exchange effects on R_2 in the two-site system as measured by the Carr-Purcell sequence. Theoretical curves are calculated from Equation 9 of the text assuming a low capacity pool containing 3% of the solvent and a chemical shift of 50 Hz across the membrane.

These somewhat cumbersome analytical expressions yield to a particularly simple form in the fast-exchange, fast-pulsing region, which is defined by $(1/2\ \tau \gg \delta,\ R_{2,1} = R_{2,2} \gg t_{cp}$:

$$R_2 - R_2^o = P_1 P_2 \delta^2 \tau [1 - (2\tau/t_{cp}) \tanh(t_{cp}/2\tau)] \qquad (9)$$

In the limit of *very* fast exchange, this expression further simplifies to cw NMR result, Equation 6. This agreement is expected in the limit of very long pulse spacing, $\tau/t_{cp} \rightarrow 0$, where the refocusing effect of the 180° pulses disappears.

These mathematical results illustrate the power of the Call-Purcell experiment relative to cw NMR for the analysis of two-site experiments. As pointed out above, the pulsed experiment determines the mean exchange lifetime unambiguously from the functional dependence of $R_2 - R_2^o$ on a single instrumental parameter, t_{cp}. Unlike experiments based on linewidth measurements, prior information concerning the chemical shift difference or the fractional site populations is not required.

The theoretical dependence of R_2 - R_2^o on t_{cp} predicted by the approximate expression, Equation 9, for various values of τ is shown in Figure 1. The P_2 and δ values selected for the calculation (0.03 and 50 Hz, respectively) are generally readily achievable in organelle suspensions. It is evident from an inspection of these results that τ determines the inflection point of the $R_2 - R_2^o$ vs. t_{cp} curve, and that τ values greater than about 100 μsec produce readily detectable increments in the apparent R_2. In general terms, the magnitude of the R_2 increment scales with the square of the chemical shift difference between the two sites and with the first power of the population of the low-capacity site.

C. Spin-Lattice Decay in Pulsed Experiments

Experiments in the previous two sections monitor chemical exchange effects through their influence on the transverse NMR relaxation time, T_2, either measured directly by pulsed methods, or indirectly, from the linewidth in a high-resolution NMR experiment. Chemical exchange effects can also be monitored through effects on spin-lattice decay rate, R_1. In a typical experimental situation, two solvent pools are characterized by significantly different relaxation properties (e.g., one site may be doped by a paramagnetic relaxation agent) and are coupled physically by the presence of chemical exchange reactions. When chemical exchange processes are slow or absent, the measured spin-lattice decay is a simple sum, weighted by the appropriate fractional populations, of the exponential decay processes that characterize the two sites:

$$M_z(t) = M^{o}_{z1}[1 - 2\exp(-R_{1,1}t)] + M^{o}_{z2}[1 - 2\exp(-R_{1,2t})] \quad (10)$$

In the limit of very rapid chemical exchange, the resonant nuclei sample, both environments on a time scale that is very short compared to T_1, and the two exponentials coalesce to a single exchange-averaged value, weighted by the appropriate fractional populations, of spin-lattice decay rates in the two sites. At intermediate exchange rates, the observed spin-lattice decay is influenced by the chemical exchange kinetics. Throughout the intermediate exchange situation, the decay kinetics remain a simple sum of two exponentials, although the apparent relaxation times depend on kinetic, as well as relaxation, parameters.

Experimental determination of T_1 normally involves the inversion-recovery technique: an initial 180° pulse inverts the spin populations and a subesquent 90° sampling pulse is applied after a variable delay τ to monitor the return of M_z to thermal equilibrium. This technique can be repeated with progressively incremented τ values to define the spin-lattice decay, or more efficiently, the decay can be recorded by a sequence of (90°-τ-180°-τ-90°) pulse triplets, which sample and restore the magnetization vector to its position along Z.[34,35] The solution of Equation 2 corresponding to the inversion-recovery method involves solving the coupled equations for M_z with $H_1 = 0$ under the initial condition $M_z\ (t = 0) = -M_z\ (t = \infty)$. The general two-site solution is a sum of exponentials with decay constants:

$$\begin{aligned} R_{\pm} = {} & (1/2)[R_{1,1} + \tau_1^{-1} + R_{1,2} + \tau_2^{-1} \\ & \pm (1/2)[(R_{1,1} - R_{1,2})^2 + (\tau_1^{-1} + \tau_2^{-1})^2 \\ & - 2R_{1,1}(\tau_2^{-1} - \tau_1^{-1}) - 2R_{1,2}(\tau_1^{-1}\tau_2^{-1})]^{1/2} \end{aligned} \quad (11)$$

(The corresponding equation in Reference 15 is in error.) In the two-site case, the NMR signal is invariably described as a sum of two exponentials, which, when water exchange is relatively slow ($\tau_{1,2} \gg T_{1,2}$), are identified closely with the internal and external water spaces, with $R_{\pm}$ describing the more rapidly relaxing environment. In the fast-exchange region, the lines coalesce to a single peak with $R_1 \approx P_1R_1 + P_2R_2$. Figure 2 contains theoretical curves showing the effect of τ_2 on R_1. These curves were calculated assuming that the small water pool contains 3% of the solvent.

The choice of an efficient and appropriate technique for investigating solvent transport in a particular biological system depends on which exchange situation applies. Previous studies of whole cells have normally been carried out under conditions of slow exchange. The required gradient in magnetic properties across the membrane has been achieved in several studies by the addition of high concentrations (typically 40 m*M*) of Mn(II), an impermeant paramagnetic relaxation reagent. Under these conditions (i.e., when the high-capacity site is strongly relaxing), T_1 measurements do not provide an appropriate probe of slow chemical exchange. The high-capactiy (external) solvent pool produces most of the observable NMR signal, and its spins are very efficiently relaxed by the dopant. Slow chemical exchange processes which transfer water into the small, weakly relaxing internal solvent pool exert only a small incremental effect on the relaxation properties of the observed signal. In contrast, the resonance of the weakly relaxing, internal water pool is very sensitive to chemical

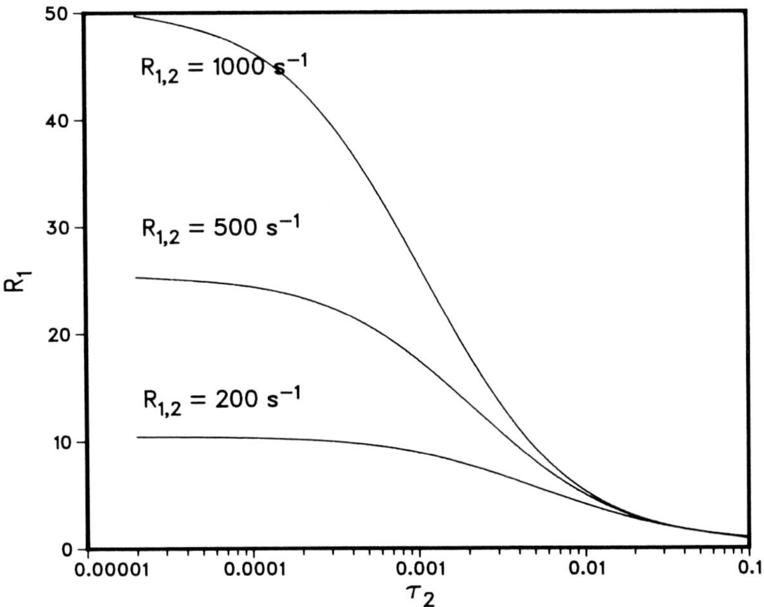

FIGURE 2. Chemical exchange effects on the solvent R_1 due to two-site chemical exchange involving an internally doped low capacity site containing 5% of the solvent. $R_{1,2}$ is the relaxation rate in the doped environment, and τ_2 is the internal residence time of water. $R_{1,1,}$ was taken to be 0.5 sec^{-1}.

exchange processes which transfer spins to the strongly relaxing external environment. Thus, in the slow-exchange situation with external doping, the choice of experimental technique revolves on how best to observe the weak internal water resonance in the presence of a much larger external resonance. In this case, the spin-echo T_2 experiment is the method of choice, since the doped external signal can be strongly suppressed by transverse decay at long 90 to 180° pulse spacings. Thus, the spin-echo method, using an external, impermeant dopant, has proven useful in analyzing slow (on the NMR time scale) water transport into erythrocytes.[5,7,10,11]

In suspensions of organelles, water transport is normally in the fast-exchange region, where external doping is inappropriate. This difference in kinetics results from the fact that the internal water space is relatively small, resulting in a short water residence time within the organelle and more nearly equal T_2s in the internal and external water resonances. In this situation, the distinction between internal and external resonances is weak and chemical exchange effects are much harder to detect. An alternative experimental procedure that is effective for rapid transport is to incorporate the dopant inside the intact organelle. In this case, the external water signal, which is large and easily observed, is very sensitive to chemical exchange processes that couple it to the strongly relaxing internal site. This procedure is well suited to studies of artificial vesicles which can be prepared, for example, by sonication in the presence of dopant and then cleansed of external ions by chelation, centrifugation, or filtration. Natural organelles provide a greater challenge with respect to the doping procedure. One method that has proven effective in studies of chromaffin granule suspensions is described in Section III.

III. EXPERIMENTAL RESULTS

A. Water Transport Across the Chromaffin Granule Membrane

Chromaffin granules are the storage site of catecholamines in the adrenal medulla. These

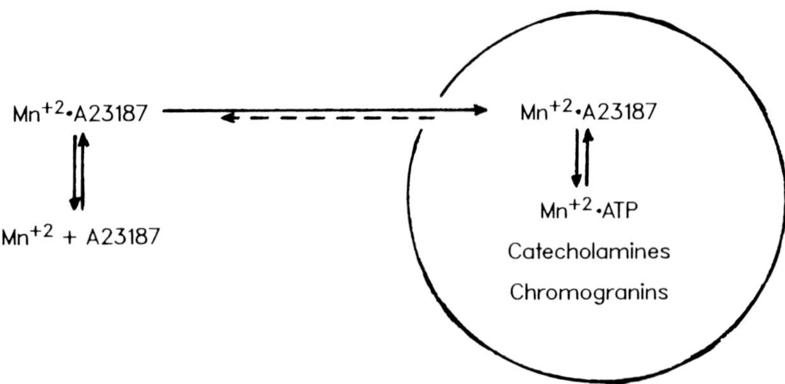

FIGURE 3. Ionophore-mediated doping of chromaffin granules.

organelles are osmotically tight, spheroidal vesicles,[36] with an average *in situ* diameter of 115 nm.[37] Chromaffin granules store the catecholamine neurohormones, epinephrine and norepinephrine, at very high internal concentration (0.5 *M* and 0.1 *M*, respectively), and deliver these hormones to the blood as a trigger of the body's response to stress. In addition to the catecholamine hormones, chromaffin granules contain large concentrations (~0.15 *M*) of nucleotides, primarily in the form of ATP, and chromogranins, which are a mixture of soluble, largely random-coil polypeptides. Chromaffin granules are close structural and functional analogs of neurotransmitter storage organelles and have been studied widely as model systems in this regard.[38,39]

The water permeability of the chromaffin granule membrane has been studied in detail using spin-lattice relaxation techniques described in Section I.C above. These techniques rely on a gradient in magnetic properties across the vesicle membrane, which, for this system, was achieved by ionophore-mediated incorporation of Mn^{+2} into the chromaffin granule interior. The paramagnetic doping process is illustrated schematically in Figure 3. Added Mn^{2+} is very slowly permeant toward the chromaffin granule membrane at 25°C in the absence of ionophore.[40,41] Addition of micromolar concentrations of the divalent cation-specific ionophore A23187 facilitates the entry of Mn^{2+} into the chromaffin granule lumen, where the metal ion binds rather tightly to the phosphate esters of ATP. The time course of Mn^{2+} entry and its subsequent binding to ATP has been monitored spectroscopically by means of the paramagnetic broadening produced by Mn(II) in the ^{31}P-ATP resonances.[41] When incorporated by chromaffin granules at 3°C, Mn^{2+} is held within the matrix for many hours in a fashion that prohibits rapid equilibration with the external aqueous phase. Incorporated Mn^{2+} is more labile at 37°C and can be removed slowly by strong chelators (e.g., EDTA) over a period of several minutes.

The effect of Mn(II) doping on the spin-lattice relaxation rate (R_1) of water protons is shown in Figure 4. Addition of 100 μM Mn^{2+} to a suspension of isolated chromaffin granules in the absence of ionophore produces a large R_1 increment, approximately fourfold, over that of the diamagnetic suspension. This value is stable over a time course of tens of minutes, during which period Mn^{2+} does not permeate appreciably into the chromaffin granule interior. Addition of A23187 to a concentration of 20 μM leads to a progressive loading of the chromaffin granule interior. This process can be followed conveniently by measurements of R_1, which falls as Mn^{2+} leaves the external aqueous phase and is sequestered inside the chromaffin granule lumen. Sequestered Mn^{2+} has a much smaller influence, at 10°C, on the solvent relaxation properties than does external Mn^{2+} because solvent exchange at this temperature is relatively slow, and the main pool of solvent protons does not efficiently sample the paramagnetic environment. Release of the sequested Mn^{2+} through intentional

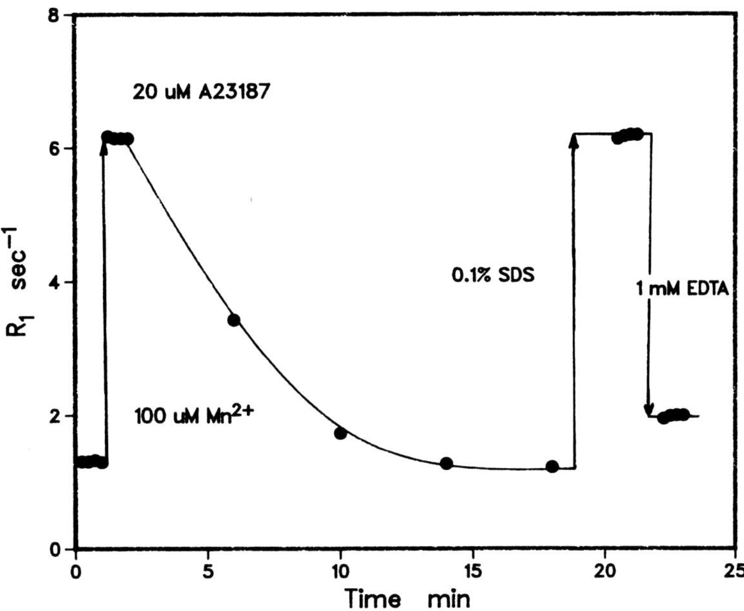

FIGURE 4. Time course of the solvent R_1 in chromaffin granule suspensions during the doping process at 10°C.

lysis of the chromaffin granules (treatment with detergent or several freeze/thaw cycles) causes an increase in R_1. (Mn^{2+} released by lysis is significantly less efficient as a relaxation reagent than the same concentration of exogenous manganese in a suspension of intact chromaffin granules; this decreased relaxation efficiency is due to chelating effects of nucleotides released by the freeze/thaw treatment.) Addition of a strong chelator (EDTA) following, but not prior to, lysis effectively suppresses the paramagnetic R_1 enhancement.

These results are characteristic of Mn^{2+} incorporation experiments at 10°C. At this temperature, water exchange across the chromaffin granule membrane is slow on the NMR time scale, and Mn^{2+} uptake decreases the measured R_1, reflecting the sequestration of manganese by the internal environment. A parallel experiment conducted at 37°C gives qualitatively different results (Figure 5). At the higher temperature, Mn^{2+} incorporation leads to an elevation of R_1, reflective primarily of more rapid solvent exchange which permits external water to sample the internal doped environment more efficiently. Perhaps surprisingly, Mn^{2+} incorporation enhances R_1 over values observed before addition of ionophore, i.e., ''sequestration'' inside chromaffin granules produces an increase in Mn(II) relaxation efficiency. This effect results from the fact that Mn^{2+} binding in the chromaffin granule matrix sharply increases its molar relaxivity relative to values corresponding to the external medium. The theoretical basis for this effect is well understood: Mn(II) immobilization by phosphate esters lengthens the ion's reorientational correlation time, which in turn produces a proportionate enhancement of its molar relaxivity toward the solvent.[16,42,43]

The temperature dependence of R_1 in a suspension of Mn^{2+}-doped chromaffin granules is shown by the solid curve in Figure 6. R_1 increases steeply with temperatures above 5°C, passing through a maximum near 35°C. This behavior is in pronounced contrast with the monotonic decrease in R_1 that is characteristic of chromaffin granule suspensions without incorporated Mn(II), either in the presence and absence of external Mn^{2+}. The strong positive temperature dependence of R_1 is a consequence of the doping procedure and has its origin in the solvent exchange processes. The observed functional form of R_1 vs. T^{-1} is qualitatively very similar to that of theoretical plots of R_1 vs. τ_2 shown in Figure 2.

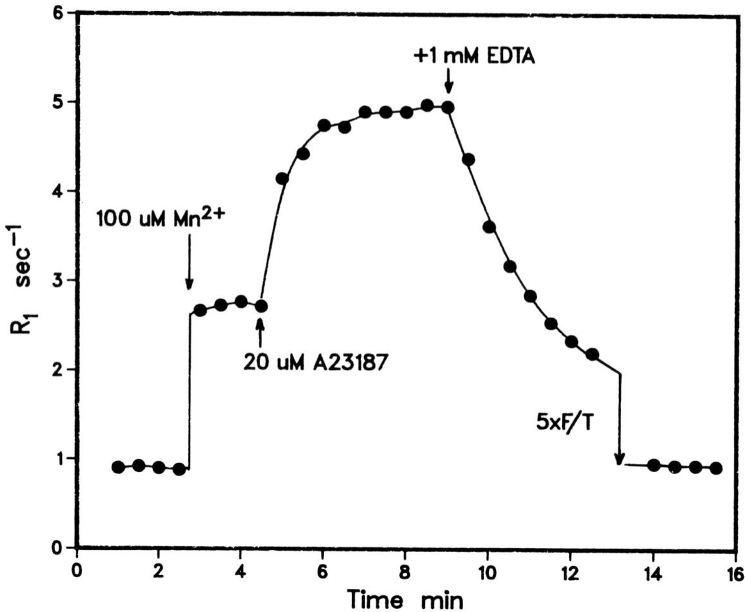

FIGURE 5. Time course of the solvent R_1 in chromaffin granule suspensions during the doping process at 37°C.

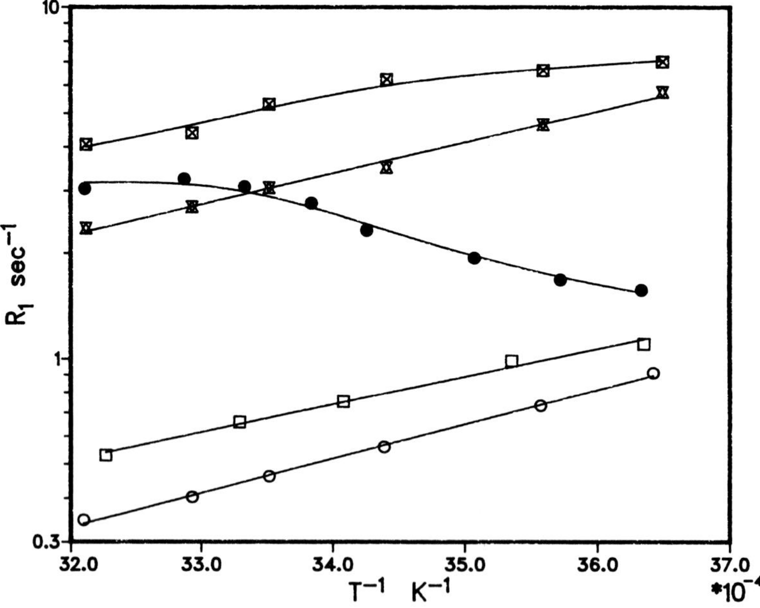

FIGURE 6. Temperature dependence of the solvent R_1 of doped chromaffin granule suspension (●). Also shown are data for the buffer (○), for chromaffin granule suspensions containing Mn^{2+} but without ionophore (■), for the doped suspension following several freeze-thaw cycles (⧗), and for isolated, washed chromaffin granule membranes (□).

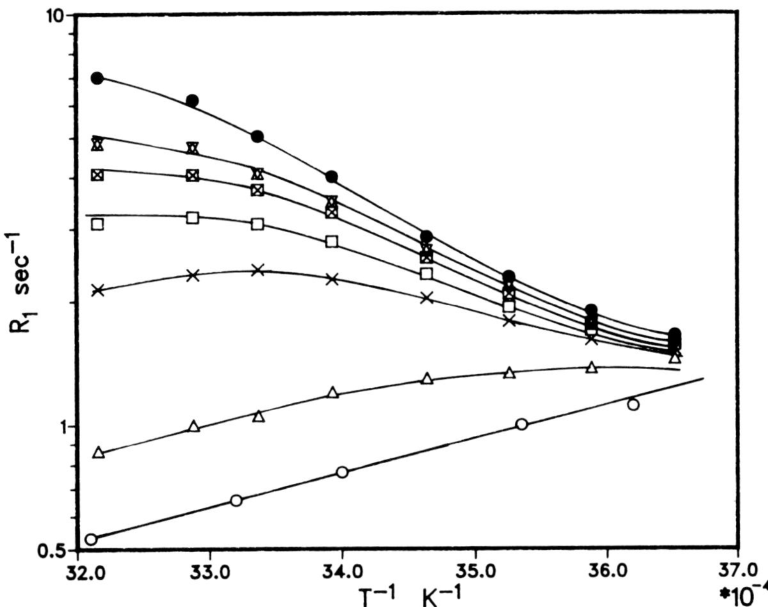

FIGURE 7. Temperature dependence of the solvent R_1 of doped chromaffin granule suspensions. Concentrations of added Mn^{2+} are (◯) 400 μ*M*, (⧗) 200 μ*M*, (⊠) 150 μ*M*, (□) 100 μ*M*, (x) 50 μ*M*. Also shown are data for undoped chromaffin granules △) and for the buffer containing washed chromaffin granule membranes (◯).

The sensitivity of the solvent T_1 to the kinetics of water transport provides a basis for measurements of the water permeability coefficient using the theoretical analysis of Section I.C. Observations of the temperature dependence of R_1 at various concentrations of relaxation reagent strongly overdetermine the kinetic parameter, τ, that describes two-site chemical exchange. The most accurate method for analyzing the data is an iterative multiparameter least-squares fit of the data to Equation 11 at various temperatures and Mn^{2+} concentrations. The result of such an analysis is shown in Figure 7. The solid lines illustrate a fit derived from the internal and external relaxation rates, $R_{1,2}$ and $R_{1,1}$, the mean exchange lifetime in the internal phase, τ_2, and the temperature dependence of these quantities expressed in Arrhenius form. Clearly, the simple two-site analysis provides an excellent description of water transport in the chromaffin granule suspensions.

One noteworthy point about the data of Figure 7 is that native chromaffin granules produce a sufficient increment of the intragranular solvent R_1 to produce readily observable chemical exchange effects in the T_1 relaxation data, even in the absence of added paramagnetic dopant. Thus, the kinetics of solvent transport have been determined in this system in the absence of perturbing divalent cations or ionophores. The permeation properties of the membrane are essentially unaltered by the doping process, a result that provides an important control for this kind of study. Possible effects of high Mn^{2+} concentrations on the water permeability of erythrocytes has been discussed as a possible source of error in earlier NMR studies on erythrocytes,[11] where the Mn^{2+} concentration in the external buffer was maintained in the range 25 to 70 m*M*.[5,6,10] However, control experiments have indicated that such effects are unimportant in erythrocyte suspensions,[11] as is true for the chromaffin granule suspensions discussed here.

Table 1 summarizes available information concerning the water permeability of the chromaffin granule membrane and provides a brief comparative summary of results in other model and natural systems. The water permeation coefficient of the chromaffin granule membrane is near the lower end of values characteristic of model membranes, in the form

Table 1
KINETIC PARAMETERS CHARACTERIZING WATER TRANSPORT IN ORGANELLES, CELLS, AND MODEL VESICLES

Organelle	T (°C)	τ_2 (msec)	P_w ($m \cdot sec^{-1}$)	E_A (kJ/mol)
Chromaffin granule	25	7.0	$2.3 (10^{-6})$	14.7
	2	68	$2.4 (10^{-7})$	
	36	2.6	$0.9(10^{-6})$	
Thylakoid membrane	25	1.14 ± 0.08	$4.4(10^{-6})$[a]	6.3
	3	2.76 ± 0.4	$1.8(10^{-6})$[a]	
Erythrocyte membrane[b]	25		$3.0(10^{-5})$	7.1
Chlorella plasma membrane[c]	20		$2.1(10^{-5})$	
Egg lecithin	25		$2.9(10^{-5})$[d]	10.5
Dipalmitoyl lecithin	25		$3.4(10^{-6})$[d]	19.3
	25		$2.6(10^{-6})$[e]	15

[a] Estimated value assuming a lamellar structure with a loculus of width 5 nm; edge effects were neglected.
[b] Average of values reported in References 5—8, 10, and 11.
[c] Reference 9.
[d] Reference 14.
[e] Reference 12.

either of planar bilayers or of vesicles. The erythrocyte membrane appears to be nearly an order of magnitude more permeant than the chromaffin granule membrane. This high permeability has previously been attributed to the presence of hydrophilic channels,[4] the existence of which is supported by comparative measurements of the hydraulic and tracer water permeability coefficients of the erythrocyte membrane. An interesting chemical observation that tends to support the existence of hydrophilic channels is the reported inhibitory effect on water transport of *p*-chloromercuribenzoate,[6,10] a covalent sulfhydryl-modifying reagent. While water channels may be present in the erythrocyte membrane, the available data strongly argue against their presence in chromaffin granules.

B. Water Transport Across the Chloroplast Thylakoid Membrane

A second natural organelle that has been subjected to water permeability studies is the thylakoid membrane of chloroplasts. Thylakoid membranes form substructures of surpassing complexity within algal cells and within the chloroplasts of higher plants, where they contain the pigments and electron-transport apparatus of the light reactions of photosynthesis. The topology of these structures is still not completely understood, although the electron microscopic work in serial section of Paolillo and co-workers[44-48] has provided considerable structural insight. Basically, the thylakoids are organized as layered stacks of flattened, osmotically tight vesicles (the grana stacks), in which are embedded the pigments and electron transport proteins which catalyze the following reactions:

$$4h\nu + 2H_2O \rightarrow O_2 + 4H^+ + 4e^- \quad \text{(photosystem II)}$$

$$4e^- + 2NADP^+ + 2H^+ \rightarrow 2NADPH \quad \text{(photosystem I)}$$

$$3H^+ + ADP + P_i = ATP \quad \text{(energy transducing complex)}$$

These reactions are driven energetically by photo-excited chlorophyll and are accompanied by the vectorial movement of protons inwardly across the membrane. The pH gradient thus

FIGURE 8. Schematic structure of thylakoid membranes in the grana stacks.

formed is dissipated by leakage through the energy transducing complex, where it provides free energy for the phosphorylation of ADP through chemiosmotic coupling.[49]

The internal aqueous space of thylakoids does not contain a well-defined impermeant osmoticum, but rather exists in vivo in a collapsed form shown schematically in Figure 8. The lumen of intact thylakoids contains a variety of simple semipermeable monovalent and divalent ions, particularly H^+, Na^+, K^+, Mg^{2+}, and Cl^-, which move in response to membrane potentials generated by the light-driven electron transport reactions.[50] Chemiosmotic coupling of the pH gradient, formed as a result of electron transport, to the synthesis of ATP, which is a primary storage form of free energy, requires that the thylakoid membrane in vivo forms an osmotic boundary which prevents diffusive dissipation of the light-driven pH gradients for periods of many minutes.

The water permeability of the thylakoid membrane has been investigated[15] using T_2 measurements, both in high resolution studies (Section I.A) and in studies of the pulse-spacing dependence of the Carr-Purcell sequence (Section I.B). The thylakoid membranes used in these studies were present in suspension as isolated, osmotically intact grana stacks. The observation of water transport effects on T_2 requires the imposition of a chemical shift and/or relaxation step across the membrane. This was produced by doping the external medium with lanthanide ion, dysprosium Dy^{+3}. Dysprosium produces large molar increments in the solvent proton chemical shift (0.10 to 0.12 ppm/m*M*[15]*) and is virtually impermeant toward the thylakoid membrane. In order to increase its concentration in the region of neutral pH, the ion was added as the ethylenediamine (en) complex. $Dy^{3+}\cdot en_3$ is soluble to the extent of several millimolar, a concentration that produces large and easily measurable effects in T_2.

A potentially complicating aspect of the thylakoid membrane system with respect to water permeation studies is that measurements of the internal volume and of the internal chemical shift are very difficult. Internal volume estimates based on the distribution of tritiated water and ^{14}C-labeled inulin have been reported,[51] but the use of these measurements under somewhat different experimental conditions is of questionable validity since the volume of the thylakoid lumen is quite small at physiological osmotic strength and undergoes large fractional changes in response to light-driven ion movement.[52,53] As an added complication, the chemical shift difference across the thylakoid membrane is difficult to estimate accurately. The chemical shift of the buffer alone does not provide a reliable estimate due to magnetic susceptibility effects produced by the lipid phase. For these reasons, the Carr-Purcell technique is the method of choice for water permeability studies in that the kinetic parameters can be determined independently of internal volume or chemical shift information.

Typical results of such studies are shown in Figure 9. In the presence of the dysprosium shift reagent, R_2 depends strongly on t_{cp}, increasing monotonically across the range 10^{-3}

* The $Dy\cdot en_3^{3+}$ complex produces, in addition to a large chemical shift, significant line broadening in the solvent ^1H resonance. This perturbation of both the relaxation parameters and the chemical shift introduces unnecessary complications in the theoretical analysis that could be avoided by selecting a less strongly broadening lanthanide ion, such as Eu^{3+}.

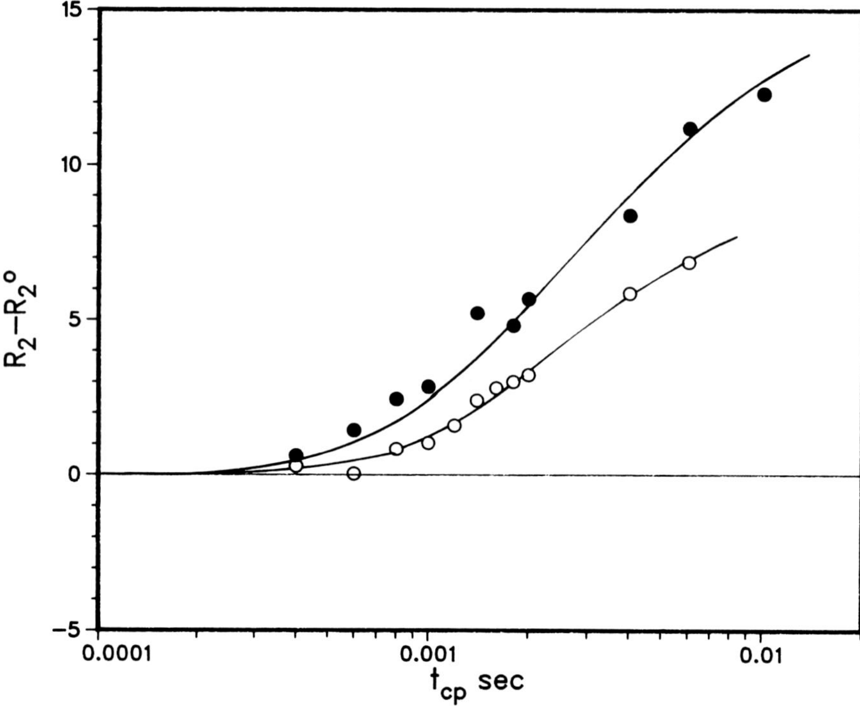

FIGURE 9. Chemical exchange effects on the solvent R_2 as measured by the Carr-Purcell sequence. The pulse-spacing dependence of R_2 is shown for two different concentrations of $Dy^{3+} \cdot (en)_3$, used as an external dopant in a suspension of thylakoid membranes; (●) ca. 4 m*M*; (○) 3 m*M*.

sec $< t_{cp} < 10^{-2}$ sec. The data follow the functional form predicted by the two-site water transport model in Figure 1. The inflection point of these curves determines the mean lifetime, $\tau \approx (^1/_2)\ t_{cp} = 1.2$ msec, of a water molecule in the thylakoid lumen. According to Equation 9, the position of the inflection point is determined only by the mean internal residence time and should be independent of the chemical shift across the membrane. This expectation is fulfilled in the results of Figure 9 (solid circles), which were obtained at a higher concentration of shift reagent. A quantitative analysis of these experiments has been carried out through a multiparameter fit to Equation 9, resulting in the solid curves. Clearly, the two-site kinetic model of water transport provides a satisfactory description of this experiment.

The temperature dependence of the water permeability has also been measured using the variation of R_2 with t_{cp} (Figure 10). With falling temperature, the inflection point in the data moves to longer t_{cp} values (i.e., longer τ), and R_2 measured in the limit of long pulse spacings increases in accord with the limiting expression, Equation 12:

$$R_2 = P_1P_2\delta^2\tau/\pi \tag{12}$$

Thus, theory predicts an intersection of the data curves measured at different temperatures, a prediction confirmed by experiment. Table 1 summarizes the kinetic parameters obtained for thylakoid suspensions.

A parallel, but much less informative, analysis can be based on the line broadening that is induced in the high-resolution water resonance by the $Dy{\cdot}en_3^{3+}$ shift reagent. In the fast-exchange region, the paramagnetic increment to the linewidth (corrected for background line broadening due to the shift reagent) is $P_1P_2\delta^2\tau/\pi$. Clearly, this approach has quantitative kinetic significance only when the internal water volume, P_2, and the transmembrane chem-

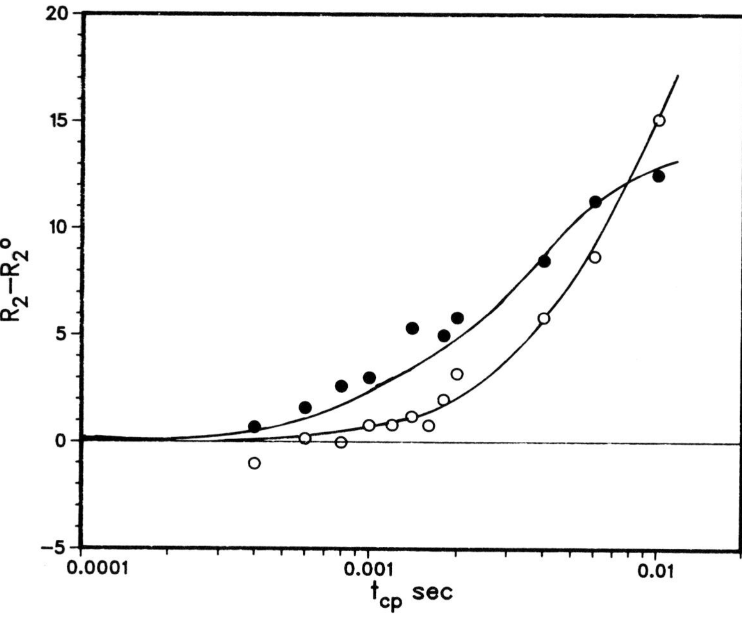

FIGURE 10. Temperature dependence of the solvent R_2 in externally doped suspensions of thylakoid membranes; (●) 25°C; (○) 3°C.

ical shift, δ, are known accurately, which is not true for thylakoid suspensions. Nevertheless, this alternate approach provides a valuable test of the self-consistency of the spin-echo method.[15]

IV. MULTI-SITE ANALYSIS

The two-site methods described in the previous sections have successfully been used to monitor the chemical exchange reactions in which water is transported across the membranes of isolated biological organelles. It should be recognized, however, that in real systems of physiological interest the two-site analysis is always an approximation, the validity of which should be examined carefully. Contamination is always a concern in preparations of physiological origin, but especially so in NMR relaxation studies, where contaminating organelles may have effects out of proportion to their population due to size differences or uneven doping. Even in ostensibly pure physiological preparations, the suspension is more accurately described in terms of distributions which reflect the sample heterogeneity with respect to the relevant NMR parameters. Distributions in size, in the numbers of pores or water channels (if these exist) per organelle, or in the membrane properties, for example between immature and mature organelles of a given type, may well be substantial. Another type of heterogeneity, which is not usually recognized in the literature, involves the permeance of a membrane to a paramagnetic ion added as a shift or relaxation reagent. The paramagnetic ion enters cells by processes, which are usually not well understood but which presumably involve the mediation of a carrier molecule. In such cases, the internal concentration of shift reagent is probably heterogeneous, reflecting a distribution in the "leakiness" of the biological membrane. This latter type of heterogeneity can be controlled to some degree by the use of ionophores to provide a uniform degree of leakiness (as, for example, in the chromaffin granule studies described above).

Nevertheless, all previous investigations of biological material have been subject to unavoidable uncertainties due to sample heterogeneity, and the results invariably have the limitations inherent in a weighted average over an unspecified distribution. While no attempt

Table 2
FRACTIONAL TRANSFER PROBABILITIES FOR A HYPOTHETICAL THREE-SITE SYSTEM

has yet been made to characterize the relevant distribution functions experimentally, the theoretical basis for such an investigation exists and would appear to be a promising source of information on the heterogeneity of biological systems. Even in studies where the sample heterogeneity is not a primary focus of the research, some consideration of the limitations of the two-site approach is normally of interest, if only to deal with effects of contamination.

The theory of multi-site chemical exchange was originally developed to describe complex chemical equilibria. Specific systems which have received detailed attention include water exchange into the coordination sphere of dissolved paramagnetic ions,[28] equilibria involving the stepwise formation of complex metal halides,[54] and the $Cl^- + Cl_2 = Cl_3^-$ equilibrium.[55] The latter analysis is most general and will be followed here. The application of the theory to organellar suspensions presumes that different sites possess distinguishable magnetic properties, either due to the characteristics of the natural system itself or due to the addition of a paramagnetic reagent.

The modified Bloch equations for two-site chemical exchange can be adapted to the multi-site situation by the inclusion of a sum of terms that describes chemical exchange events which transfer a spin, j, into any of the various connected sites, i. The net change in magnetization in the j site due to transfers with site i can be written

$$(dG_j/dt) = -G_j p_{ji}/\tau_j + G_i p_{ij}/\tau_i \tag{13}$$

p_{ji} is a fractional transfer probability for a $j \rightarrow i$ event, and τ_j is the mean lifetime of a spin in site j. p_{ji} is the probability, expressed fractionally, that a chemical exchange event out of site j results in the transfer of a spin to site i; clearly, $\Sigma p_{ji} = 1$. Assignment of the p_{ji} is closely related to the stoichiometry of the system. A practical example, based on a three-site system in which the fractional spin populations are 0.90, 0.08, and 0.02, is shown in Table 2.

The motion of the steady state xy magnetization vector at the i site, in the rotating frame of reference, can be written in the normal manner (Equation 2) by including a sum of terms of the form (Equation 13) describing all chemical exchange processes leading into, or out of, site j:

$$-G^j[R_{2j} - \iota(\omega - \omega_j) + \tau_j^{-1}] + \sum_{i \neq j} G^i p_{ij}/\tau_i = -\iota\gamma H_1 M_o^j \tag{14}$$

By appropriate arrangement of terms, the modified Bloch equations for n sites consist of a set of n coupled vector equations

$$\begin{aligned}
\alpha_{11}G_1 + \alpha_{12}G_2 + \cdots \alpha_{1i}G_i + \cdots \alpha_{1n}G_n &= \iota\gamma H_1 M_o^1 \\
\alpha_{21}G_1 + \alpha_{22}G_2 + \cdots \alpha_{2i}G_i + \cdots \alpha_{2n}G_n &= \iota\gamma H_1 M_o^2 \\
\cdot \quad \cdot \quad \cdot \quad \cdot \quad & \cdot \\
\cdot \quad \cdot \quad \cdot \quad \cdot \quad & \cdot \\
\alpha_{n1}G_1 + \alpha_{n2}G_2 + \cdots \alpha_{ni}G_i + \cdots \alpha_{nn}G_n &= \iota\gamma H_1 M_o^n
\end{aligned} \tag{15}$$

The diagonal coefficients, α_{jj}, are composed of an imaginary part describing the Larmor precessional motion of the spins in the jth site plus a sum of real terms describing magnetization loss in the jth site due to relaxation, with the rate constant R_{2j}, and due to chemical exchange processes which transfer spins out of the jth site:

$$\alpha_{jj} = -(R_{2j} - \iota(\omega - \omega_j) + \tau_j^{-1} \tag{16}$$

Off-diagonal terms are real and describe the gain of transverse magnetization into site j due to transfer out of site i, appropriately weighted by the fractional transfer probabilities,

$$\alpha_{ij} = p_{ij}/\tau_i \tag{17}$$

The observable spectral information (R_{2j}, $\Delta\omega_j$) is obtained by solving Equation 15 for the lineshape function $G(\omega) = \Sigma G_j(\omega)$. In the special case where chemical exchange is rapid enough to exchange-average the observed resonances to a single Lorentzian peak, or alternatively, when a single site strongly dominates the observed signal, $G(\omega)$ can be written as a complex function of Lorentzian form, $G(\omega) = U(\omega) + \iota v(\omega)$, the real and imaginary parts of which describe the dispersion and absorption mode NMR signals, which are of the form

$$u \,\alpha\, (\omega - \omega_o)T_2/[1 + (\omega - \omega_o)^2T_2^2], \quad \text{and} \quad v \,\alpha\, T_2/[1 + (\omega - \omega_o)^2T_2^2 \tag{18}$$

Useful analytical expressions for multi-site cases have been given for several cases of practical interest.[28,54,55] One model that leads to simple expressions in closed form and is of particular relevance to water transport studies, occurs when a single site contains most of the resonant spins, i.e., $M_1^o \gg M_i^o$ for all $i \neq 1$. In this case, the observed resonance is describable as a single Lorentzian due to spins in site 1. Using the shortened notation $\tau_{ji} = p_{ji}/\tau_j$, and the condition $p_{1i}/\tau_1 \ll p_{i1}/\tau_i$, it is readily shown that

$$R_2 = R_{2,1} + \sum \tau_{1i}^{-1} \frac{[R_{2,i}^2 + R_{2,i}\tau_{i1} + \Delta\omega_{i]}^2}{[(R_{2i} + \tau_{i1}^{-1})^2 + \Delta\omega_i^2]} \tag{19}$$

$$\Delta\omega_1 = -\sum (\Delta\omega_i)[\tau_{1i}\tau_{i1}(R_{2i} + \tau_{i1}^{-1})^2 + \Delta\omega_i^2 \tag{20}$$

These expressions, while originally written for the three-site case,[28] are applicable to the general multi-site chemical systems where a single site contains most of the magnetization. Analytical expressions appropriate to other specific cases of chemical interest have also appeared in the literature.

A general solution of the coupled Equations 15 has been obtained in a form suitable for numerical evaluation through the use of matrix methods,

$$\boldsymbol{\alpha} \cdot G = \iota\gamma H_1 M_o \tag{21}$$

where $\alpha = \mathbf{R} + \iota\mathbf{I}$, $G = u + \iota v$. The formal solution of Equation 21 for the absorption mode signal is written $v(\omega) = \mathrm{Im}[(\mathbf{R} + \iota\mathbf{I})^{-1}\iota\gamma H_1 M_o]$. A significant simplification of Equation 21 is obtained by separating the real and imaginary parts of G,

$$-\begin{bmatrix} \mathbf{R} & \mathbf{I} \\ \mathbf{I} & \mathbf{R} \end{bmatrix} \Big[\; = \gamma H_1 \begin{matrix} M_o \\ \mathbf{O} \end{matrix} \tag{22}$$

where **O** is the null matrix and the square matrix is of order 2n. An explicit solution for the ith element of v gives

$$v_i = \gamma H_1 \sum \frac{\det \operatorname{cof} \begin{bmatrix} \mathbf{R} & \mathbf{I} \\ \mathbf{I} & \mathbf{R} \end{bmatrix}_{ij}}{\det \begin{bmatrix} \mathbf{R} & \mathbf{I} \\ -\mathbf{I} & \mathbf{R} \end{bmatrix}} \tag{23}$$

In the common situation where the spins exchange between a single site, 1, of large capacity and several other sites, i, of much smaller capacity, Equation 23 leads to a considerable simplification. In this case, contributions to the signal other than that of G_1 are negligible, and the absorption mode signal can be written

$$v(\omega) \approx v_1 = \gamma H_1 M_o^1 \frac{\det \operatorname{cof} \begin{bmatrix} \mathbf{R} & \mathbf{I} \\ -\mathbf{I} & \mathbf{R} \end{bmatrix}_{11}}{\det \begin{bmatrix} \mathbf{R} & \mathbf{I} \\ -\mathbf{I} & \mathbf{R} \end{bmatrix}} \tag{24}$$

This form does not readily provide physical insight into the kinetic system as do analytical expressions such as Equations 19 and 20. However, the matrix solution is unrestricted in its application to liquid systems described by the Bloch equations, and, while somewhat cumbersome, it yields readily to numerical analysis.

V. CONCLUSIONS

This review has attempted to illustrate the major NMR methods that are suitable for water permeability studies in suspensions of intact subcellular organelles. Previous studies using whole cells, particularly erythrocytes, have been conducted in the slow-exchange region, in most cases using high concentrations of Mn(II) as an impermeant external relaxation reagent. This method is essentially inapplicable to the study of organelles, where, due to the much smaller internal volumes, fast exchange conditions almost invariably apply. Fortunately, several alternative NMR approaches are available for this situation and have been outlined in Section I. The utility of these approaches has been examined in two specific cases involving respectively suspensions of chromaffin granules and thylakoid membranes, and found to be generally satisfactory. In the opinion of the author, the approach of greatest versatility for future studies is the Carr-Purcell technique. Through systematic variation of the 180° pulse spacing, kinetic parameters over a kinetic window of some four orders of magnitude or more can be determined, even in a complete absence of information concerning the physical structure of the system or of the magnetic gradients that produce the chemical exchange effects. The instrumental degree of freedom provided by the ability to vary t_{cp} is indeed valuable in analyses of chemical exchange kinetics.

While the data in Figure 9 and in Reference 15 clearly illustrate the power of the Carr-Purcell technique, they only partially delineate its range of applicability. In the fast exchange region, the sensitivity of T_2 to chemical exchange effects depends on the square of the chemical shift across the membrane. The chemical shift scales with magnetic field strength. Investigations conducted at superconducting magnetic field strengths (corresponding to ^1H frequencies of 360 or 500 MHz using current-generation, high-resolution spectrometers), would achieve molar chemical shifts a factor six to eight times greater than that used in the study described above. At these elevated field strengths, the kinetic window of the C-P experiment could be extended by more than an order of magnitude at the short end. This extended window could be used to study chemical exchange processes in more permeant vesicles, or equivalently from an NMR standpoint, vesicles with smaller internal volumes.

Alternatively, the additional kinetic sensitivity provided by high magnetic fields permits the use of lower concentrations of shift reagent. In the limiting situation, kinetic analyses of organelles might be possible in the absence of paramagnetic dopants entirely, using the natural chemical shift gradient across the vesicle membrane. The realization of this possibility, which has been achieved in the slower exchange situation represented by erythrocytes,[6,8] depends on the magnitude of the *in situ* chemical shift or relaxation gradient in the organelle of interest. It is interesting in this regard that the natural relaxivity of the chromaffin granule lumen produced easily observable chemical exchange effects in the T_1 experiments of Figure 7, thereby permitting a permeability determination in the absence of added paramagnetic ions.

Areas requiring further theoretical attention in studies of complex biological systems involve the effects of heterogeneity, of structural complexity, and of contamination in preparations of physiological origin. The theoretical basis for such an analysis is available from the theory, outlined in Section III, of multi-site chemical exchange in fast chemical equilibria. Further attention to this area is needed, however. The theory outlined above describes lineshape phenomena rigorously but does not exhaust the options available to the experimentalist, particularly through the use of the Carr-Purcell sequence. A very general formalism describing the Carr-Purcell experiment has been developed[33] that is suitable for descriptions of multi-site chemical exchange effects, although the theory in its present form requires a rather complex analysis of data.

There is reason to expect that the versatility exhibited by the C-P experiment in the two-site analysis will be mirrored in multi-site situations. In the presence of a single major pool of spins (the buffer phase), the C-P experiment will exhibit a superposition of sigmoidal features, like those in Figure 1, corresponding to the various water pools participating in exchange. Evidence of sample heterogeneity should be evident as a broadening in functional form of the sigmoidal dependence of R_2 on t_{cp} (in fact, the data on thylakoid membranes (Figure 9) are not inconsistent with the presence of effects of this type). Future studies in this area would benefit from additional theoretical work and will require data of higher quality than has yet been published to characterize accurately the pulse-spacing dependence of T_2.

REFERENCES

1. **Dick, D. A. T.,** *Cell Water,* Butterworths, London, 1966, chap. 6.
2. **Sha'afi, R. I., Rich, G. T., Sidel, V. W., Bossert, W., and Solomon, A. K.,** The effect of the unstirred layer on human red cell water permeability, *J. Gen. Physiol.,* 50, 1377, 1967.
3. **Vieira, F. L., Sha'afi, R. I., and Solomon, A. K.,** The state of water in human and dog red cell membranes, *J. Gen. Physiol.,* 55, 451, 1970.
4. **Macey, R. I., Daran, D. M., and Farmer, R. E. L.,** Properties of water channels in human red cells, in *Biomembranes,* Vol. 3, Kreuzer, F., and Slegers, J. F. G., Eds., Plenum Press, New York, 1973, 331.

5. **Conlon, T. and Outhred, R.**, Water diffusion permeability of erythrocytes using an NMR technique, *Biochim. Biophys. Acta,* 288, 354, 1972.
6. **Andrasko, J.**, Water diffusion permeability of human erythrocytes studied by a pulsed gradient NMR technique, *Biochim. Biophys. Acta,* 428, 304, 1976.
7. **Fabry, N. E. and Eisenstadt, M.**, Water exchange between red cells and plasma: measurement by nuclear magnetic relaxation, *Biophys. J.,* 15, 1101, 1975.
8. **Shporer, M. and Civan, M. M.**, NMR study of ^{17}O from $H_2{}^{17}O$ in human erythrocytes, *Biochim. Biophys. Acta,* 385, 81, 1975.
9. **Stout, D. G., Steponkis, P. L., Bustard, L. D., and Cotts, R. M.**, Water permeability of Chlorella cell membranes by nuclear magnetic resonance, *Plant Physiol. (Bethesda),* 62, 146, 1978.
10. **Conlon, T. and Outhred, R.**, The temperature dependence of erythrocyte water diffusion permeability, *Biochim. Biophys. Acta,* 511, 408, 1978.
11. **Morariu, V. V. and Benga, G.**, Evaluation of a nuclear magnetic resonance technique for the study of water exchange through erythrocyte membranes in normal and pathological subjects, *Biochim. Biophys. Acta,* 469, 301, 1977.
12. **Andrasko, J. and Forsen, S.**, NMR study of rapid water diffusion across lipid bilayers in dipalmitoyl lecithin vesicles, *Biochem. Biophys. Res. Commun.,* 60, 813, 1974.
13. **Haran, N. and Shporer, M.**, Study of water permeability through phospholipid vesicle membranes by ^{17}O NMR, *Biochim. Biophys. Acta,* 426, 638, 1976.
14. **Lipschitz-Farber, C. and Degani, H.**, Kinetics of water diffusion across phospholipid membranes, *Biochim. Biophys. Acta,* 600, 291, 1980.
15. **Sharp, R. R. and Yocum, C. F.**, The kinetics of water exchange across the chloroplast membrane, *Biochim. Biophys. Acta,* 592, 169, 1980.
16. **Sharp, R. R. and Sen, R.**, Water permeability of the chromaffin granule membrane, *Biophys. J.,* 40, 17, 1982.
17. **Fukushima, E. and Roeder, S. B. W.**, *Experimental Pulse NMR,* Addison-Wesley, Reading, Mass., 1981.
18. **Harris, R. K.**, *Nuclear Magnetic Resonance Spectroscopy — A Physicochemical View,* Pitman, London, 1983.
19. **Farrar, T. C. and Becker, E. D.**, *Pulse and Fourier Transform NMR,* Academic Press, New York, 1971.
20. **Whittam, R.**, *Transport and Diffusion in Red Blood Cells,* Edward Arnold Publishing, London, 1964.
21. **Cooke, R. and Wien, R.**, The state of water in muscle tissue as determined by proton nuclear magnetic resonance, *Biophys. J.,* 11, 1002, 1971.
22. **Packer, K. J.**, The effects of diffusion through locally inhomogeneous magnetic fields on transverse nuclear spin relaxation in heterogeneous systems. Proton transverse relaxation in striated muscle tissue, *J. Magn. Reson.,* 9, 438, 1973.
23. **Glasel, J. A. and Lee, K. H.**, On the interpretation of water relaxation times in heterogeneous systems, *J. Am. Chem. Soc.,* 96, 970, 1975.
24. **Woessner, D. E., Snowdon, B. S., Jr., and Meyer, G. H.**, Tetrahedral model for pulsed nuclear magentic resonance transverse relaxation: application to the clay-water system, *J. Colloid Interface Sci.,* 34, 43, 1970.
25. **McConnell, H. M.**, Reaction rates by nuclear magnetic resonance, *J. Chem. Phys.,* 28, 430, 1958.
26. **Johnson, C. S., Jr.**, Chemical rate processes and magnetic resonance, *Adv. Magn. Reson.,* 1, 33, 1966.
27. **Rogers, M. T. and Woodbrey, J. C.**, Proton magnetic resonance study of hindered internal rotation in some substituted N,N-dimethylamides, *J. Phys. Chem.,* 66, 540, 1962.
28. **Swift, T. J. and Connick, R. E.**, NMR-relaxation mechanisms of ^{17}O in aqueous solutions of paramagnetic cations and the lifetime of water molecules in the first coordination sphere, *J. Chem. Phys.,* 37, 307, 1962.
29. **Carr, H. Y. and Purcell, E. M.**, Effects of diffusion on free precession in the nuclear magnetic resonance experiment, *Phys. Rev.,* 94, 630, 1954.
30. **Bloom, M., Reeves, L. W., and Wells, E. J.**, Spin-echos and chemical exchange, *J. Chem. Phys.,* 42, 1615, 1965.
31. **Allerhand, A. and Gutowsky, H. S.**, Spin-echo NMR studies of chemical exchange. I. Some general aspects, *J. Chem. Phys.,* 41, 2115, 1964.
32. **Allerhand, A. and Gutowsky, H. S.**, Spin-echo studies of chemical exchange. II. Closed formulas for two sites, *J. Chem. Phys.,* 42, 1587, 1965.
33. **Gutowsky, H. S., Vold, R. L., and Wells, E. J.**, Theory of chemical exchange effects in magnetic resonance, *J. Chem. Phys.,* 43, 4107, 1965.
34. **Morris, S. J., Schultens, H. A., and Schober, R.**, An osmometer model for changes in the buoyant density of chromaffin granules, *Biophys. J.,* 20, 33, 1977.
35. **Coupland, R. E.**, Determining sizes and distribution of sizes of spherical bodies such as chromaffin granules in tissue section, *Nature (London) New Biol.,* 217, 384.

36. **Winkler, H.,** The composition of adrenal chromaffin granules: an assessment of controversial results, *Neuroscience,* 1, 65, 1976.
37. **Winkler, H. and Westhead, E.,** The molecular organization of adrenal chromaffin granules, *Neuroscience,* 5, 1803, 1980.
38. **Daniels, A. J., Johnson, L. N., and Williams, R. J. P.,** Uptake of manganese by chromaffin granules in vitro, *J. Neurochem.,* 33, 923, 1979.
39. **Sharp, R. R. and Yoon, P. S.,** unpublished data, 1985.
40. **Bloembergen, N. and Morgan, L. O.,** Proton relaxation times in paramagnetic solutions. Effects of Electron spin relaxation, *J. Chem. Phys.,* 34, 842, 1961.
41. **Sharp, R. R. and Yocum, C. F.,** Field-dispersion profiles of the proton spin-lattice relaxation rate in chloroplast suspensions. Effect of manganese extraction by EDTA, Tris, and hydroxylamine, *Biochim. Biophys. Acta,* 592, 185, 1980.
42. **Paollilo, D. J., Jr. and Falk, R. H.,** The ultrastructure of grana in mesophyll plastids of *Zea mays, Am. J. Bot.,* 53, 173, 1966.
43. **Paollilo, D. J., Jr., MacKay, N. C., and Graffius, J. R.,** The structure of grana in flowering plants, *Am. J. Bot.,* 56, 344, 1969.
44. **Paollilo, D. J., Jr., Falk, R. H., and Reighard, J. A.,** The effect of chemical fixation on the fretwork of chloroplasts, *Trans. Am. Microsc. Soc.,* 86, 225, 1967.
45. **Paollilo, D. J., Jr. and Reighard, J. A.,** On the relationship between mature structure and ontogeny in the grana of chloroplasts, *Can. J. Bot.,* 45, 773, 1967.
46. **Kreutz, W.,** X-ray structure research on the photosynthetic membrane, *Adv. Bot. Res.,* 3, 53, 1970.
47. **Neumann, J. and Jagendorf, A. T.,** Light-induced pH changes related to phosphorylation by chloroplasts, *Arch. Biochem. Biophys.,* 107, 109, 1964.
48. **Hind, G., Nakatani, H. Y., and Izawa, S.,** Light-dependent redistribution of ions in suspensions of chloroplast thylakoid membranes, *Proc. Natl. Acad. Sci. U.S.A.,* 71, 1484, 1974.
49. **Gaensslen, R. E. and McCarty, R. E.,** Amine uptake in chloroplasts, *Arch. Biochem. Biophys.,* 147, 55, 1971.
50. **Deamer, D. W., Crofts, A. R., and Packer, L.,** Mechanisms of light-induced structural changes in chloroplasts. I. Light-scattering increments and ultrastructural changes mediated by proton transport, *Biochim. Biophys. Acta,* 131, 81, 1967.
51. **Crofts, A. R., Deamer, D. W., and Packer, L.,** Mechanisms of light-induced structural change in chloroplasts. II. The role of ion movements in volume changes, *Biochim. Biophys. Acta,* 131, 97, 1967.
52. **Hertz, H. G.,** Austauschgeschwindigkeiten brom- und jodhaltiger Ionen in wassriger Losung aus der Linienbreite der magnetischen Kernresonanz, *Z. Elektrochem.,* 65, 36, 1961.
53. **Hall, C., Kydon, D. W., Richards, R. E., and Sharp, R. R.,** Multi-site chemical exchange effects in the n.m.r. spectra of quadrupolar nuclei: the aqueous Cl^- + Cl_2 system as an example, *Proc. R. Soc. London Ser. A.,* 318, 119, 170.

Chapter 2

OSMOTIC AND DIFFUSIONAL WATER PERMEABILITY IN RED CELLS: METHODS AND INTERPRETATIONS

R. I. Macey and J. Brahm

TABLE OF CONTENTS

I. INTRODUCTION

Water permeates biological membranes through several routes. These include the lipid bilayer, possibly ion channels, and, in many instances, nonspecific "porous" pathways which occur for example in blood capillaries, epithelial shunts, and gap junctions. In addition, water appears to move through specific water channels, at least in the red cell membrane, in the toad bladder, and presumably the distal portions of the renal nephron. These channels have not been isolated, seen, or localized. Their existence is inferred largely from permeability measurements.

The purpose of this article is to review the methods and interpretations of water permeability measurements. Although considerable progress has been made in toad bladder studies, interpretations are sometimes complicated by the fact that the preparation consists of two or more nonidentical membranes in series.[1,2] On the other hand, the red cell membrane is particularly attractive because of its simplicity compared to other biological membranes, and because more is known about its structure and chemical composition than any other animal cell membrane. Accordingly, we confine this article to water channels in the red cell. (For related reviews on methods see References 3 and 4.)

There are two types of permeability measurements which have been used extensively in red cell studies: (1) P_f, the osmotic (or filtration) permeability coefficient with a dimension of cm/sec, is obtained from measurements of volume flow induced by an osmotic gradient and (2) P_d, the diffusional permeability coefficient also with dimension cm/sec, is measured by exchange of labeled water at equilibrium (*vide infra*). The most striking characteristic of water channels is that these two measurements do not agree. In fact, the discrepancy $P_f > P_d$ is generally taken as the major criterion for the existence of channels. The basis for this discrepancy is attributed to a fundamental difference in diffusional and osmotic flow through a channel. In the former case, movement occurs through random and essentially independent molecular motion, while osmotic flow in a channel occurs by hydraulic movement of the entire fluid within the channel. The hydraulic character of osmotic flow can be predicted by simply examining the boundary conditions at the mouth of a channel. As Mauro has shown, the discontinuity in solute concentration must be balanced by an equivalent discontinuity in pressure in order for the chemical potential of water to remain continuous across the boundary.[5] The logical consequence is that actual hydrostatic pressure gradients develop through the channel.

II. CELL WATER AND CELL VOLUME

A. Cell Water

In the past there has been considerable dispute on the existence of structured or polarized intracellular water that could exclude solutes. Current beliefs, based on substantial evidence no longer entertain the possibility that these factors play a significant role in the osmotic balance of normal red cells.[6-9] Both nonelectrolytes and the major ions Na^+, K^+, and Cl^- are evenly distributed throughout the cell water.[6-9] Further, Freedman and Hoffman have shown that Na^+ and K^+ binding to Hb (hemoglobin) is insignificant (although K^+/Na^+ selectivity = 1.3 in shrunken cells) and that mean activity coefficients of the major salts NaCl and KCl are equal in the intra- and extracellular solutions.[9] Thus, to a good approximation the water and osmotically active solute contents of the red cell interior appear to be well behaved and predictable in terms of elementary physical chemistry. The major exception to this generalization is offered by hemoglobin whose osmotic coefficient, Φ_{Hb} shows an unusually large concentration dependence which may be represented by the following virial equation in both solution and in cells:

$$\Phi_{Hb} = 1 + 0.0645[Hb] + 0.0258[Hb]^2 \tag{1}$$

where [Hb] is measured in millimolal.[9,10] For typical concentration of Hb in isotonic solutions, the osmotic coefficient is about 2.6, resulting in a contribution of approximately 17 mOsm or about 6% of the internal tonicity.

Although there has been dispute over the best method for estimating total water volume in red cells, the magnitude of the disagreement is small; the volume reported in the literature ranges from 65 to 72% of the isotonic volume.[7,11-14] The actual variation is probably much less since the isotonic volume is somewhat arbitrary and depends on the osmotic pressure of the solution that is considered isotonic as well as the temperature and pH. These latter two variables are important because they alter the ionization of weak electrolytes (particularly Hb) which influences the distribution of permeating salts.[15]

B. Cell Volume and Osmotic Pressure

The relation between cell volume and osmotic pressure is required in all determinations of P_f. If n_i represents the total number of osmotically active particles of species i at any particular time, and if ϕ_i denotes its osmotic coefficient then the osmolality (= osmotic pressure/RT) inside the cell, Π_{in}, is given by

$$\Pi_{in} = \left(RT \sum \Phi_i n_i\right)/(V - \alpha) \tag{2}$$

where V is the cell volume, and α represents the volume of the cell solids so that $V - \alpha$ is simply the volume of cell water. When cells equilibrate with a solution whose osmotic pressure is Π_{out}, we have $\Pi_{in} = \Pi_{out}$. Solving Equation 2 for V, we arrive at the expression

$$V = \alpha + \left(RT \sum \Phi_i n_i\right)/\Pi_{out} \tag{3}$$

In earlier papers it was assumed that $RT\Sigma\Pi_i n_i$ is constant and, after equilibrating cells with different solutions V was plotted against $1/\Pi_{out}$. The anticipated linear plot is obtained, but the intercept from the plot designated as β, does not agree with the value of α as obtained from direct measurements of cell water. The empirical equation is

$$V = \beta + g/\Pi_{out} \tag{4}$$

where g is a constant. The most obvious sources of discrepancy between Equations 3 and 4 stem from the assumption that $RT\Sigma\phi_i n_i$ is constant. We have seen that ϕ_{Hb} is concentration dependent, and it has also been shown that small shifts of electrolytes may also occur particularly in highly buffered extracellular media which will force shifts in intracellular pH.[7] Both of these effects act in a direction to reduce the difference between α and β, but the discrepancy has never been totally resolved.[15] In any case, current practice is to regard the linear plot of V against $1/\Pi_{out}$ as suggested by Equation 4, as an accurate and useful empirical equation. The constant g can be obtained by referring to standard isotonic conditions. Letting V_{iso} and Π_{iso} denote the isotonic cell volume and isotonic osmotic pressure respectively, and substituting these values into Equation 4 we obtain $g = \Pi_{iso}(V_{iso} - \beta)$. Using this value of g and introducing the following new variables normalized to isotonic conditions:

$$v = V/V_{iso},\ \pi = \Pi/\Pi_{iso},\ b = \beta/V_{iso} \tag{5}$$

we have

$$v = b + (1 - b)/\pi \tag{6}$$

or alternatively solving for π

$$\pi = (1 - b)/(v - b) \tag{6a}$$

as a simple expression for the cell volume in terms of osmotic pressure. Like α, values of β (or b) depend on a precise specification of "isotonic conditions" and it is dependent on the buffering capacity of the external solution as well as pH.[7] Osmolality of normal human blood plasma is approximately 285 mOsm.[14] Using 285 mOsm as our isotonic standard, values reported for b range from 0.41 to as high as 0.48.[7,11] Contrast this with values of a $= \alpha/V_{iso}$ which range from 0.28 and 0.35.[7,11,14] In red cell ghosts, the values of a and b are both close to zero.

III. OSMOTIC WATER PERMEABILITY

A. Kinetics

Osmotic permeability is measured by subjecting cells to a sudden change in the osmotic pressure of the suspending medium and recording the subsequent change in cell volume. Interpretations are simplified by the fact that hydrostatic pressure gradients are insignificant when compared to osmotic gradients[16] and because volume changes occur through changes in shape with little or no change in cell surface area. If j represents the molar water flow into the cell, and A the surface area, then

$$j = P_f A(\Pi_{in} - \Pi_{out}) \tag{7}$$

To convert j to volume flow, multiply both sides of Equation 7 by V_w, the molar volume of water (~18 cm^3/mol). $j \cdot V_w$ then becomes dV/dt, and introducing the normalized variables from Equation 5 we have

$$dv/dt = (P_f V_w A/V_{iso})\, \Pi_{iso}\, (\pi_{in} - \pi_{out}) \tag{8}$$

Using Equation 6a with $\pi = \pi_{in}$ we can rewrite Equation 8 in the form

$$dv/dt = k_f[(1 - b)/(v - b) - \pi_{out}] \tag{9}$$

where

$$k_f = (P_f V_w A/V_{iso})\, \Pi_{iso} \tag{10}$$

Permeability experiments begin with cells equilibrated in solutions with $\pi_{out} = \pi_o$ so that by Equation 6 the corresponding initial v_o is given by

$$v_o = b + (1 - b)/\pi_o \tag{11}$$

Then π_{out} is suddenly changed to a new value π_∞ and the cells swell (or shrink) to a new volume

$$v_\infty = b + (1 - b)/\pi_\infty \tag{12}$$

Solving Equation 12 for π_∞ and letting $\pi_{out} = \pi_\infty$, we finally arrive at

$$dv/dt = k_f(1 - b)[1/(v - b) - 1/(v_\infty - b)] \tag{13}$$

Table 1
SOME USEFUL PARAMETERS FOR DESCRIPTION OF WATER PERMEABILITY OF RED CELL MEMBRANES

			Ref.
Isotonic cell volume	V_{iso}	87×10^{-12} cm^3	
Water fraction of isotonic cells	$(1 - \alpha)/V_{iso}$	0.65—0.72	7,13,14
Cell solid fraction of isotonic cells	$a = \alpha/V_{iso}$	0.28—0.35	7,11,14
Isotonic cell water volume	$(1 - a)V_{iso}$	56—63×10^{-12} cm^3	
Osmotic "dead volume"	$b = \beta/V_{iso}$	0.41—0.48	7,11,14
Osmotic "active volume"	$(1 - b)V_{iso}$	45—51×10^{-12} cm^3	
Cell membrane area	A	1.42×10^{-6} cm^2	12

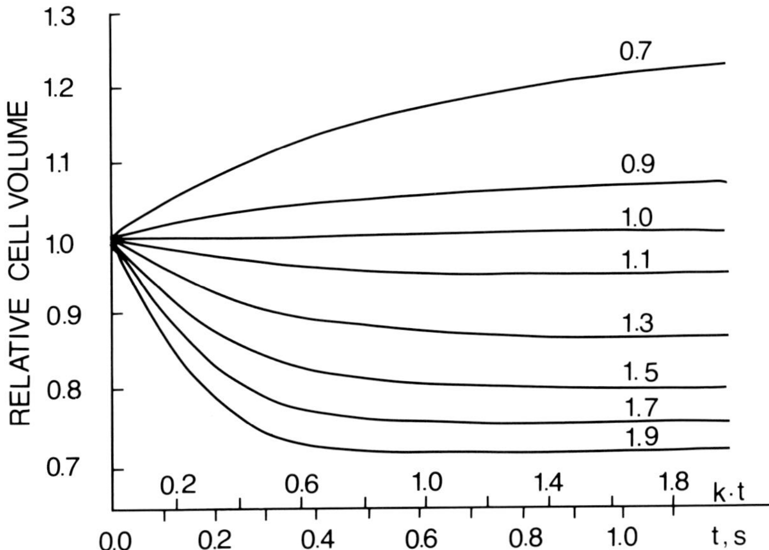

FIGURE 1. Numerical solutions to Equation 9 with various values of π_{out} beginning with $\pi_{out} = 0.7$ (upper curve) and progressing in increments of 0.2 until $\pi_{out} = 1.9$ (lower curve). The horizontal axis represents time in dimensionless units $k \cdot t$ that are converted to real time units by dividing numbers on scale by k, the rate coefficient. Using $P_f = 0.02$ cm/sec together with nominal values of the other parameters (see Table 1) $k = 1.64$ sec^{-1}.

Equation 13 can be integrated with $v = v_o$ at $t = 0$ to give:

$$(v - v_o) + (v_\infty - b) \ln[(v - v_\infty)/(v_o - v_\infty)] =$$
$$-k_f[(1 - b)/(v_\infty - b)] = -k_f\pi_\infty t \quad (14)$$

If the volume is measured as a function of time, the left-hand side of Equation 14 can be calculated and plotted against t. The slope of the predicted straight line can then be used to calculate P_f from Equation 10 together with parameters listed in Table 1. Numerical solutions to Equations 9, 13, or 14 are illsutrated in Figure 1.

A useful alternative is to design experiments where volume changes are small so that Equation 13 can be linearized (see below for advantages).[17] In this case we may expand the right-hand side of Equation 13 in a Taylor's series about the point $v = v_\infty$ to obtain

Table 2
PERCENTAGE OF ERRORS INVOLVED IN THE LINEAR APPROXIMATION

Π_{out}/Π_{iso}	% Error	Relative cell volume
0.75	6.61	1.19
0.80	5.33	1.15
0.85	4.06	1.10
0.90	2.77	1.06
0.95	1.40	1.03
1.00	0.00	1.00
1.05	−1.39	0.97
1.10	−2.70	0.95
1.15	−3.97	0.92
1.20	−5.22	0.90
1.25	−6.48	0.88
1.30	−7.66	0.87
1.35	−8.80	0.85
1.40	−9.87	0.83
1.45	−11.10	0.82
1.50	−12.10	0.81

Note: Percentage error encountered by curve-fitting data with the exponential approximation in Equation 18 to extract an estimate of P_f from the time constant given by Equation 18, or equivalently from the rate constant k_f given by Equation 10. Numerical solutions to Equation 9 were used to simulate data, and semi-log plots of simulated data were used to calculate k_f by linear regression. Since k_f is proportional to P_f, the error in the estimated value of k_f (as compared to the "true" value used in the numerical solutions) is the same error that would be obtained for P_f. In the last portions of an exponential curve the deviations of "real" data from the asymptote are difficult to distinguish from noise. Accordingly, simulated data were discarded from the semilog plots when the deviation from the asymptote was less than 5% of the initial amplitude.

$$dv/dt = k_f[(1 - b)/(v_\infty - b)^2](v - v_\infty) + \cdots \tag{15}$$

The solution to Equation 15 is given by

$$v = v_\infty + (v_o - v_\infty)e^{-t/\tau_f} \tag{16}$$

where

$$\tau_f = (v_\infty - b)^2/[k_f(1 - b)] \tag{17}$$

Using Equations 12, 14, and 5, Equation 17 can be rewritten in a number of useful forms, e.g.,

$$\begin{aligned}\tau_f &= V_{iso}(V_\infty - b)/(P_f V_w A \Pi_\infty) \\ &= (V_{iso}(1 - b)/P_f V_w A)[1/(\Pi_{iso}\pi_\infty^2)]\end{aligned} \tag{18}$$

Errors involved in the linear approximation (Equation 15) are shown in Table 2. In general they are small (i.e., < ±5%) provided volume changes are within ±10%.

B. Measurements

Rapid changes in cell volume in response to osmotic perturbations have generally been followed by changes in light scattering or transmission. The measurement depends on the difference in refractive index between cells and the suspending solution. As cells shrink in a dilute suspension, their internal contents become more concentrated, increasing their index of refraction with a resulting increase in light scattering (decrease in light transmission). Opposite changes occur as cells swell. Empirical calibrations of light scattering (or transmission) vs. cell volume are easily obtained under steady conditions. However, a number of problems arise related to the requirements for rapid mixing and for extracting meaningful signals immediately following the mix.

Rapid mixing is usually obtained through use of a stopped-flow apparatus although a simple rapid injection of a small volume of anisotonic solution into a cuvette of rapidly stirring cells can be adequate and offers some advantages in certain cases.[17,18-22] Performance criteria for conventional stopped-flow mixing and optical recording of homogeneous solutions have been discussed in detail by Gibson.[23] With heterogeneous solutions containing discoid particles like red cells, a number of new artifacts are encountered.[20,22,24] These become apparent in initial mixing transients, drifting baselines, and in the loss of scattering signal as cells settle.[20]

Of these, initial mixing transients are most serious because they are large and require about 300 msec to decay, a time which is comparable to the relaxation time in a typical osmotic perturbation. Examination of the stopped-flow mixing chamber shows that cells are aligned in vortices during an injection and that the light transmission signal is dependent on the linear velocity of flow. Although the light scattering signal is highly dependent on the geometry of the optical path so that it is difficult to generalize from one device to the next, it is not surprising to find large scattering changes when flow is stopped and cells are allowed to revert to a more random orientation. Further, the faster the injection velocity the larger the alignment artifact, while slow injections may compromise mixing. These initial mixing transients disappear when spherical polystyrene spheres are substituted for cells and they are substantially reduced when cells are preswollen into a more spherical shape or when the cells are pretreated with lecithin to create spherocytes.[20,22] Since lecithin treatment does not appear to change P_f, Mlekoday et al. recommend lecithin pretreatment for actual P_f measurement.[22]

Slowly drifting baseline artifacts are much smaller, but more variable and not always present. Generally they are in the direction of increased scattering (i.e., opposite to the direction of cell settling). Their origin is poorly understood and they assume significance when they interfere with obtaining accurate measurements of the final asymptotic cell volume. These slower artifacts are largely eliminated when instead of using a stopped-flow device, the experimental solution is rapidly injected into a cuvette where cells are continuously stirred.

In any case, initial mixing artifacts as well as drifting baselines and cell settling can be observed by rapidly mixing cells with the same solution that they are suspended in so that no volume change takes place. It is common practice to record these as control traces and to subtract them from experimental curves which presumably contain the same artifacts superimposed on the experimental signal. Caution is required in this procedure because the recorded signal is sensitive to many variables; simply changing the temperature, for example, will cause changes in mixing artifacts. Careful attention to these problems is mandatory, particularly when attempting to demonstrate small changes in P_f.

The relation between cell volume and light scattering (or transmission) is dependent on differences in the refractive index between cells and suspending solution and on the concentration of cells. Empirical calibrations are obtained by equilibrating cells in media of different π and either measuring the cell volume directly or calculating it with the aid of

Equation 11. For precise results over a large range of volume, corrections should be made for differences in refractive indices of each suspension medium.[24] For small volume changes, these corrections are not significant and the relation between signal and volume change is linear. This linearity has been exploited by Mlekoday et al.[22] They used Equations 11 and 12 to calculate values of v_o and v_∞ and then compared them with measured photocell signals to estimate the calibration constants (an iterative curve-fitting routine is then used to improve the accuracy). This procedure has the advantage that each kinetic curve is the source of its own calibration.

A convenient alternative is to confine measurements to small volume changes and to take advantage of the exponential approximations. In this case, no calibration is necessary. As shown by Equation 17, P_f is obtained from τ_f, which can be taken directly from raw data. Further, since τ_f is independent of the signal amplitude, data analysis can proceed at any arbitrary point along the exponential decay. In addition to subtracting control (isovolume) traces from experimental curves, we have found it expedient to discard the first 250 msec of the trace to remove all remnants of the initial mixing transient. An additional advantage of the exponential case is that it is insensitive to poor initial mixing, the distortions show up in the amplitude of the decay, but not in the time constant.[25] Finally, for precise results it is worth noting in Equation 17 that a plot of t vs. $1/\pi_\infty^2$ yields a convenient straight line whose slope can be used to calculate P_f. The disadvantages of the small perturbation approach are that: (1) the exponential is an approximation to the true curve and (2) that signal retrieval is more difficult with small changes in volume. The first objection is not particularly significant as shown by the results of an error analysis illustrated in Table 2. These results were obtained by using Equation 9 to simulate data which were then used in semi-log plots to calculate τ_f. Errors were obtained by using τ_f to calculate k_f (which is proportional to P_f) and comparing these values to the k_f values used to generate the original data. It can be seen that if the volume change is constrained to 10% (corresponding to osmotic perturbations from $\pi = 1$ to $\pi = 0.85$ for swelling, or to $\pi = 1.2$ for shrinking) the error will be within 5%. On the other hand, the small signal can present a problem particularly when there is significant baseline drift which will compromise the analysis and yield values of P_f that are too low.

IV. DIFFUSIONAL WATER PERMEABILITY

At steady-state conditions, no net movement of water takes place in the system. Water molecules do, however, cross the red cell membrane due to the irregular thermic motion of all water molecules (Brownian motion). Thus, no measurable force moves the water molecules from one side to the other side of the membrane. Further, the accidental movement of a water molecule across the membrane remains unrecognized unless the molecule is specifically labeled so that the event can be detected.

Assuming that the membrane does not discriminate between labeled and nonlabeled water molecules (e.g., no isotope effect), one can determine a rate coefficient (k_d, sec^{-1}) for the exchange of labeled molecules for nonlabeled molecules on the other side of the membrane, and calculate a permeability coefficient for water by:

$$P_d = k_d (V - \alpha)/A = (0.69/T_d) (V - \alpha)/A = 1/\tau_d (V - \alpha)/A \qquad (19)$$

where T_d and $1/\tau_d$ are the half time and the time constant of the exchange process, respectively, and (V − α/A is the ratio of cell water volume to cell membrane area. Note that k_d here represents the rate coefficient for the ''unidirectional'' diffusional flux of labeled water from a minor to an ''infinitely'' large volume. If such experimental conditions are not obtainable, k_d must be adjusted. Since $C_o^{H_2O} \simeq C_i^{H_2O}$, the correction is $k_d^{measured} \cdot V_2/(V_1 +$

Table 3
OSMOTIC (P_f) AND DIFFUSIONAL (P_d) WATER PERMEABILITY IN HUMAN RED CELLS AT ROOM TEMPERATURE[37]

cm/sec·10³							
P_f				P_d			
Influx		Efflux		Influx		Efflux	
Volumetric methods				Continuous flow tube methods			
12.8			(45)	5.3	(52)	2.4	(12)
	17.5		(46)	3.4	(53,54)		
13.5			(19)	2.8	(39*)		
22.8		16.1	(17)	NMR methods			
20.0		16.7	(20)	4.0	(55,56)		
	18.3		(47)	3.9	(57,58*)		
32.4		21.5	(48)	3.7	(43)		
		22.1	(49)	2.6	(59*)		
	24.8		(24)	2.5	(60*)		
	19.8		(22)	2.4	(61,62)		
		22.0	(43)	2.1	(63*)		
		19.0	(50)	2.0	(64*)		
		18.0	(51)				

Note: * denotes that the permeability coefficient is calculated from a published time constant assuming that the ratio $(V - \alpha)/A$ is 4.3 $\cdot 10^{-4}$ cm as is the ratio for "normal" erythrocytes under isotonic conditions.[12,44]

$V_2) = k_d^{\text{"true"}}$ for labeled water leaving the compartment with water volume V_1 into the compartment with water volume V_2.

Two approaches have been used to label water molecules for measuring the diffusion permeability: (1) incorporation of tritium (^3H) into the water molecule in substitution for ordinary hydrogen ions (^1H) and (2) orienting the water molecules in a magnetic field, which orders the molecules as long as the field is applied.

A. Radioactive Isotope Studies

There are principally two ways to determine diffusional water permeability, either as influx or as efflux of tracer molecules. Most studies have determined the rate of ^3H_2O uptake, while only one study has determined the rate of efflux (cf. Table 3). The rate with which water permeates the red cell membrane is so large that it is necessary to apply special devices with high time resolutions for the measurements. Such a device is the continuous-flow tube method, which was originally introduced by Hartridge and Roughton to study kinetics of binding of oxygen and carbon dioxide to red cell suspensions.[26]

The principle of the flow tube method is to convert time to distance.[27] A sample of isolated and packed red cells is continuously mixed in a mixing chamber with a test solution, and the cell suspension thus formed flows down an observation pipe. If the flows of cells and solution into the mixing chamber are kept constant, the flow down the pipe is also constant. Each distance along the pipe, thus, is convertible to time if the linear velocity of flow (e.g. cm/sec) is known. Though simple in principle and rather simple in construction the flow tube method compiles a list of sources of errors which should be considered. A detailed description of the continuous-flow tube method and its applicability has recently been published.[4] In the present context, it suffices to say that the method in the version developed recently is well suited to measure transport processes in isolated cells with half times as short as about 3 msec, which expressed as a permeability coefficient for human red cells with a physiological water volume of about 60 μm^3, is about 10^{-2} cm/sec.

The method can be used to measure either influx or efflux of tracer. It should be noted, however, that the kinetics of tracer equilibration in influx experiments generally require a high hematocrit in the cell suspension, which increases the degree of hemolysis. The very dilute cell suspension used for tracer efflux experiments greatly reduces hemolysis, but at the same time limits the performance of the methods to efflux measurements only.

B. Nuclear Magnetic Resonance Studies

In brief the NMR method relies on the fact that nuclear magnetic moments in disordered water molecules become oriented in a magnetic field. Application of a radiofrequency pulse reorients the moment, and this reorientation serves as a label which can be detected. Following the pulse, the label decays exponentially with time. The relaxation time is shortened considerably (the label is quenched) when paramagnetic Mn ions are present in the solution. Since the red cell membrane is relatively impermeable to Mn, addition of Mn to a cell suspension results in quenching of the extracellular phase only. Application of a pulse to a red cell suspension containing Mn labels both intra- and extracellular water. Following the pulse, label disappears very quickly from the extracellular spaces because of Mn quenching, and thereafter the disappearance of label is almost entirely due to water leaving the cells for the Mn-rich suspension medium. The time constant for water exchange can be extracted after correcting for spontaneous decay of label within the cells and for any inefficiency of the Mn quench. Though in principle this method is also usable for influx measurements with Mn in the intracellular water phase, it has been used solely for efflux measurements (Table 3). The advantage of the method is that the number of measurements is almost unrestrained, and that the cells are not exposed to a possible damaging shear of the membrane during the measurement.

C. Comparison of Data Obtained with Isotopic and NMR Methods

Table 3 summarizes the results of studies in which diffusional water permeability was determined by means of either methods. In all but one study with the continuous-flow tube method, the permeability of water was determined as tracer influx, while the NMR method has been used for efflux measurements only. It is noteworthy that as years have passed, P_d determined by the continuous-flow tube method has decreased while P_d determined by means of the NMR method has increased (cf. Table 3).[28] The reason for this discrepancy is not known.

V. OSMOTIC AND DIFFUSIONAL PERMEABILITY: RESULTS AND INTERPRETATIONS

A. Ratio of Time Constants

Although it can be seen from Table 3 that $P_f > P_d$, typical relaxation times for the two types of measurements are reversed with $T_f \gg T_d$. The basis for this apparent paradoxical behavior lies in the fact that P_f is defined for water flow in terms of a solute (osmolar) gradient while P_d is defined for water flow in terms of its own gradient. This discrepancy can be clarified by comparing Equations 18 and 19. If we neglect the small difference between v − a and v − b (or between V − α and V − β), then

$$\tau_f/\tau_d = (P_d/P_f)(1/V_w\Pi_\infty) \quad (20)$$

Assuming dilute solutions with n_s moles of solute and n_w moles of water, $\Pi = n_s/(n_w V_w)$ so that Equation 20 can be rewritten as

$$\tau_f/\tau_d = (P_d/P_f)(n_w/n_s) \tag{21}$$

It follows that under physiological conditions P_f would have to be about 185 (n_w/n_s) times larger than P_d in order for τ_d to equal τ_f. The time constant τ_d in a diffusion experiment is a measure of the time required to dissipate a concentration gradient of water; each time 185 water molecules move across the membrane, it dissipates a gradient comprised of 185 molecules. In contrast, τ_f is a measure of the time required to dissipate a solute gradient. In this case, each time 185 water molecules move across the membrane, it dissipates a gradient of only one solute molecule.

B. Ratio of Permeabilities: Pores or Channels

Current interpretations of the discrepancy between P_f and P_d begin with the empirical observations that this discrepancy also exists in artificial membranes that are known to be porous, but vanishes in membranes (e.g., lipid bilayers) that are believed to be nonporous. Accordingly, when this discrepancy is significant (and can be shown to occur in the absence of unstirred layers), pores or channels are suspect. The basis for the discrepancy becomes apparent in extreme cases. First, consider a pore that is sufficiently large to apply continuum mechanics to the fluid within it. Using a right circular cylinder as a model, together with the fact that osmotic and hydraulic flows are equivalent, it follows from Poiseuille's law that P_f will be proportional to r^4 the fourth power of the radius. On the other hand, in a diffusional experiment, the flux is proportional to area, or to r^2. Thus, the ratio P_f/P_d, being proportional to r^2, reflects the geometry of the pore. This fact has been exploited by Solomon in a number of papers on red cell permeability.[29] Its applicability to pores of molecular dimensions has been discussed in detail by Levitt who presents an interesting simulation of hydraulic flow based on hard spheres colliding with each other and with the walls of a channel with $r = 3.2$ Å.[30]

At the other extreme, consider a narrow pore or channel that constrains water molecules to move through it in single file; i.e., the channel radius is less than the diameter of a water molecule so that two water molecules cannot exist side by side. In this case, it can be shown that $P_f/P_d = n$, the number of water molecules within the channel.[31-34] Here again, the ratio reflects the geometry of the channel, but its interpretation is quite different.

Our current view is that water channels in red cells correspond to the narrow channel described above. Evidence for this position is based on a large number of observations with transport inhibitors that indicate that solute and water permeability are independent. On the other hand, Solomon et al. view the channel as a more generalized pore which is capable of transporting a variety of solutes.[35] Pros and cons of these positions have been discussed at length and will not be reviewed here.[29,36-37]

C. Blocking the Channels

Both diffusional and osmotic water permeability in human red cells can be reversibly depressed by mercurial reagents such as PCMBS (*p*-chloromercuribenzenesulfonate).[12,38-41] These reagents appear to act by blocking water channels, because the water transport in red cells treated with saturating doses of PCMBS (e.g., 2 m*M*) cannot be distinguished from water transport in corresponding lecithin-cholesterol bilayers.[39] In particular, after treatment with PCMBS (1) both P_f and P_d are reduced to a value which agrees with corresponding lecithin-cholesterol bilayers, (2) the activation energy for water transport is raised to ~55 kJ/mol which also agrees with lipid bilayers, and most importantly, (3) the ratio $P_f/P_d \simeq 1$. Since the evidence for water channels disappears, we conclude the PCMBS reaction with the membrane (probably SH groups) results in channel blockage. This view is strengthened by comparing the above results on human cells with corresponding experiments on chicken red cells which do not appear to have channels (i.e., they have a low lipid bilayer-type water permeability, $P_f = P_d$, and a high activation energy $\simeq$ 55 kJ/mol).[17,42] Water transport in chicken cells is not changed by PCMBS. This result also suggests that PCMBS has no effect on lipid bilayer permeability.

Assuming that PCMBS closes water channels and does not affect the bilayer, it can be used to decompose the gross water permeability into two component parts: the channel permeability, p, and the background bilayer permeability, q. Since these permeabilities are in parallel, the measured permeability P is simply the sum of the two. Writing these quantities as a function of PCMBS concentration c, we have

$$P(c) = p(c) + q \tag{22}$$

When c is large $p \simeq O$ so that q can be estimated by $q = P(\infty)$, where ∞ denotes a saturating dose ($\geq$2 m*M*) of PCMBS. It follows that permeability of the channel can be obtained from the measured permeabilities as

$$p(c) = P(c) - P(\infty) \tag{23}$$

For osmotic flow $P_f(\infty) \simeq 0.1 \cdot P_f(O)$, so that in the untreated human red cell approximately 90% of the osmotic flow takes place through channels with the remaining 10% going through the bilayer. In the case of diffusional flow, the difference between channel and bilayer is much smaller as $P_d(\infty) \simeq 0.4 \cdot P_d(O)$, and only ~ 60% of the diffusional flow goes through channels. In either case, it is important to subtract off the bilayer permeability before ascribing measured properties to the channel. By taking this precaution, Moura et al. have been able to show that the ratio

$$p(c)_f/p(c)_d = [P(c)_f - P(\infty)_f]/[P(c)_d - P(\infty)_d] = n \simeq 11$$

remains constant (independent of c) as more and more water channels are inhibited by increasing concentrations of PCMBS.[43] This implies that PCMBS action on water channels is all or none. When the reagent reacts with a channel it closes it completely. The amount of water (or geometry) of each open channel remains constant and it is only the *number* of open channels that diminishes with PCMBS inhibition. Although it does not exclude the possibility of wider pores, this result seems to be a natural consequence of single file channels. If we adopt this interpretation, then it follows that there are approximately 11 single-file water molecules per channel. Other recent estimates of this number are 14 to 15.[28,36-37]

When using PCMBS for water transport studies, it is important to account for the fact that these reagents promote cation leakages which lead to an increase of cell volume over a period of time. Although these volume changes are too slow to interfere with accurate determinations of τ_f or τ_d, the reaction with PCMBS with water channels is equally slow (taking about 40 min for completion, depending on dosage). It follows that the initial cell volume used in the τ_f or τ_d determinations can be variable and can lead to faulty interpretations of the relevant permeability. Proper caution requires either a direct determination of cell volume *at the time of measurement,* or use of an extracellular incubation medium which will minimize these changes. We have found that a slightly hypertonic medium containing 140 m*M* KCl, 10 m*M* NaCl, 27 m*M* sucrose, and 5 m*M* Hepes buffer will reduce the shifts of Na^+ and K^+, and stabilize cell volume for over 100 min.

VI. CONCLUDING REMARKS

The fact that osmotic water permeability is larger than diffusional water permeability in human red cells as demonstrated repeatedly by different methods and techniques emphasizes the concept of pores in the human erythrocyte membrane. The structural basis for such pores is not known though it is tempting to assume that integral membrane proteins contribute to the formation of such pores. Recent binding studies with labeled inhibitors of water transport suggest that integral membrane proteins located in bands 3 and 4.5 in SDS-gel electrophoresis create the structural basis for $P_f > P_d$.[41,65] Whether band 3 proteins which definitely transport inorganic anions, are involved in water transport is, however, still a matter of debate as is the specificity of the presumed pores.[12,25,28,29,35-38,42]

ACKNOWLEDGMENT

Supported by NIH grants GM-18819 and HL-37593.

REFERENCES

1. **Levine, S. D., Jacoby, M., and Finkelstein, A.,** The water permeability of toad urinary bladder. I. Permeability of barriers in series with the luminal membrane, *J. Gen. Physiol.,* 83, 529, 1984.
2. **Levine, S. D., Jacoby, M., and Finkelstein, A.,** The water permeability of toad urinary bladder. II. The value of P_f/P_d(w) for the antidiuretic hormone-induced water permeation pathway, *J. Gen. Physiol.,* 83, 543, 1984.
3. **Macey, R. I. and Moura, T.,** Water channels, in *Methods in Enzymology,* Packer, L., Ed., Academic Press, Orlando, Fla., 1986, 598.
4. **Brahm, J.,** Fast transport techniques for measurements of anion, urea and glucose transport in the red blood cell and ghost systems, in *Methods in Enzymology,* Fleischer, B. and Fleischer, S., Eds., Academic Press, Orlando, Fla., in press, 1988.
5. **Mauro, A.,** Nature of solvent transfer in osmosis, *Science,* 126, 252, 1957.
6. **Miller, D. M.,** Sugar uptake as a function of cell volume in human erythrocytes, *J. Physiol.,* 170, 219, 1964.
7. **Cook, J. S.,** Nonsolvent water in human erythrocytes, *J. Gen. Physiol.,* 50, 1311, 1967.
8. **Gary-Bobo, C. M.,** Nonsolvent water in human erythrocytes and hemoglobin solutions, *J. Gen. Physiol.,* 50, 2547, 1967.
9. **Freedman, J. C. and Hoffman, J. F.,** Ionic and osmotic equilibria of human red blood cells treated with nystatin. *J. Gen. Physiol.,* 74, 157, 1979.
10. **Adair, G. S.,** A theory of partial osmotic pressures and membrane equilibria, with special reference to the application of Dalton's law to haemoglobin solutions in the presence of salts, *Proc. R. Soc. London Ser. A,* 120, 573, 1928.
11. **Lefevre, P. G.,** The osmotically functional water content of the human erythrocyte, *J. Gen. Physiol.,* 47, 585, 1964.
12. **Brahm, J.,** Diffusional water permeability of human erythrocytes and their ghosts, *J. Gen. Physiol.,* 79, 791, 1982.
13. **Brahm, J.,** Temperature-dependent changes of chloride transport kinetics in human red cells, *J. Gen. Physiol.,* 70, 383, 1977.
14. **Savitz, D., Sidel, V. W., and Solomon, A. K.,** Osmotic properties of human red cells, *J. Gen. Physiol.,* 48, 79, 1964.
15. **Dalmark, M.,** Chloride and water distribution in human red cells, *J. Physiol.,* 250, 65, 1975.
16. **Rand, R. P. and Burton, A. C.,** Mechanical properties of the red cell membrane. I. Membrane stiffness and intracellular pressure, *Biophys. J.,* 4, 115, 1964.
17. **Farmer, R. E. L. and Macey, R. I.,** Pertubation of red cell volume: rectification of osmotic flow, *Biochim. Biophys. Acta,* 196, 53, 1970.
18. **Sha'afi, R. I., Rich, G. T., Sidel, V. W., Bossert, W., and Solomon, A. K.,** The effect of the unstirred layer on human red cell water permeability, *J. Gen. Physiol.,* 50, 1377, 1967.

19. **Sirs, J. A.,** The rate of osmotic influx of water by flexible and inflexible erythrocytes, *J. Physiol.*, 205, 147, 1969.
20. **Blum, R. M. and Forster, R. E.,** The water permeability of erythrocytes, *Biochim. Biophys. Acta*, 203, 410, 1970.
21. **Owen, J. D. and Eyring, E. M.,** Reflection coefficients of permeant molecules in human red cell suspensions, *J. Gen. Physiol.*, 66m 251, 1975.
22. **Mlekoday, H. J., Moore, R., and Levitt, D.,** Osmotic water permeability of the human red cell. Dependence on direction of water flow and cell volume, *J. Gen. Physiol.*, 81, 212, 1983.
23. **Gibson, Q. H.,** Rapid mixing: stopped flow, in *Methods in Enzymology*, Vol. 16, Kustin, K., Ed., Academic Press, New York, 1969, 187.
24. **Terwilliger, T. C. and Solomon, A. K.,** Osmotic water permeability of human red cells, *J. Gen. Physiol.*, 77, 549, 1981.
25. **Macey, R. I.,** Transport of water and nonelectrolytes across red cell membranes, in *Membrane Transport in Biology*, Giebisch, G., Tosteson, D. C., and Ussing, H. H., Eds., Springer-Verlag, Berlin, 1979, 1.
26. **Hartridge, H. and Roughton, F. J. W.,** A method of measuring the velocity of very rapid chemical reactions, *Proc. R. Soc. London Ser. A*, 104, 376, 1923.
27. **Tosteson, D. C.,** Halide transport in red blood cells, *Acta Physiol. Scand.*, 46, 19, 1959.
28. **Galey, W. R. and Brahm, J.,** The failure of hydrodynamic analysis to define pore size in cell membranes, *Biochim. Biophys. Acta*, 818, 425, 1985.
29. **Solomon, A. K.,** Characterization of biological membranes by equivalent pores, *J. Gen. Physiol.*, 51, 355s, 1968.
30. **Levitt, D. G.,** Kinetics of diffusion and convection in 3.2-Å pores. Exact solution by computer simulation, *Biophys. J.*, 13, 186, 1973.
31. **Lea, E. J. A.,** Permeation through long narrow pores, *J. Theor. Biol.*, 5, 102, 1963.
32. **Dick, D. A. T.,** *Cell Water*, Bittar, E. E., Ed., London, 1966, 102.
33. **Levitt, D. G.,** A new theory of transport for cell membrane pores. I. General theory and application to red cells, *Biochim. Biophys. Acta*, 373, 115, 1974.
34. **Finkelstein, A. and Rosenberg, R. A.,** Single-file transport: Implications for ion and water movement through gramicidin a channels, in *Membrane Transport Processes*, Vol. 3, Stevens, C. F. and Tsien, R. W., Eds., Raven Press, New York, 1979, 73.
35. **Solomon, A. K., Chasan, B., Dix, J. A., Lukacovic, M. F., Toon, M. R., and Verkman, A. S.,** The aqueous pore in the red cell membrane: band 3 as a channel for anions, cations, nonelectrolytes and water, *Ann. N. Y. Acad. Sci.*, 414, 97, 1983.
36. **Brahm, J.,** Urea permeability of human red cells, *J. Gen. Physiol.*, 82, 1, 1983.
37. **Brahm, J.,** Water transport through the red cell membrane, *Period. Biol.*, 85, 109, 1983.
38. **Macey, R. I. and Farmer, R. E. L.,** Inhibition of water and solute permeability in human red cells, *Biochim. Biophys. Acta*, 211, 104, 1970.
39. **Macey, R. I., Karen, D. M., and Farmer, R. E. L.,** Properties of water channels in human red cells, in *Biomembranes*, Kreuzer, F. and Slegers, J. F. G., Eds., Plenum Press, New York, 1972, 331.
40. **Benga, Gh., Pop, V. I., Popescu, O., Ionescu, M., and Mihele, V.,** Water exchange through erythrocyte membranes: nuclear magnetic resonance studies on the effects of inhibitors and of chemical modification of human membranes, *J. Membr. Biol.*, 76, 129, 1983.
41. **Benga, Gh., Popescu, O., and Pop, V. I.,** *p*-(chloromercuri)-benzenesulfonate binding by membrane proteins and the inhibition of water transport in human erythrocytes, *Biochemistry*, 25, 1535, 1986.
42. **Brahm, J. and Wieth, J. O.,** Separate pathways for urea and water, and for chloride in chicken erythrocytes, *J. Physiol.*, 266, 727, 1977.
43. **Moura, T. F., Macey, R. I., Chien, D. Y., Karan, D., and Santos, H.,** Thermodynamics of all-or-none water channel closure in red cells, *J. Membrane Biol.*, 81, 105, 1984.
44. **Wieth, J. O., Funder, J., Gunn, R. B., and Brahm, J.,** Passive transport pathways for chloride and urea through the red cell membrane, in *Comparative Biochemistry and Physiology of Transport*, Bolis, L., Bloch, K., Luria, S. E., and Lynen, F., Eds., North-Holland, Amsterdam, 1974, 317.
45. **Sidel, V. W. and Solomon, A. K.,** Entrance of water into human red cells under an osmotic pressure gradient, *J. Gen. Physiol.*, 41, 243, 1957.
46. **Rich, G. T., Sha'afi, R. I., Romualdez, A., and Solomon, A. K.,** Effect of osmolality on the hydraulic permeability coefficient of red cells, *J. Gen. Physiol.*, 52, 941, 1968.
47. **Colombe, B. W. and Macey, R. I.,** Effects of calcium on potassium and water transport in human erythrocyte ghosts, *Biochim. Biophys. Acta*, 363, 266, 1974.
48. **Galey, W. R.,** Determination of human erythrocyte membrane hydraulic conductivity, *J. Membrane Sci.*, 4, 41, 1978.
49. **Levin, S. W., Levin, R. L., and Solomon, A. K.,** Improved stop-flow apparatus to measure permeability of human red cells and ghosts, *J. Biochem. Biophys. Meth.*, 3, 255, 1980.

50. **Chasan, B., Lukacovic, M. F., Toon, M. R., and Solomon, A. K.,** Effect of thiourea on PCMBS inhibition of osmotic water transport in human red cells, *Biochim. Biophys. Acta,* 778, 195, 1984.
51. **Dix, J. A., Ausiello, D. A., Jung, C. Y., and Verkman, A. S.,** Target analysis studies of red cell water and urea transport, *Biochim. Biophys. Acta,* 821, 243, 1985.
52. **Paganelli, C. V. and Solomon, A. K.,** The rate of exchange of tritiated water across the human red cell membrane, *J. Gen. Physiol.,* 41, 259, 1957.
53. **Barton, T. C. and Brown, D. A. J.,** Water permeability of the fetal erythrocyte, *J. Gen. Physiol.,* 47, 839, 1964.
54. **Vieira, F. L., Sha'afi, R. I., and Solomon, A. K.,** The state of water in human and dog red cell membranes, *J. Gen. Physiol.,* 55, 451, 1970.
55. **Dix, J. A. and Solomon, A. K.,** Role of membrane proteins and lipids in water diffusion across red cell membranes, *Biochim. Biophys. Acta,* 773, 219, 1984.
56. **Benga, Gh., Borza, V., Popescu, O., Pop, V. I., and Muresan, A.,** Water exchange through erythrocyte membranes: nuclear magnetic resonance studies on resealed ghosts compared to human erythrocytes, *J. Membr. Biol.,* 89, 127, 1986.
57. **Conlon, T. and Outhred, R.,** The temperature dependence of erythrocyte water diffusion permeability, *Biochim. Biophys. Acta,* 511, 408, 1978.
58. **Morariu, V. V., Ionescu, M. S., Frangopol, M., Grosescu, R., Lupu, M., and Frangopol, P. T.,** Nuclear magnetic resonance investigation of human erythrocytes in the presence of manganese ions. Evidence for a thermal transition, *Biochim. Biophys. Acta,* 815, 189, 1985.
59. **Shporer, M. and Civan, M. M.,** NMR study of ^{17}O in human erythrocytes, *Biochim. Biophys. Acta,* 385, 81, 1975.
60. **Andrasko, J.,** Water diffusion permeability of human erythrocytes studied by a pulsed gradient NMR technique, *Biochim. Biophys. Acta,* 428, 304, 1976.
61. **Fabry, M. E. and Eisenstadt, M.,** Water exchange between red cells and plasma. Measurements by nuclear magnetic relaxation, *Biophys. J.,* 15, 1101, 1975.
62. **Chien, D. Y. and Macey, R. I.,** Diffusional water permeability of red cells. Independence on osmolality, *Biochim. Biophys. Acta,* 464, 45, 1977.
63. **Ashley, D. L. and Goldstein, J. H.,** In vitro erythrocyte water transport in Duchenne muscular dystrophy: an NMR investigation, *Neurology,* 33, 1206, 1983.
64. **Pirkle, J. L., Ashley, D. L., and Goldstein, J. H.,** Pulse nuclear magnetic resonance measurements of water exchange across the erythrocyte membrane employing a low Mn concentration, *Biophys. J.,* 25, 389, 1979.
65. **Benga, Gh. and Popescu, O.,** Water permeability in human erythrocytes: identification of membrane proteins involved in water transport, *Eur. J. Cell Biol.,* 41, 252, 1986.

Chapter 3

MEMBRANE PROTEINS INVOLVED IN THE WATER PERMEABILITY OF HUMAN ERYTHROCYTES: BINDING OF *p*-CHLOROMERCURIBENZENE SULFONATE TO MEMBRANE PROTEINS CORRELATED WITH NUCLEAR MAGNETIC RESONANCE MEASUREMENTS

Gheorghe Benga

TABLE OF CONTENTS

I. INTRODUCTION

Because of its relatively simple structure the red blood cell has been for many years a favorite object for studying water permeability. There are two basic strategies for measuring water exchange through the erythrocyte membrane: (1) nonstationary methods and (2) stationary methods.

The nonstationary methods involve subjecting the cells to an osmotic gradient that creates a net flux of water in one direction or the other, depending on whether the cells swell or shrink. The membrane is subjected to stress, and the cell will eventually hemolyze. Much information about water transport through erythrocyte membranes was obtained between 1930 and 1950 by these methods. The information on osmotic permeability of erythrocyte membrane has been reviewed by Sha'afi.[1]

In the case of stationary methods, the diffusion movement of water is measured and therefore there is no net flux of water through the membrane. The cells remain in their normal state, which is often considered an advantage over nonstationary methods. The stationary methods can be classified in two groups: (1) radiotracer methods and (2) nuclear magnetic resonance (NMR) methods. A review on diffusional water permeability has recently been published by Morariu and Benga.[2]

Two mechanisms for the permeation of water through erythrocyte membranes have been proposed.[1] One model is based on the concept that the molecular motion of the hydrocarbon chains of membrane lipids generates structural defects through which water permeates.[3] The second model assumes the presence of aqueous membrane channels or "pores" assembled from membrane-integral proteins which span the human red cell membrane.[1] The first model is thought to represent water flux through the lipid bilayer accounting for about 10% of the total flux observed in red cell membranes. Functional evidence for the existence of water channels or "pores" located in membrane proteins has come from comparisons with lipid bilayers.[4] The activation energy for the diffusional and osmotic permeabilities of water in red blood cells (4 to 6 kcal/mol) is much lower than the values for water permeation through lipid bilayers (11 to 14 kcal/mol) and the water permeability of red blood cells is much higher than the corresponding permeabilities of lipid bilayers. In addition, the osmotic permeability is several times higher than the diffusional permeability,[5] while in artificial bilayers these values are equal.[6] Despite these observations, the proteins in the erythrocyte membranes that may accommodate the water pathways have not been identified to any degree of certainty in previous experiments.[4]

A new approach to the study of transport processes in the erythrocyte membrane has been the use of chemical probes.[7] This has allowed the identification of one major protein of the membrane, the "band 3" protein, as being involved in anion transport, based on the selective binding to this protein of a radioactively labeled inhibitor of anion transport.[8] The name of the protein is derived from its location in polyacrylamide gels following sodium dodecyl sufate (SDS) gel electrophoresis.

An important characteristic of the water permeability of erythrocytes is its inhibition by sulfhydryl-binding mercurial reagents.[9-13] Macey and Farmer[9] found that when *p*-chloromercuribenzene sulfonate (PCMBS) reached maximum effectiveness (after an incubation time of 10 to 20 min), it inhibited osmotic permeability by about 90% and diffusional permeability by about 50%. The organic mercurials were found not only to decrease water permeability, but also to elevate activation energy and reduce the ratio of osmotic to diffusional water permeability to unity. Water transport properties of red blood cells under these conditions are hardly distinguishable from lipid bilayers. A straightforward interpretation is that mercurials react with the sulfhydryl groups of proteins associated with water channels, resulting in the closure of the channels.[4] Consequently, the labeling of red blood cell membrane proteins with mercurials, under conditions of inhibition of water diffusion, and blocking the nonspecific SH-groups by noninhibitory sulfhydryl reagents, would allow the identification of those membrane-spanning proteins associated with the water channels. Previous labeling experiments[14-16] have not correlated the binding of sulfhydryl reagents to the effect on the inhibition of water transport, so that evidence from radioactive tagging with inhibitors was considered to be equivocal at best.[4] The suggestion[14-17] that band 3 protein contains the water channel had to be demonstrated.[4]

The aim of our chapter is to review our work in this area,[11,12,18-24] including recent experiments that allowed us to identify the proteins involved in the water permeability of human erythrocytes.[25-27]

II. METHODOLOGY

A. Blood Sample Preparations

The NMR labeling experiments have been performed on human red blood cells or resealed ghosts. Human blood was obtained by venipuncture in heparinized tubes, the erythrocytes were isolated by centrifugation, and washed three times in 166 m*M* NaCl. For the preparation of resealed (pink) ghosts, the procedure of Bodemann and Passow as described by Wood and Passow[28] has been used, while for the white ghosts that described by Bjerrum has been used.[29] Finally, the erythrocytes or the ghosts were suspended in 150 m*M* NaCl, 5.5 m*M* glucose, 5 m*M* Hepes (pH 7.4), and 0.5% bovine serum albumin (BSA) at a cytocrit of 50%.

For polyacrylamide gel electrophoresis (PAGE) analysis, purified erythrocyte membranes (open ghosts) were prepared by mixing rapidly 1 mℓ of red blood cells into 20 mℓ of cold 5 m*M* sodium phosphate buffer (pH 8.0), followed by the preparation procedure of Dodge et al.[30] modified by Fairbanks et al.[31] After three or four washes in the above phosphate solution, followed each time by centrifugation at 20,000 *g* for 20 min, the white pellets are suspended in the same solution to a final concentration of about 4 mg protein per mℓ.

B. NMR Measurements of Water Diffusion

The principle of the NMR method[32] of measuring water exchange across membranes is described in Reference 2. This method relies upon the characterization of a system, consisting of two compartments, A and B, by two nuclear relaxation times, T_a and T_b, of the same type of nuclei residing in each of the compartments (Figure 1). The nuclear relaxation times are parameters that characterize the return to equilibrium after a suitable radiofrequency (rf) perturbation of the nuclei in an NMR experiment.[33] Long relaxation times are often associated with nuclei that are a part of molecules with fast motion. We will next assume that we deal with the same type of molecules distributed in compartments A and B and the corresponding NMR relaxation times differ, for some reason, so that $T_a \gg T_b$. Two cases can be considered: (1) as shown in Figure 1a, there is no exchange of molecules between the two compartments and (2) as shown in Figure 1b, there is a relatively fast exchange process transferring molecules between compartments. The question is how the exchange process will affect the relaxation times of the two compartments.

In case 1, the two relaxation times, T_a and T_b, will be detected in an actual NMR experiment (Figure 1). (In fact, even if there is a very slow exchange so that the nuclei will have time to relax in each compartment, the result will be much the same.) However, in case 2, the nuclei in A will start relaxing with T_a, but will end up in compartment B where its relaxation T_b will be faster. As a result, the observable relaxation time of phase A will be T'_a, which is shortened compared to T_a (Figure 1b). The faster the exchange, the shorter T'_a. The equations describing this phenomenon have been derived by Woessner,[34] enabling the calculation of exchange times.

Let us see further how this model can be applied to an erythrocyte suspension (Figure

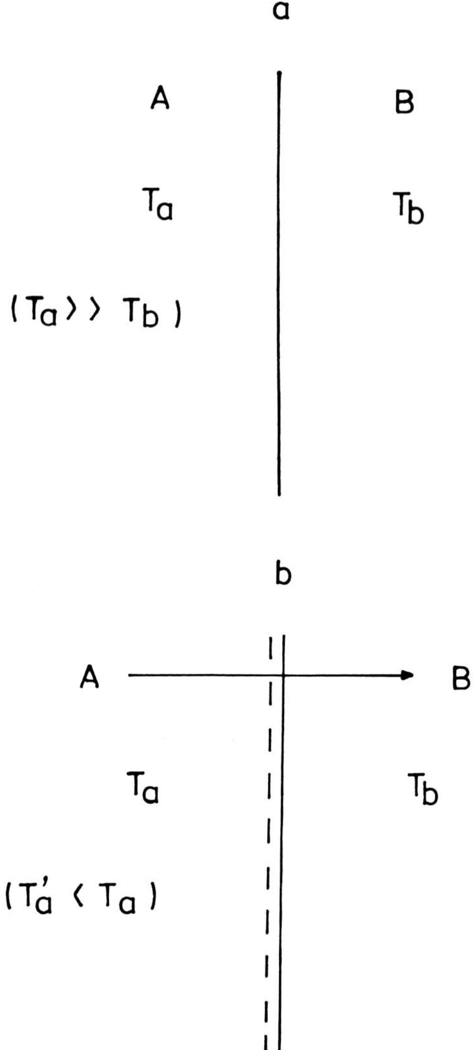

FIGURE 1. Two nuclear compartments having the same type of nucleus. (a) No exchange between compartments; the nuclei relax in each compartment with the relaxation time T_a and T_b, respectively (T_a is assumed to be much higher than T_b). (b) Fast exchange of nuclei from A to B. The observed relaxation of nuclei in compartment A, which also relax in B due to the exchange process, will be shortened compared to T_a.[2]

2). The nucleus of concern in this case is the water proton, which can reside either inside the cell (compartment A) or outside the cell (compartment B). There is an exchange of water molecules between these two compartments. However, if we perform an NMR experiment on such a system, we will only detect a single relaxation time. This is due to the fact that the relaxation times of the water protons in both compartments are not very different (of the order of hundreds of milliseconds), and the rapid exchange between compartments makes the distinction between the two compartments impossible (Figure 2A). Obviously, some way is needed of making $T_a \gg T_b$ as described above. One way of doing this is the method of paramagnetic doping. If we add a paramagnetic ion, such as manganese, to the cell suspension, then the proton relaxation time, T_b, of water molecules in the suspending solution will become much shorter by a mechanism known as electron-proton interaction (see, e.g.,

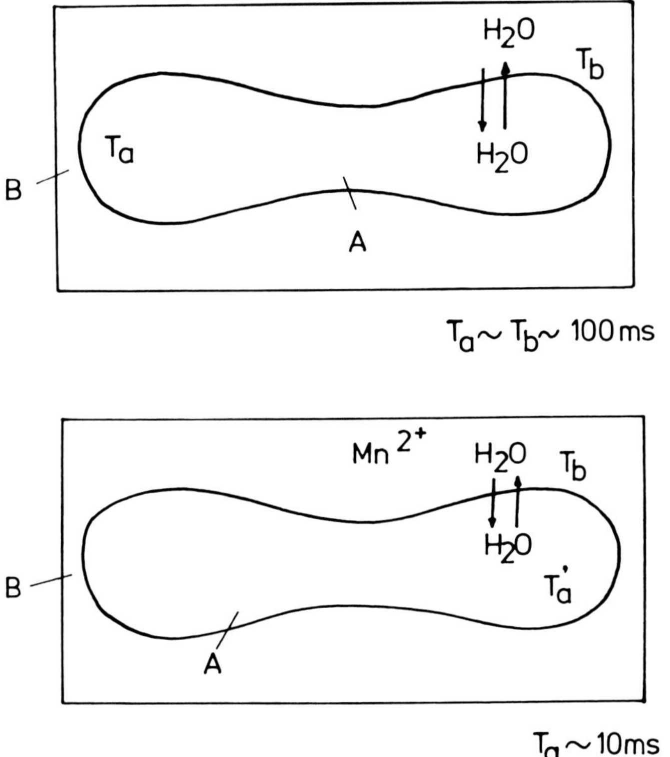

FIGURE 2. Illustration of the NMR paramagnetic doping method for measuring the water exchange through erythrocyte membranes. In each part of the figure A is the red blood cell compartment and B is the suspending solution compartment. (A) In a suspension of erythrocytes the relaxation times of water protons are similar in the two compartments ($T_a \sim T_b \sim 100$ msec) and fast exchange occurs between them. As a result, a single relaxation time is detected and therefore this experiment cannot be used for the measurement of water exchange. (B) If manganese ions are added to the suspending solution, then T_b becomes much shorter ($\sim$0.1 msec). In the absence of exchange between compartments A and B, T_a should remain unchanged ($\sim$100 msec). However, fast exchange of water occurs through the membrane, and T_a becomes T'_{2a} ($\sim$10 msec). The water exchange time can be calculated from T'_{2a} (see the text).[2]

Dwek[35] for an explanation). Then we will have a system with $T_a \gg T_b$ as described in Figure 1b, and the same type of experimental approach can be applied to erythrocytes. Of course, a prerequisite to this experiment is that the manganese ions do not penetrate the red blood cells. Fabry and Eisenstadt[36] showed that the penetration of manganese ions is hampered in the presence of albumin and we and others could not detect indeed any penetration of the Mn^{2+} into red blood cells.[19,37]

So far, the general term "nuclear relaxation time" has been used, whereas in reality there are two different relaxation time processes that can be measured: T_1, the so-called spin-lattice or longitudinal relaxation time, and T_2, the spin-spin or transversal relaxation time. Both of these relaxation times can be used for the determination of water exchange times.

The principle described above could be adapted to any other type of cell, and it is not necessarily restricted only to protons. If we refer to water, the resonance of ^{17}O in water has also been used to estimate water exchange through erythrocyte membranes.[38] In the case of water ^{17}O resonance, however, it is not necessary to use paramagnetic doping because the relaxation times in the two compartments are sufficiently different to distinguish them, and both are shorter than the exchange times.[8] The disadvantage of this method is its limited sensitivity to exchange rates over only a narrow temperature range.

The observed transversal relaxation time T'_{2a} of the erythrocytes is related to the individual relaxation times T_{2a} and T_{2b} and the lifetimes τ_a, τ_b by the following relationship:[34]

$$\frac{1}{T'_{2a}} = \frac{1}{2}\left\{\frac{1}{T_{2a}} + \frac{1}{T_{2b}} + \frac{1}{\tau_a} + \frac{1}{\tau_b} - \left[\left(\frac{1}{T_{2a}} - \frac{1}{T_{2b}} + \frac{1}{\tau_a} - \frac{1}{\tau_b}\right) + \frac{4}{\tau_a \cdot \tau_b}\right]^{1/2}\right\} \quad (1)$$

$1/\tau_a$ and $1/\tau_a$ are related to the populations of the phases by the relationship, $P_a/\tau_a = P_b/\tau_b$.

The significance of the parameters in Equation 1 is as follows: T_{2a} and T_{2b} are the water proton transversal relaxation times of the isolated cells and doped plasma, respectively; τ_a and τ_b are the lifetimes of the water proton in the corresponding compartments; P_a and P_b are the population fractions of water protons in the compartments (or nuclear phases as they are usually termed), therefore $P_a + P_b = 1$.

In order to calculate τ_a, we need to know the following parameters, which can be determined experimentally: T_a, T_b, T'_{2a}, and P_a or P_b.

The population parameter can be estimated by knowing the content of water in erythrocytes. The average fraction of intracellular volume represented by water is 0.71 to 0.72 and for plasma is 0.95 (Fabry and Eisenstadt;[36] Pirkle et al.[39]). This allows the calculation of P_a and P_b from the hematocrit value. However, a more precise knowledge of P_a should take into account the variations of the hematocrit value and hemoglobin content with age.[2,40]

Equation 1 refers to a T_2 transversal relaxation time experiment. The same expression holds for a T_1 longitudinal relaxation time.

If a relatively low concentration of manganese ions (1 to 2 m*M*) is used for doping, Equation 1 should be used for the calculation of τ_a.[39] However, if higher concentrations of manganese are used (20 to 50 m*M*), then Equation 1 is considerably simplified. This is due to the fact that $T_{2a} \gg T_{2b}$ and Equation 1 can be well approximated by the simple equation:

$$\frac{1}{T'_{2a}} = \frac{1}{T_{2a}} + \frac{1}{\tau_a} \quad (2)$$

The lifetime τ_a of the protons in the erythrocytes, which is the quantity of interest to us, was called by Conlon and Outhred[32] the water exchange time, T_e, while T_{2a} was called in our papers T_{2i} (the transverse relaxation time of the cell interior).

We have, therefore, Equation 3:

$$\frac{1}{T_e} = \frac{1}{T'_{2a}} - \frac{1}{T_{2i}} \quad (3)$$

The membrane permeability for water diffusion, P_b, is related to $1/T_e$, the cell water volume, V, and the cell surface area, A, by:

$$P_d = \frac{V}{A} \cdot \frac{1}{T_e} \quad (4)$$

Since different authors have used different values of V and A, in order to compare our results with previous ones, we have used two sets of values. On one hand, we have taken a value of 65 μm^3 for the intracellular solvent volume of erythrocytes and 86 μm^3 for that of resealed ghosts and a value of 1.42×10^{-6} for the membrane area, after Brahm.[13] These give V/A ratios of 4.58×10^{-5} cm and 6.06×10^{-5}cm for erythrocytes and ghosts, respectively. On the other hand, we used a slightly higher V/A ratio, after Dix and Solomon,[41] e.g., 5.33×10^{-5} cm for erythrocytes and the corresponding value for ghosts. In their initial work, Conlon and Outhred[32] diluted 1 mℓ of blood with 0.5 mℓ of doping solution containing 20 to 100 m*M* of manganese chloride. In later work,[42] 0.4 mℓ of whole blood was diluted with 0.9 mℓ of manganese solution (50 or 100 m*M* of manganese chloride made isotonic with sodium chloride). This was found necessary to keep the packed cell volume less than 20% in order to eliminate any dependence of relaxation time on the packed cell volume.

We have performed the measurements of T'_{2a} on washed erythrocytes or resealed ghosts suspended in 150 m*M* NaCl, 5 m*M* Hepes (pH 7.4) containing 0.5% albumin, and 5.5 m*M* glucose. The presence of albumin is important to prevent the entry of manganese ions into the cells. In our earlier experiments,[19] we have used 1 mℓ of blood or suspension of erythrocytes and 0.5 mℓ solution of 40 m*M* manganese chloride made isotonic with sodium chloride; the NMR measurements have been performed with a Bruker-Physik SXP spectrometer, at 90 MHz.

Now we are routinely using 0.2 mℓ of erythrocyte or ghost suspensions and 0.1 mℓ doping solution. The NMR measurements are now performed with an AREMI-78 spectrometer (manufactured by the Institute of Physics and Nuclear Engineering, Bucharest-Măgurele, Roumania) at a frequency of 25 MHz. The temperature is controlled to ± 0.2°C by air flow over an electrical resistance using the variable temperature unit attached to the spectrometer. The actual temperature in the sample was measured with a thermocouple connected to a microprocessor thermometer (Comark Electronics Limited, Rustington, Littlehampton, England). T'_{2a} is evaluated by the spin-echo method[33] using a computer unit coupled on line with the spectrometer.

T_{2i} was measured by the 90 to 180° method using the Carr-Purcell-Meiboom-Gill sequence,[33] on packed cells or ghosts from which the supernatant, with no added Mn, had been removed by centrifugation at $50,000 \times g$ for 60 min.

The inhibition of water diffusion across human red blood cell membranes was calculated assuming that the permeability coefficient is inversely related to T'_{2a}, according to the formula:[12]

$$\% \text{ Inhibition} = \frac{\dfrac{1}{T'_{2a}\text{ (control)}} - \dfrac{1}{T'_{2a}\text{ (sample)}}}{\dfrac{1}{T'_{2a}\text{ (control)}}} \times 100 \qquad (5)$$

C. Measurements of Binding of PCMBS to Membrane Proteins

Red blood cells or resealed ghosts, suspended at a cytocrit of 10% in 150 m*M* NaCl, 5 m*M* sodium phosphate buffer (pH 7.5), were incubated with ^{203}Hg-PCMBS under conditions as described in the legends to figures and tables. After incubation, the cells were pelleted by centrifugation and washed three times in the same medium. An aliquot of the final pellet was taken for radioactivity measurement with a gammacounter (Packard Multichannel Analyzer Model A 9012, Downers Grove, Illinois). The remaining pellet was lysed with 5 m*M* phosphate buffer (pH 8.0) and purified erythrocyte membranes were prepared as described above. An aliquot of the purified membrane was taken for radioactivity and protein concentration measurements and the remainder used for PAGE.

During PAGE, β-mercaptoethanol was omitted and 20 m*M* *N*-ethylmaleimide (NEM) was added to prevent the release of PCMBS and its subsequent binding to NEM binding sites. Membrane peptides were separated using the discontinuous SDS polyacrylamide gel system designed by Laemmli,[43] as previously described.[44] The slab gels used consisted of a running gel of 7.5% acrylamide and 5% stacking gel. The acrylamide to bis-acrylamide ratio was maintained at 36.5:1 in both the stacking and running gel. The slab gels were 16 cm × 14 cm × 0.1 cm. Running times were 1 hr at 70 V and then 3 hr at 125 V (18°C) in buffer (25 m*M* Tris, 192 m*M* glycine, and 0.1% SDS). Following electrophoresis, gels were cut into 2-mm slices and the radioactivity measured with the gammacounter. Parallel samples were stained overnight with a solution containing 0.075% Coomassie Brilliant Blue R 250, 45% methanol, and 10% acetic acid. Destaining was performed with 10% acetic acid followed by drying as previously described.[45] The proportion of various membrane peptides was calculated from densitometric scans obtained using a microdensitometer, MD 100, coupled to an automatic integrating recorder, K 201 (both manufactured by C. Zeiss, Jena, GDR).

III. CONDITIONS FOR INHIBITION OF WATER DIFFUSION IN ERYTHROCYTES AND GHOSTS

Comparative values of parameters characterizing the diffusional water permeability in erythrocytes and ghosts at various temperatures are listed in Table 1. The temperature values were chosen to enable comparison with values reported by other authors for measurements performed at various temperatures. It is obvious that for all temperatures, the values were higher in ghosts than in erythrocytes. However, when the permeability values were estimated it appeared that resealed ghosts have a permeability similar to erythrocytes. The longer values of T_e in ghosts are thus due to a higher intracellular solvent volume caused by the removal of hemoglobin.

We have also compared diffusional permeability of pink and white ghosts. No significant differences in their water permeability were found.

A variety of sulfhydryl reagents and chemical treatments have been tested for their effects on the water exchange time in erythrocytes (Figure 3). It can be seen that mercury-containing compounds are the only reagents acting as efficient inhibitors, in agreement with previous results:[9,12,13] none of the other sulfhydryl reagents, such as NEM, iodoacetamide (IAM), or 5,5′-dithiobis-2-nitro-benzoate (DTNB) either inhibited or prevented the inhibitory effect of a mercurial. This suggests that the sulfhydryl groups involved in water transport exhibit some specificity to mercurials, a finding that is important for evaluating the experiments aimed at associating water channels with specific membrane proteins using radioactive-sulfhydryl labeling methods.

When optimal conditions for inhibition are used, all mercurials: $HgCl_2$, *p*-chloromercuribenzoate (PCMB), mersalyl, PCMBS, and fluoresceinmercuric acetate (FMA) produce the same degree of inhibition, around 45%, which also corresponds to the maximal value of inhibition of the water permeability that can be obtained at 37°C. At lower temperatures, the degree of inhibition increases, reaching 87% at 6°C.[9] In all cases where FMA was present, the inhibition of water diffusion could not be reversed by a large excess of cysteine.[11]

Our observations of the irreversible inhibition of water transport by FMA provide further insight into the nature of the site reacting with the aromatic sulfhydryl reagents. A comparison of the structures of PCMBS and FMA (Figure 4) illustrates differences which may account for the greater potency of FMA and the irreversibility of its inhibition by cysteine. The FMA molecule can be divided into two almost symmetrical halves, each containing a reactive mercury atom adjacent to an aromatic ring. The inhibition by FMA suggests that there are a pair of SH groups in close proximity that react with the sulfhydryl reagents. This would offer an explanation for the greater potency of FMA and the failure of cysteine to reverse its inhibition, as FMA would bind more tightly and cysteine molecules would have to gain access to two SH groups to release the inhibitor. Such an hypothesis would suggest that to block water transport, the binding of two molecules of PCMBS may be required. From the time course of the inhibition of diffusion by PCMB and the effect of pH on this inhibition, Ashley and Goldstein[10] concluded that there were probably two SH reactive sites, one in a

Table 1
DIFFUSIONAL PERMEABILITY (P_d) OF THE HUMAN ERYTHROCYTES AND RESEALED GHOSTS[a]

Temperature (°C)	T_e (msec)	Erythrocytes P_d ($cm.sec^{-1} \times 10^3$) I	II	T_e (msec)	Resealed ghosts P_d ($cm.sec^{-1} \times 10^3$) I	II	Statistical significance of the differences T_e	P_d I	II
15	15.4 ± 1.8	2.8 ± 0.2	3.2 ± 0.3	23.4 ± 4.3	2.6 ± 0.1	3.0 ± 0.1	$p<0.001$	NS	NS
20	13.3 ± 1.3	3.4 ± 0.3	4.0 ± 0.5	19.3 ± 3.3	3.1 ± 0.1	3.6 ± 0.2	$p<0.001$	NS	NS
25	11.4 ± 1.3	4.7 ± 0.5	4.7 ± 0.5	17.0 ± 2.9	3.6 ± 2.9	4.2 ± 0.2	$p<0.001$	NS	NS
30	9.6 ± 0.8	4.8 ± 0.6	5.6 ± 0.8	13.5 ± 2.1	4.5 ± 0.2	5.2 ± 0.3	$p<0.001$	NS	NS
37	7.4 ± 0.7	6.2 ± 0.7	7.2 ± 0.9	10.4 ± 1.4	5.8 ± 0.3	6.8 ± 0.4	$p<0.001$	NS	NS

[a] The measurements have been performed on duplicate and triplicate blood sample from 12 donors as described in Section II. Results are expressed as mean ± standard deviation. The permeability was calculated from T_e using a V/A ratio of 4.58 ± 10^{-5} cm for erythrocytes and 6.06×10^{-5} cm for ghosts in column I, and the slightly higher value for the V/A ratio (e.g., 5.33×10^{-5} cm for erythrocytes) in column II. The statistical significance was calculated using unpaired *t*-test. NS = statistically not significant.

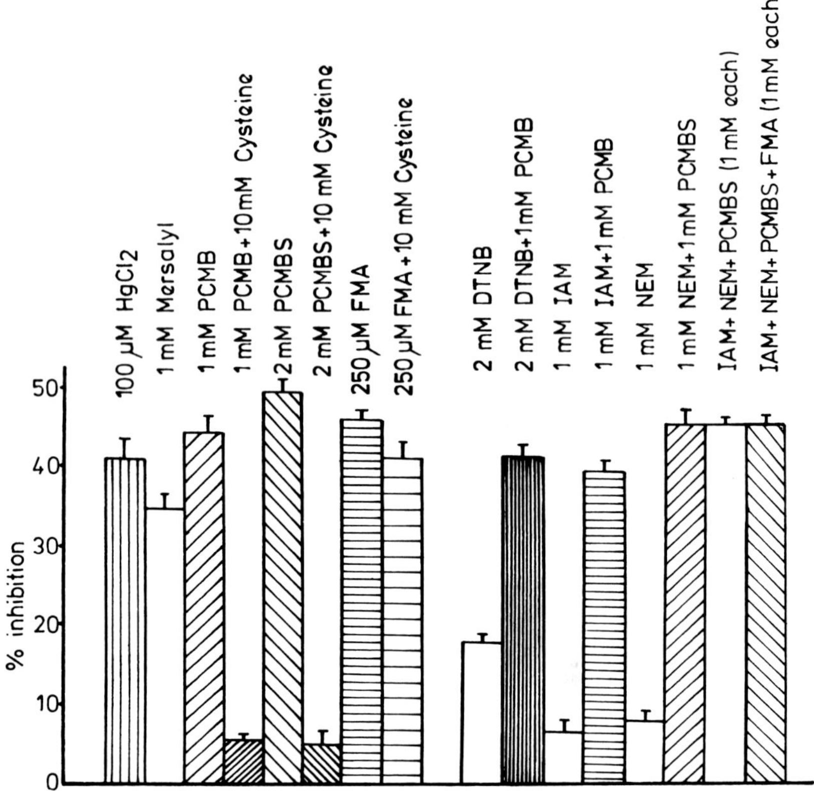

FIGURE 3. Inhibition of water diffusion in human erythrocytes by sulfhydryl reacting reagents. All incubations were performed for 60 min at 37°C at a hematocrit of 10% in 150 m*M* NaCl, 5 m*M* Hepes (pH 7.4), 5.5 m*M* glucose, except for mersalyl and cysteine, where incubations of 15 min at 37°C were used. After incubation, three washes in the same medium were performed. The results are the means (columns) and the standard deviation (bars on top of columns) for 4 to 50 determinations. Inhibition was calculated as described in Section II based on NMR measurements.

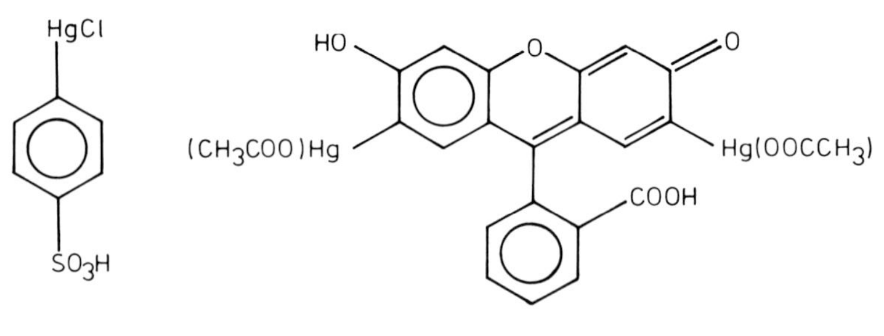

PCMBS

FMA

FIGURE 4. The structures of PCMBS and FMA.

hydrophobic environment and one in the anion channel. Whether there is any relationship between the two SH groups suggested by FMA inhibition and these two reactive sites is not yet known. An alternative explanation is that the larger aromatic ring structure of FMA produces a higher affinity for a sulfhydryl group at the active site.

The time course of development of the water diffusion inhibition by 1 m*M* PCMBS on erythrocytes at 20°C is shown in Figure 5a. It appears that inhibition develops in a two-step manner: the first rapid step is reached in 15 min and this amounts to about half of the maximal inhibition. The inhibition rate then slows, reaching maximum inhibition after 60 min. Similar kinetics of inhibition of osmotic permeability are described by Naccache and Sha'afi.[46] With longer incubation times, some hemolysis occurred, giving an apparent fall in the degree of inhibition. The inhibition is dependent not only on the temperature and time of exposure of erythrocytes to the mercurial, but also on preincubation of the cells with the noninhibitory sulfhydryl reagents. As shown in Figure 2, the maximal inhibition of water exchange could be induced at 37°C in 15 to 30 min by incubating erythrocytes with 0.2 to 0.5 m*M* PCMBS *after* a preincubation with NEM, or in 30 to 60 min with 1.0 to 2.0 m*M* PCMBS if *no* preincubation with NEM was employed. This indicates that treatment of erythrocytes with NEM prior to exposure to PCMBS results in the inhibition of water exchange occurring faster and at a lower concentration of mercurial.

The two-step manner of the development of water inhibition induced by PCMBS, in agreement with similar features of osmotic permeability described by Naccache and Sha'afi,[46] may correspond to two populations of sulfhydryl groups which participate in the control of water transport, differing in location. Some would appear to be located close to the outer surface of the membrane, being readily accessible to PCMBS, and others located deeper in the membrane interior. Different populations of membrane sulfhydryl groups have been described by other authors using various techniques.[47]

The time course of the inhibition induced by PCMBS on resealed ghosts was studied at 0 and 37°C (Figure 6). At 0°C, no significant inhibition occurred up to 30 min of incubation. At 37°C, at least 15 min of incubation were necessary for a significant inhibitory effect to occur, and the maximal inhibition was obtained in 30 min with 0.1 m*M* PCMBS, without preincubation with NEM. If a NEM preincubation is used, a nearly maximal inhibitory effect occurs in 5 min at 37°C with 0.1 m*M* PCMBS.

When the reversibility by cysteine of the PCMBS-induced inhibition of water diffusion in ghosts was studied it appeared that more than a tenfold excess of cysteine is required to remove the inhibition. This is again in contrast with studies on erythrocytes where 10 m*M* cysteine fully reversed the inhibition induced by 1 m*M* PCMBS.[12,46] It may be concluded that PCMBS is approximately ten times more potent in inhibiting water diffusion in ghosts compared to erythrocytes. This can be due to the absence of hemoglobin, which probably binds the excess of PCMBS in erythrocytes.

In order to better understand the development of water diffusion inhibition induced by PCMBS and the reason for the potentiating effects of NEM preincubation, we have studied the uptake of ^{203}Hg-PCMBS by erythrocytes and its binding to the membrane. Uptake of the PCMBS is slightly slower in cells preincubated with NEM compared to those with no exposure to NEM (Figure 7), but exposure to NEM prior to PCMBS actually enhances the binding of the inhibitor. With no NEM preincubation, PCMBS binding ranges from 3 nmol/mg protein (an amount that is bound within 2 min at 20°C, when very little inhibition of water diffusion occurs) to a constant 7 nmol PCMBS per mg protein after 60 min of incubation. In contrast, following incubation with NEM the binding of PCMBS continuously increases over the 30- to 80-min period of time explored. NEM increases the yield of binding (expressed as %PCMBS to the membranes to the total amount of PCMBS captured by the red blood cells).[27] The results indicate that the potentiating effect of NEM on PCMBS-induced inhibition of water diffusion may be a result of an enhanced binding of PCMBS to the membrane.

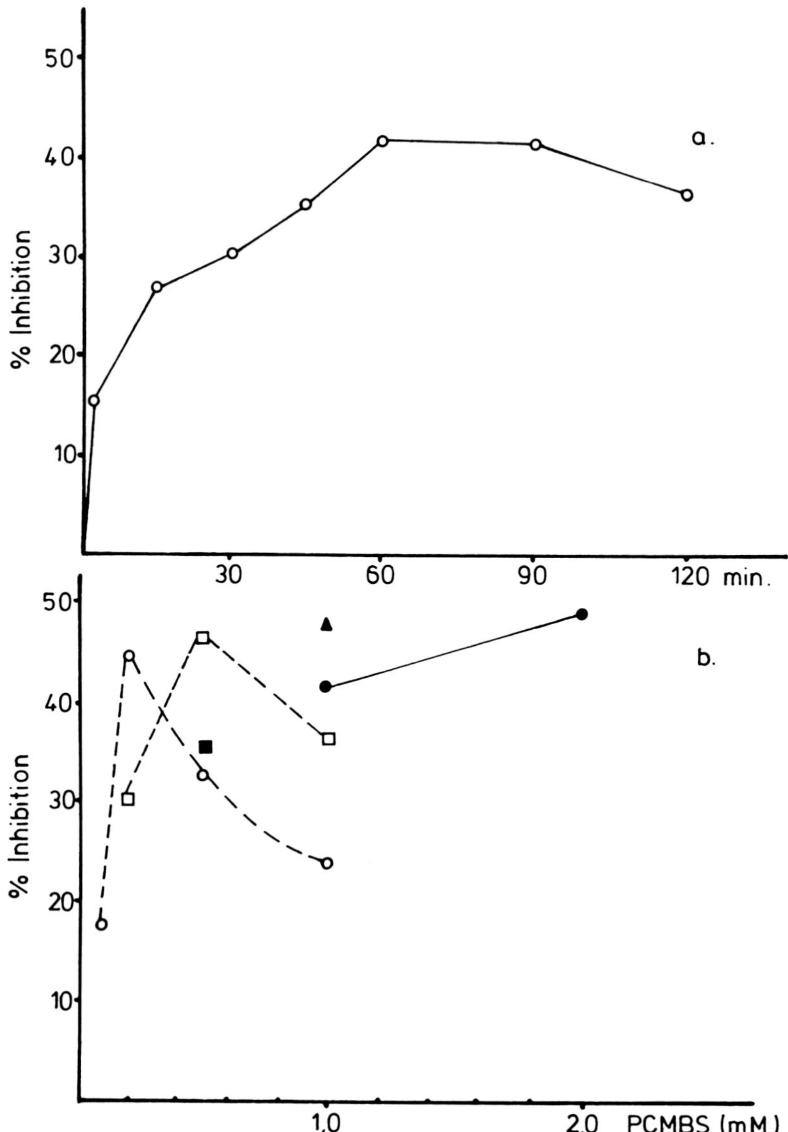

FIGURE 5. (a) The time course of inhibition of water diffusion in red blood cells induced by 1 m*M* PCMBS at 20°C. Red blood cells were incubated for various durations at a hematocrit of 10% and washed as in Figure 3. The inhibition was calculated on the basis of NMR measurements as described in Section II. (b) Dependence of the inhibition of water diffusion in erythrocytes induced by various concentrations of PCMBS at 37°C upon a preincubation with NEM, a noninhibitory SH reagent. For samples marked by empty symbols erythrocytes were preincubated for 60 min at 25°C with 2 m*M* NEM at a hematocrit of 25% in 150 m*M* NaCl, 5 m*M* sodium phosphate buffer (pH 7.5); they were then diluted with medium containing 2 m*M* NEM to a hematocrit of 10% with PCMBS to give the final concentrations indicated and incubated for 15 min (□) or 30 min (○) at 37°C. Other samples were incubated only with PCMBS (and no NEM) for 15 min (■), 30 min (●), or 60 min (▲) at 37°C. After incubation, three washes of erythrocytes in the same medium containing 2 m*M* NEM were performed and the inhibition was calculated on the basis of NMR measurements as described in Section II.

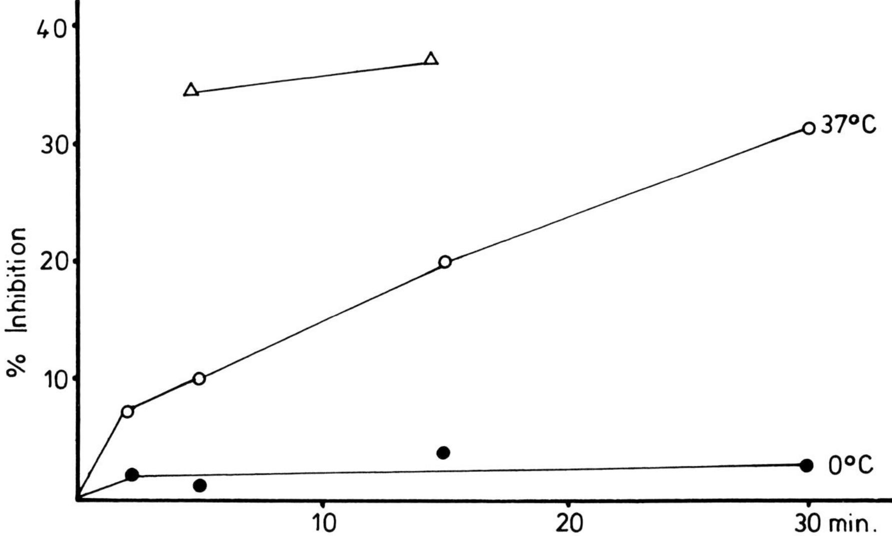

FIGURE 6. The time course of inhibition of water diffusion in resealed ghosts induced by 0.1 m*M* PCMBS at 0°C (●—●) and 37°C without (○—○) and with (△—△) a preincubation with 1 m*M* NEM. Resealed ghosts were incubated at a cytocrit of 10% with 0.1 m*M* PCMBS in 150 m*M* NaCl, 5 m*M* Hepes (pH 7.4), and 5.5 m*M* glucose.

The actual binding of PCMBS to the membrane appears to be strongly dependent not only on the duration, but also upon the temperature of incubation of the erythrocytes and resealed ghosts (Figure 8). For erythrocytes an incubation time of 15 min with 0.5 m*M* PCMBS results in a binding ranging from 3 nmol/mg protein (at 0°C) to 19 nmol/mg of protein (at 37°C), while for an incubation time of 30 min, the amount of PCMBS is higher and increases with temperature from 7 nmol/mg of protein at 20°C to 11 nmol at 27°C and 27 nmol/mg of protein at 37°C. For longer incubation times at 37°C, the binding further increases, reaching 37 nmol/mg of protein in 90 min.

The total amount of sulfhydryl groups in erythrocytes titrated by mercury has been reported as 4.1×10^{-15} nmol per cell.[48,49] Out of these, membrane sulfhydryl groups constitute $< 5\%$ with an equal quantity attributable to reduced glutathione. The remaining groups are predominantly associated with hemoglobin.[49] Using a conversion factor of 1 mg protein being equivalent to 1.1×10^9 cells,[48,49] the amount of membrane sulfhydryl groups found by us turned out to be 165 nmol/mg protein. Alternatively, using the conversion factors of Dodge et al.[30] or Lepke et al.[50] of 1 mg protein equivalent to 1.41×10^9 cells or 1.95×10^9 cells, respectively, then the sulfhydryl group concentrations in the membrane would be 129 nmol/mg or 93 nmol/mg protein. Reported values for this concentration vary from 89 to 144 nmol/mg protein.[47]

PCMBS reacts with only 25% of these mercurial-titratable groups in hemoglobin-free ghosts[51] which is in close agreement with the value of 37 nmol/mg protein found in the present experiments. Sutherland et al.[49] report that the binding of PCMBS to intact erythrocytes is associated with an inhibition of glucose uptake, a loss of K^+, and decreased osmotic fragility. The effect of mercurials on K^+ loss of inhibition of glucose uptake increases with mercurial concentration to a maximum and then declines similar to the effect of PCMBS on the inhibition of water diffusion reported here. The "recovery" has been attributed[49] to a desorption of mercury from the membrane due to competition with soluble thiol substances (hemoglobin, glutathione) released into the medium by ruptured cells. Indeed we noticed a degree of hemolysis occurring in parallel with the decrease in inhibition of water diffusion at longer than optimal incubation times or higher concentrations of PCMBS.

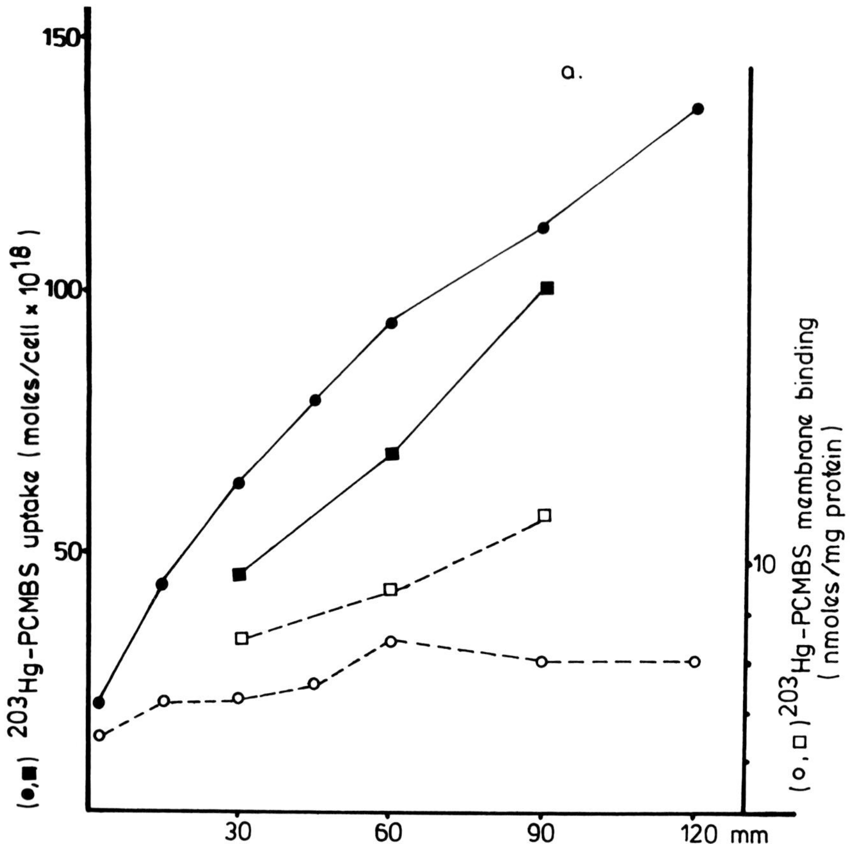

FIGURE 7. The time course of the uptake of ^{203}Hg-PCMBS by red blood cells (solid lines) and the binding to the membrane proteins (dotted lines) after incubations at 20°C with 1 m*M* ^{203}Hg-PCMBS without (○,●) or after a preincubation at 20°C for 60 min with 1 m*M* NEM (□,■).

When PCMBS is used on erythrocyte samples preincubated with NEM, inhibition of water diffusion occurs when only approximately 10 nmol of PCMBS are bound per mg protein. Subtracting the amount of PCMBS bound at 0°C in 15 min, when no inhibition of water diffusion occurs, the minimum number of sulfhydryl groups apparently involved in water diffusion would be 7 nmol/mg protein.

IV. IDENTIFICATION OF MEMBRANE PROTEINS INVOLVED IN THE WATER PERMEABILITY OF HUMAN ERYTHROCYTES

A problem in using radioactive probes to identify transport components is that of recognizing a relatively small number of specific transport sites against a large background of nonspecific sites. Accordingly, in the present investigation, procedures were designed to minimize the number of nonspecific sites. A relatively high concentration of NEM was used for preincubation of erythrocytes and resealed ghosts before and during the treatment with ^{203}Hg-PCMBS. NEM was also present in the washing medium after incubation with mercurial. To avoid displacing the mercurial during electrophoresis, disulfide-reducing agents were omitted from the sample while NEM was added. This avoided the possibility that some thiol groups, not capable of reacting with either reagent in the absence of detergent undergo some exchange on adding SDS.[52]

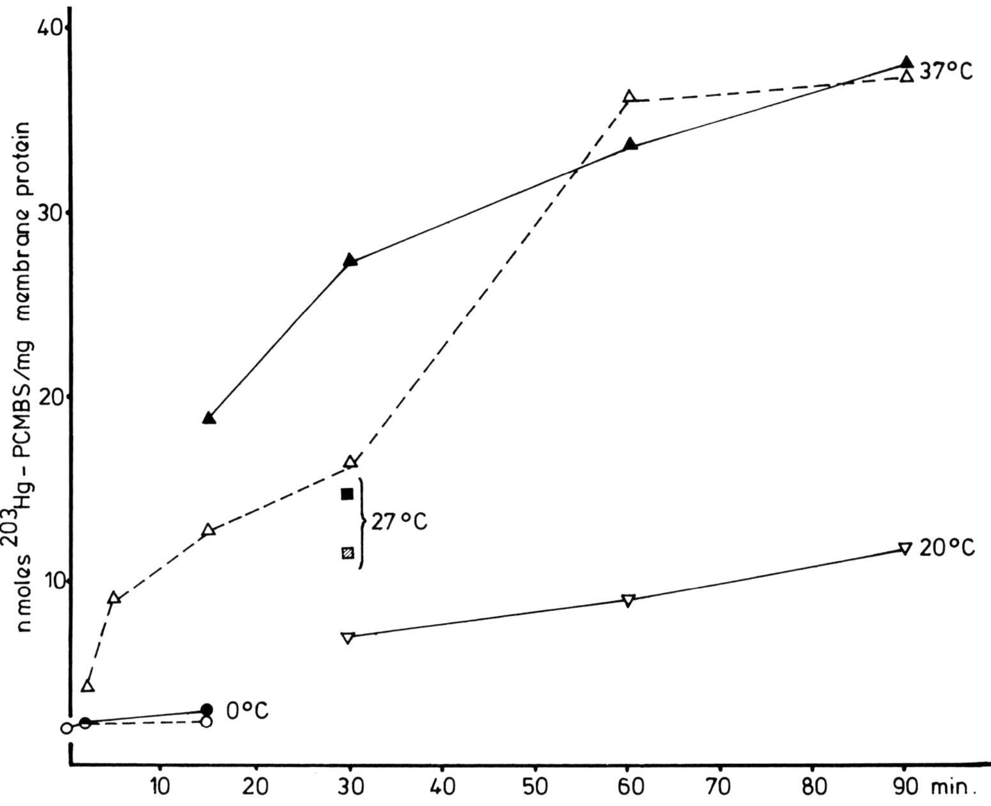

FIGURE 8. The binding of ^{203}Hg-PCMBS to erythrocyte membranes in various conditions of incubation. Washed erythrocytes or resealed ghosts were suspended in a wash medium (150 m*M* NaCl, 5 m*M* phosphate buffer, and pH 7.5) containing 2 m*M* NEM at a cytocrit of 25% and incubated for 60 min at 25°C. They were then diluted with the same medium containing NEM to a cytocrit of 10% with ^{203}Hg-PCMBS added to erythrocytes to give final concentrations of 0.5 m*M* (▨ ,▲,●) and 1 m*M* (■,▽) and to resealed ghosts of 0.1 m*M* (○,△) and incubated at the temperatures indicated. After completion of the incubation, resealed ghosts and erythrocytes were washed three times in 20 volumes of 150 m*M* NaCl, 5 m*M* sodium phosphate (pH 7.5), and 2 m*M* NEM, by centrifugation at 8000 and 2000 *g*, respectively, for 20 min at 4°C. Purified membranes were prepared from the intact erythrocytes and resealed ghosts to remove ^{203}Hg-PCMBS that may have bound to hemoglobin and other cytoplasmic components. Other details are described in Section II.

Taking these precautions it was interesting to find that under conditions of inhibition of water diffusion the ^{203}Hg-PCMBS labeling pattern of membrane proteins revealed significant binding of the inhibitor only to polypeptides migrating as band 3, band 4.2, and band 4.5. Such a pattern of labeling obtained with resealed ghosts incubated at 37°C for 5 min is shown in Figure 9. A similar pattern was obtained with erythrocytes incubated at 37°C for 15 min.[26]

These conditions are the same as those under which maximal inhibition of water diffusion occurs and which a *minimal* amount of ^{203}Hg-PCMBS is bound per milligram of membrane protein.

The distribution of radioactivity in the various polypeptide fractions of resealed ghosts labeled with ^{203}Hg-PCMBS, under conditions to block nonspecific binding with NEM, is presented in Table 2. Assuming values of molecular weights of 95,000 for band 3 and 55,000 for band 4.5,[53] it was possible to estimate the amount of PCMBS bound per mole of polypeptide. Under conditions of inhibition of water diffusion, when the major part of the radioactivity is distributed in band 3 and 4.5, we find approximately 1 mol PCMBS bound per mole of polypeptide.

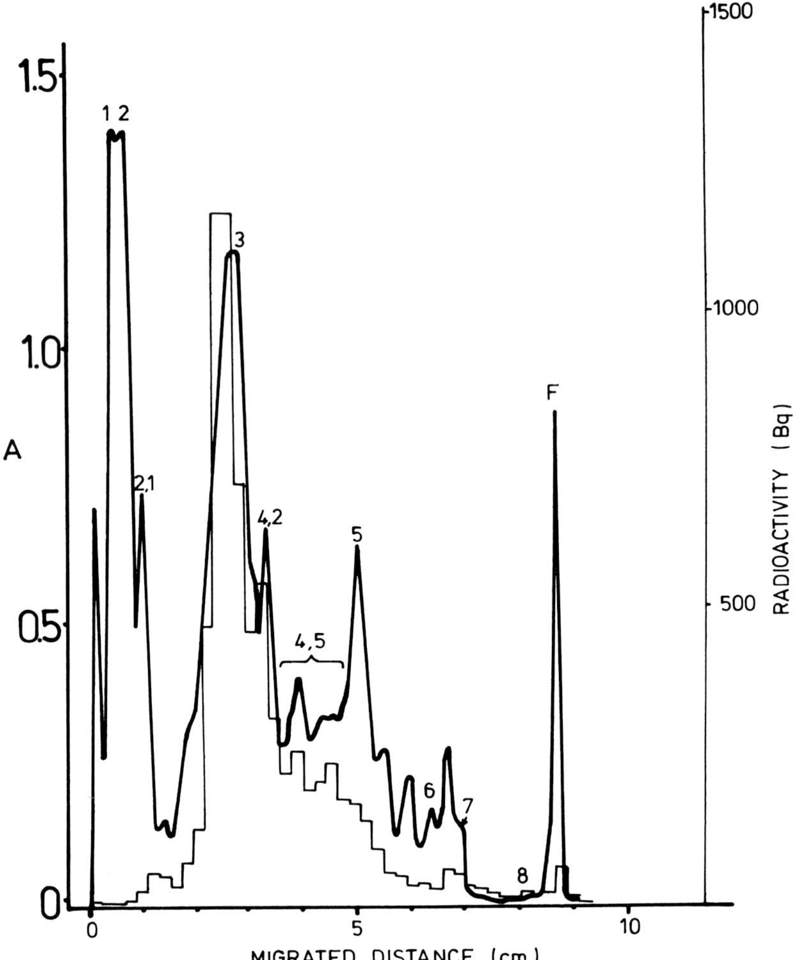

FIGURE 9. The binding of ^{203}Hg-PCMBS to proteins in resealed ghosts resolved by polyacrylamide gel electrophoresis. After a preincubation with 2 m*M* NEM for 60 min at 25°C resealed ghosts were incubated in the same medium containing 0.1 m*M* ^{203}Hg-PCMBS for 5 min at 37°C or 15 min at 0°C. Purified membranes were prepared as described in the legend to Figure 5. Membrane polypeptides were separated by electrophoresis, gels were cut into 2-mm slices and the radioactivity measured as described in Section II. Densitometric scans of Coomassie blue-stained gels are shown by the continuous tracing. The radioactivity, illustrated by the bar graphs, is the difference for the incubation of resealed ghosts for 5 min at 37°C (i.e., when a maximal inhibition of water diffusion occurs) and for 15 min at 0°C (when no inhibition is noticed). The nomenclature derived is used to identify membrane proteins, and F represents the migration of the tracking dye. (From Fairbanks, G., Steck, T. L., and Wallach, D. F. H., *Biochemistry,* 10, 2606, 1971. With permission.)

Brown et al.[14] were among the first to report the results of labeling experiments using ^{14}C-DTNB after preincubation of the cells with NEM and IAM. A binding of ^{14}C-DTNB to the band 3 protein was found and they suggested that band 3 is involved in water transport, on the assumption that DTNB is an inhibitor of this process. Although Naccache and Sha'afi[46] report an inhibition of osmotic permeability by DTNB, work in other laboratories as well as our findings, show no inhibition of water diffusion by DTNB. Later work by Sha'afi and Feinstein[15] presented evidence for selective labeling of band 3 with ^{14}C-PCMBS after preincubation of erythrocytes with IAM, NEM, and mersalyl, compounds that were considered not to inhibit water transport. However, as shown by us, mersalyl is a strong inhibitor of

Table 2
DISTRIBUTION OF PROTEIN AND SH GROUPS AMONG MEMBRANE POLYPEPTIDES OF RESEALED GHOSTS INCUBATED WITH NEM AND ^{203}Hg-PCMBS UNDER CONDITIONS OF NO INHIBITION (A) OR MAXIMAL INHIBITION (B) OF WATER DIFFUSION

Band number	% Protein		% Radioactivity		Number of SH groups per protein molecule	
	A	B	A	B	A	B
1 + 2	23.0	21.0	1.5	0.7	0.02	0.06
2.1	10.7	8.9	1.0	1.9	0.02	0.35
3	28.9	27.2	24.3	46.8	0.08	1.38
4.1	2.9	2.5	4.7	6.2	0.14	1.66
4.2	4.5	6.0	3.6	11.2	0.06	1.14
4.5	9.3	10.7	19.2	19.9	0.12	0.86
5	8.0	9.0	8.8	3.7	0.05	0.15
6	1.2	1.3	4.7	1.2	0.14	0.26
7	1.1	1.0	8.7	2.4	0.23	0.61
8	0.5	0.3	3.6	0.5	0.18	0.35

Note: Resealed ghosts were incubated for 60 min at 25°C with 2 m*M* NEM in 150 m*M* NaCl, 5 m*M* phosphate buffer, pH 7.5, at a cytocrit of 25%. They were then diluted with the same medium containing NEM to a cytocrit of 10% with ^{203}Hg-PCMBS added to give a final concentration of 0.1 m*M*. After an incubation of 5 min at 0°C (A) or 37°C (B), the ghosts were washed three times in 20 volumes of 150 m*M* NaCl, 5 m*M* sodium phosphate (pH 7.5), and 2 m*M* NEM, by centrifugation at 8000 *g* for 10 min at 4°C. Purified membranes were prepared to remove ^{203}Hg-PCMBS that may have bound to hemoglobin and other cytoplasmic components and fractionated by polyacrylamide gel electrophoresis. Other details are described in Section II. Total binding of ^{203}Hg-PCMBS was 1.11 nmol/mg of protein in experiment A and 8.37 nmol/mg of protein in experiment B.

diffusional water permeability. More recently, Solomon et al.[16] have reported the localization of radiolabeled PCMBS on band 3 following incubation with human erythrocyte ghosts at 0°C for 2 min. In the present experiments, we also find that a small amount of PCMBS binds under these conditions, but longer incubation times or higher temperatures are needed before binding can be quantitatively associated with any inhibition of water diffusion.

The data presented above clearly show that when conditions of maximal inhibition of water diffusion are chosen, together with minimal PCMBS binding to the membrane, the mercurial binds to band 3 and the polypeptides in band 4.5. These findings strongly point to the proteins in either or both of these bands to be associated with water channels in erythrocytes. Using an average conversion factor of 1 mg protein corresponding to 1.5×10^9 cells, the 7 nmol/mg protein of sulfhydryl groups associated with water diffusion would correspond to approximately 2.8×10^6 sulfhydryl groups per cell.

The reactivity of sulfhydryl groups in band 3 protein has been studied by a number of authors,[16,54-56] and there seems to be a general agreement that this protein has six groups, five of which are reactive to NEM, and the sixth reactive only to mercurials such as PCMBS. Indeed, in the present work, under conditions where all the NEM reactive sulfhydryl groups in band 3 are blocked, 1 mol of PCMBS binds per mole of band 3. The number of these monomers per cell is considered to be 1.2×10^6.[57] Consequently, out of the 2.8×10^6 sulfhydryl groups per cell involved in inhibition of water diffusion, approximately half are binding to proteins other than band 3. The other sulfhydryl groups appear to be associated with band 4.2 and 4.5. The former is an extrinsic protein located in the cytoplasmic region of band 3[53] and is reported to be highly reactive to sulfhydryl reagents.[56] Under the conditions of inhibition of water diffusion used in our work, about 1 mol of PCMBS binds per mol of protein. Since this protein occurs as 0.23×10^6 copies per cell, an equivalent number of sulfhydryl groups could be accounted for in PCMBS binding. Band 4.1, another extrinsic protein in contact with the intrinsic membrane domain,[53] occurs at the same frequency and also seems to bind 1 mol PCMBS per mol of protein. However, this binding could be a reflection of its close proximity on the gels to both bands 3 and 4.2. The remaining PCMBS binding protein is band 4.5 which occurs at about 0.9×10^6 copies per cell. Again, under conditions of inhibition of water diffusion, this band binds approximately 1 mol PCMBS per mol protein. It can be seen, therefore, that there is good agreement between the total number of sulfhydryl groups reactive to ^{203}Hg-PCMBS when water diffusion is inhibited and the number of copies of the individual polypeptides binding the reagent.

V. CONCLUSION

It could be assumed that only intrinsic proteins crossing the lipid bilayer would provide a transmembrane pathway for facilitated water diffusion. The labeling of bands 4.1 and 4.2 may be explained mainly by their association with band 3, the main route of PCMBS permeation into the cell.[13,58] The present results clearly associate this protein, together with the polypeptides migrating in the region of band 4.5, as the intrinsic protein components involved in water channels. Band 4.5 has previously been identified with other transport functions, notably the transport of glucose.[59,60] Reconstitution studies using purified bands 3 and 4.5 incorporated into liposomes may be useful in order to clarify further their role in water transport. A better understanding of this transport process in human erythrocytes is warranted by reports that water permeability is decreased in erythrocytes from individuals with epilepsy,[18] Duchenne muscular dystrophy,[24] and in red blood cells of the McLeod phenotype.[61]

ACKNOWLEDGMENTS

The collaboration of Drs. Victoria Borza, Ana Muresan, Adriana Hodârnău (Medical and Pharmaceutical Institute Cluj-Napoca, Roumania), Ross P. Holmes (Burnsides Research Laboratory, University of Illinois at Urbana-Champaign), and John M. Wrigglesworth (Department of Biochemistry, King's College, U.K.) in part of the work reviewed above, as well as the financial support of the Ministry of Education and Teaching, Academy of Medical Sciences (Romania) and The Wellcome Trust (U.K) are gratefully acknowledged.

REFERENCES

1. **Sha'afi, R. I.,** Permeability for water and other polar molecules, in *Membrane Transport,* Bonting, S. L. and De Pont, J. J. H. H. M., Eds., Elsevier/North Holland, Amsterdam, 1981, 29.
2. **Morariu, V. V. and Benga, Gh.,** Water diffusion through erythrocyte membranes in normal and pathological subjects: nuclear magnetic resonance investigations, in *Membrane Processes. Molecular Biology and Medical Applications,* Benga, Gh., Baum, H., and Kummerow, F. A., Eds., Springer-Verlag, Berlin, 1984, 121.
3. **Träuble, H.,** The movement of molecules across lipid membranes: a molecular theory, *J. Membr. Biol.,* 4, 193, 1971.
4. **Macey, R. I.,** Transport of water and urea in red blood cells, *Am. J. Physiol.,* 246, 195, 1984.
5. **Vieira, F. L., Sha'afi, R. I., and Solomon, A. K.,** The state of water in human and dog red blood cell membranes, *J. Gen. Physiol.,* 55, 451, 1970.

6. **Cass, A. and Finkelstein, A.,** Water permeability of thin lipid membranes, *J. Gen. Physiol.*, 50, 1765, 1967.
7. **Cabantchik, Z. I., Knauf, P. A., and Rothstein, A.,** The anion transport system of the red blood cell. The role of membrane protein evaluated by the use of "probes", *Biochim. Biophys. Acta*, 515, 239, 1978.
8. **Cabantchik, Z. I. and Rothstein, A.,** Membrane proteins related to anion permeability of human red blood cells. I. Localization of disulfonic stilbene binding sites in proteins involved in permeation, *J. Membr. Biol.*, 15, 207, 1974.
9. **Macey, R. I. and Farmer, R. E. L.,** Inhibition of water and solute permeability in human red cells, *Biochim. Biophys. Acta*, 211, 104,1970.
10. **Ashley, D. L. and Goldstein, J. H.,** Time dependence of the effect of *p*-chloromercuribenzoate on erythrocyte water permeability: a pulsed nuclear magnetic resonance study, *J. Membr. Biol.*, 61, 199, 1981.
11. **Benga, Gh., Pop, V. I., Ionescu, M., Holmes, R. P., and Popescu, O.,** Irreversible inhibition of water diffusion through erythrocyte membranes by fluoresceinmercuric acetate, *Cell Biol. Int. Rep.*, 6, 775, 1982.
12. **Benga, Gh., Pop, V. I., Popescu, O., Ionescu, M., and Mihele, V.,** Water exchange through erythrocyte membranes: nuclear magnetic resonance studies of the effects of inhibitors and of chemical modification of human membranes, *J. Membr. Biol.*, 76, 129, 1983.
13. **Brahm, J.,** Diffusional water permeability of human erythrocytes and their ghosts, *J. Gen. Physiol.*, 79, 791, 1982.
14. **Brown, P. A., Feinstein, M. B., and Sha'afi, R. I.,** Membrane proteins related to water transport in human erythrocytes, *Nature (London)*, 254, 523, 1975.
15. **Sha'afi, R. I. and Feinstein, M. D.,** Membrane water channels and SH-groups, in *Membrane Toxicity*, Miller, M. W. and Shamoo, A. E., Eds., Plenum Press, New York, 1977, 67.
16. **Solomon, A. K., Chasan, B., Dix, J. A., Lukacovic, M. F., Toon, M. R., and Verkman, A. S.,** The aqueous pore in the red cell membrane: band 3 as a channel for anions, cations, nonelectrolytes and water, in *Biomembranes and Cell Function*, Kummerow, F. A., Benga, Gh., and Holmes, R. P., Eds., Ann. N. Y. Acad. Sci., 414, 97, 1983.
17. **Lukacovic, M. F., Verkman, A. S., Dix, J. A., and Solomon, A. K.,** Specific interaction of a water transport inhibitor, PCMBS with band 3 in red blood cell membranes, *Biochim. Biophys. Acta*, 778, 253, 1984.
18. **Benga, Gh. and Morariu, V. V.,** Membrane defect affecting water permeability in human epilepsy, *Nature (London)*, 265, 636, 1977.
19. **Morariu, V. V. and Benga, Gh.,** Evaluation of a nuclear magnetic resonance technique for the study of water exchange through erythrocyte membranes in normal and pathological subjects, *Biochim. Biophys. Acta*, 469, 301, 1977.
20. **Morariu, V. V., Pop, V. I., Popescu, O., and Benga, Gh.,** Effects of temperature and pH on the water exchange through erythrocyte membranes: evidence for state transitions, *J. Membr. Biol.*, 62, 1, 1981.
21. **Benga, Gh., Popescu, O., Holmes, R. P., and Pop, V. I.,** NMR studies on the mechanism of water diffusion through human erythrocyte membranes, *Bull. Magn. Resonance*, 5, 265, 1983.
22. **Benga, Gh., Popescu, O., Pop, V. I., Holmes, R. P., Pavel, T., and Ionescu, M.,** Modifications of human erythrocyte membranes and their effect on water permeability studied by a nuclear magnetic resonance technique, in *Water and Ions in Biological Systems*, Pullmann, A., Vasilescu, V., and Packer, L., Eds., Plenum Press, New York, 1985, 303.
23. **Benga, Gh., Popescu, O., and Pop, V. I.,** Water exchange through erythrocyte membranes: *p*-chloromercuribenzene sulfonate inhibition of water diffusion in ghosts studied by a nuclear magnetic resonance technique, *Biosci. Rep.*, 5, 223, 1985.
24. **Serbu, A. M., Marian, A., Popescu, O., Pop, V. I., Borza, V., Benga, I., and Benga, Gh.,** Decreased water permeability of erythrocyte membranes in patients with Duchenne muscular dystrophy, *Muscle & Nerve*, 9, 243, 1986.
25. **Benga, Gh., Borza, V., Popescu, O., Pop, V. I., and Muresan, A.,** Water exchange through erythrocyte membranes: nuclear magnetic resonance studies on resealed ghosts compared to human erythrocytes, *J. Membr. Biol.*, 89, 127, 1986.
26. **Benga, Gh., Popescu, O., Pop, V. I., and Holmes, R. P.,** *p*-chloromercuribenzene sulfonate binding by membrane proteins and the inhibition of water transport in human erythrocytes, *Biochemistry*, 25, 1535, 1986.
27. **Benga, Gh., Popescu, O., Borza, V., Pop, V. I., Muresan, A., Mocsy, I., Brain, A., and Wrigglesworth, J.,** Water permeability of human erythrocytes: identification of membrane proteins involved in water transport, *Eur. J. Cell. Biol.*, 41, 252, 1986.
28. **Wood, P. G. and Passow, H.,** Techniques for the modification of the intracellular composition of red blood cells, in *Techniques in Cellular Physiology*, Part 1, Munksgaard, Copenhagen, 1981, 1.
29. **Bjerrum, P. J.,** Hemoglobin-depleted human erythrocyte ghosts: characterization of morphology and transport functions, *J. Membr. Biol.*, 48, 43, 1979.

30. **Dodge, J. T., Mitchell, C., and Hanahan, D. J.,** The preparation of haemoglobin free ghosts of human erythrocytes, *Arch. Biochem. Biophys.*, 100, 119, 1963.
31. **Fairbanks, G., Steck, T. L., and Wallach, D. F. H.,** Electrophoretic analysis of the major polypeptides of the human erythrocyte membrane, *Biochemistry*, 10, 2606, 1971.
32. **Conlon, T. and Outhred, R.,** Water diffusion permeability of erythrocytes using an NMR technique, *Biochim. Biophys. Acta*, 288, 354, 1972.
33. **Farrar, Th. C. and Becker, E. D.,** *Pulse and Fourier Transform NMR*, Academic Press, New York, 1971.
34. **Woessner, D. E.,** Temperature dependence of nuclear-transfer and spin-relaxation phenomena of water adsorbed on silica gel, *J. Chem. Phys.*, 39, 2783, 1963.
35. **Dwek, R. A.,** *Nuclear Magnetic Resonance in Biochemistry: Applications to Enzyme Systems*, Clarendon Press, Oxford, 1973.
36. **Fabry, M. E. and Eisenstadt, M.,** Water exchange between red cells and plasma. Measurement by nuclear magnetic relaxation, *Biophys. J.*, 15, 1101, 1975.
37. **Yoon, S. C., Toon, M. R., and Solomon, A. K.,** Relation between red cell anion exchange and water transport, *Biochim. Biophys. Acta*, 778, 385, 1984.
38. **Shporer, R. and Civan, M. M.,** NMR study of ^{17}O from $H_2{}^{17}O$ in human erythrocytes, *Biochim. Biophys. Acta*, 358, 81, 1975.
39. **Pirkle, J. L., Ashley, D L., and Goldstein, J. H.,** Pulse nuclear magnetic resonance measurements of water exchange across the erythrocyte membrane employing a low Mn concentration, *Biophys. J.*, 25, 389, 1979.
40. **Vîlcu, Al.,** *Eritrocitul*, Editura Medicală, Bucuresti, Roumania, 1977.
41. **Dix, J. A. and Solomon, A. K.,** Role of membrane proteins and lipids in water diffusion across red cell membranes, *Biochim. Biophys. Acta*, 773, 219, 1984.
42. **Conlon, T. and Outhred, R.,** The temperature dependence of erythrocyte water diffusion permeability, *Biochim. Biophys. Acta*, 511, 408, 1978.
43. **Laemmli, U. K.,** Cleavage of structural proteins during the assembly of the head of bacteriophage T_4, *Nature (London)*, 227, 680, 1970.
44. **Benga, Gh., Ionescu, M., Popescu, O., and Pop, V. I.,** Effect of chlorpromazine on proteins in human erythrocyte membranes as inferred by spin labeling and biochemical analyses, *Mol. Pharmacol.*, 23, 771, 1983.
45. **Popescu, O.,** A simple method for drying polyacrylamide slab gels using glycerol and gelatin, *Electrophoresis*, 4, 432, 1983.
46. **Naccache, P. and Sha'afi, R. I.,** Patterns of nonelectrolyte permeability in human red blood cell membrane, *J. Gen. Physiol.*, 62, 714, 1973.
47. **Abbott, R. E. and Schächter, D.,** Impermeant maleimides. Oriented probes of erythrocyte membrane proteins, *J. Biol. Chem.*, 251, 7176, 1976.
48. **Rothstein, A.,** Mercurials and red cell membranes, in *The Function of Red Blood Cells: Erythrocyte Pathobiology*, Alan R. Liss, New York, 1981, 105.
49. **Sutherland, R. M., Rothstein, A., and Weed, R. I.,** Erythrocyte membrane sulfhydryl groups and cation permeability, *J. Cell Physiol.*, 69, 185, 1967.
50. **Lepke, S., Fasold, H., Pring, M., and Passow, H.,** A study of the relationships between inhibition of anion exchange and binding to the red blood cell membrane of 4,4′-diisothiocyanostilbene-2,2′-disulfonic acid (DIDS) and its dihydro derivative (H_2DIDS), *J. Membr. Biol.*, 29, 147, 1976.
51. **Vansteveninck, J., Weed, R. I., and Rothstein, A.,** Localization of erythrocyte membrane sulfhydryl groups essential for glucose transport, *J. Gen. Physiol.*, 848, 617, 1965.
52. **Ralston, G. B. and Crisp, E. A.,** The action of organic mercurials on the erythrocyte membrane, *Biochim. Biophys. Acta*, 649, 98, 1981.
53. **Haestm, C. W. M.,** Interactions between membrane skeleton proteins and the intrinsic domain of the erythrocyte membrane, *Biochim. Biophys. Acta*, 694, 331, 1982.
54. **Ramjeesingh, M., Gaarn, A., and Rothstein, A.,** The locations of the three cysteine residues in the primary structure of the intrinsic segments of band 3 protein, and implications concerning the arrangement of band 3 protein in the bilayer, *Biochim. Biophys. Acta*, 729, 150, 1983.
55. **Rao, A.,** Disposition of the band 3 polypeptide in the human erythrocyte membrane. The reactive sulfhydryl groups, *J. Biol. Chem.*, 254, 3503, 1979.
56. **Rao, A. and Reithmeier, R. A. F.,** Reactive sulfhydryl groups of the band 3 polypeptide from human erythrocyte membranes. Location in the primary structure, *J. Biol. Chem.*, 254, 6144, 1979.
57. **Weinstein, R. S., Khodadad, J. K., and Steck, T. L.,** The band 3 protein intramembrane particle of the human red blood cell, in *Membrane Transport in Erythrocytes*, Lassen, U. V., Ussing, H. H., and Wieth, J. O., Eds., Munksgaard Copenhagen, 1980, 35.
58. **Macey, R. I., Karan, D. M., and Farmer, R. E. L.,** Properties of water channels in human red cells, in *Biomembranes*, Kreuzer, F. and Slegers, J. F. G., Eds., Plenum Press, New York, 1972, 331.

59. **Jones, M. N. and Nickson, J. K.,** Monosaccharide transport proteins of the human erythrocyte membrane, *Biochim. Biophys. Acta,* 650, 1, 1981.
60. **Wheeler, T. J. and Hinkle, P. C.,** Kinetic properties of the reconstituted glucose transporter from human erythrocytes, *J. Biol. Chem.,* 256, 8907, 1981.
61. **Galey, W. R., Evan, A. P., Van Nice, P. S., Dail, W. G., Wimer, B. M., and Cooper R. A.,** Morphology and physiology of the McLeod erythrocyte. I. Scanning electron microscopy and electrolyte and water transport properties, *Vox Sang.,* 3, 1, 1977.

Chapter 4

BULK DIFFUSION METHODS FOR MEASURING WATER PERMEABILITY OF BIOLOGICAL MEMBRANES

Beate Klösgen, Hansjürgen Schönert, and Bernhard Deuticke

TABLE OF CONTENTS

I. INTRODUCTION

The measurement of rapid solute exchange through biological membranes by conventional techniques becomes increasingly difficult in the range of high permeabilities.[1] This is particularly true when the system investigated consists of rather small cells with little enclosed volume as is the case with the red blood cell. In such systems the equilibration of solutes with high permeabilities occurs within seconds or less.[1-4] In addition, most methods used to measure the rapid transport of highly permeable solutes such as water or other small nonelectrolytes suffer from the problem of unstirred layers which may affect the transfer rate to a nonnegligible extent.[1,5-8] A review of the different experimental methods and theoretical approaches is given by Stein[9,10] and, especially concerning water transport in erythrocytes, by Sha'afi and Gary-Bobo[11] and Forster.[12]

Both of the handicaps mentioned above, solute equilibration too fast to be measured and uncertainties concerning the rate limiting steps can be overcome by a bulk diffusion method that was originally proposed by Redwood et al.[13] This approach combines the classical capillary tube technique[14-18] for measuring diffusion coefficients in homogeneous systems with a radioactive tracer method applied to a cell suspension. Instead of following the approach to equilibrium of a labeled solute moving from an intra- to an extracellular compartment, or vice versa, this technique involves the measurement of the diffusion of labeled solutes out of a column of packed and preequilibrated cells where the solute passes through intra- and extracellular compartments as well as through the membranes separating these compartments. This results in rather slow rate coefficients from which the membrane permeabilities can be calculated on the basis of a number of assumptions and correlated data. The detailed theoretical evaluation accounts for the diffusion of solute through both the intra- and extracellular pathways and thus is not affected by unstirred layer problems. The original "differential" procedure[13] consisted of measuring the concentration profile by slicing the capillaries and measuring piece by piece.

Another possibility consists in a continuous monitoring of the concentration change of a tracer which diffuses through a solution; this has been demonstrated by Passiniemi et al.[19] for aqueous electrolyte solutions. However, this procedure requires labeled solutes with high radioactive decay energies in order to overcome quenching effects and it fails when tracers with low energy radioactivity have to be used. Thus, it is not applicable to the measurement of the diffusion of solutes that can only be H^3- or C^{14}-labeled as is the case for water and a large number of solutes of biological relevance.

Osberghaus et al.[20] succeeded in simplifying the original technique of Redwood et al.[13] by evaluating the space-averaged concentration decrease within the whole capillary. This so-called "integral" measuring method has been previously applied to diffusion measurements in homogeneous solutions by Anderson et al.[14] and by Wang et al.[15-18] However, the method still required very high cell packing and suffered from an increasing lack of experimental precision as the membrane permeability decreased to about $1 \cdot 10^{-3}$ cm/sec or below. So a complete reconsideration[21] of both experimental and theoretical aspects has been performed which led to a new and more general formulation of the theoretical concept for the derivation of membrane permeability coefficients of highly permeable solutes from their bulk diffusion coefficients in a suspension of packed cells.

As a test, the permeation of water across human red blood cell membranes was investigated. As yet there is a wide range of results reported for the water transport both from measurement of diffusive (P_d) and hydraulic (P_f) transport parameters. The fact that in erythrocytes the hydraulic permeability is about three to five (Hansson Mild and Løvtrup[22]) times higher than the diffusive one points to the contribution of an aqueous (protein-related) pathway.[23] It may be inhibited by SH-reagents like *p*-chloro-mercuri-phenyl-sulfonate (PCMBS).[24,25] At maximal inhibition by these reagents, both diffusive and hydraulic permeabilities almost completely coincide and there is still a rate of water transport. This has been explained by a nonporous lipid pathway (''high resistance pathway'', in parallel to the ''low-resistance aqueous pathway''). When the temperature dependence of water transport is measured before and after inhibition by mercurials, an increase of the activation energy is observed towards values obtained for pure lipid membranes.[26] This finding supports the lipid nature of the high-resistance pathway. This is further substantiated by the good agreement of some diffusive permeabilities found for red blood cells and those from the investigation of artificial lipid systems.[26-30]

The improved version of the capillary method that is going to be introduced here is well adapted to the measurement of the high permeability of water and even higher transfer rates. So, the water transport system has been chosen as a test object. The experimental and theoretical conditions of the technique differ markedly from those of other methods. Thus, the results may be regarded as independent values. Furthermore, the new method allows a wide range of applications with respect to the permeabilities, to the packing density of the cell suspension, and to the cell types that may be investigated.

II. THEORETICAL ANALYSIS

The system we consider consists of a macroscopically homogeneous suspension of (biological) cells that are irregularly oriented in either direction. On a microscopic scale, there is something like an elementary unit of the suspension which consists of the biological cell itself together with its aqueous environment, the cell membrane being the boundary between the extracellular and the intracellular compartment. As pointed out by Colton et al.[31] the diffusion characteristics of this subsystem depend on the diffusion coefficients D_1 and D_2 of the two compartments, respectively, and on the diffusive permeability coefficient P_d of the membrane itself, weighted by some coefficients that follow from the geometric model chosen to depict the cell suspension.

Now, if the cell suspension is regarded as a bulk liquid, any diffusion by the solute may be described by the measurable quantity of an effective diffusion coefficient D_{eff} that is unambiguously correlated to the subsystem by a special function

$$D_{eff} = f\,(P_d, D_1, D_2, V_{rel}, L_2, \text{model}) \tag{1}$$

that will be derived below.

The symbols herein stand for:

1. D_{eff}, the effective diffusion coefficient in a cell suspension as described by Crank[32] that is measured by monitoring a solute's diffusion when regarding the cell suspension as a homogeneous liquid.
2. D_1, the diffusion coefficient of the solute in the extracellular diffusion medium.
3. D_2, the diffusion coefficient of the solute in the intracellular space.
4. P_d, the diffusive membrane permeability coefficient for the solute.
5. V_{rel}, the extracellular volume relative to the volume of the whole suspension.
6. L_2, the length of the cell measured in the diffusion direction.

The parameters D_{eff}, D_1, D_2, and V_{rel} are obtained in independent experiments. The geometric parameter L_2 is derived from the mean cell volume $\overline{V}_{cell}$. L_2 is calculated on the basis of assumptions on the cell size and shape (cube, quader, ellipsoid, disk, cylinder). It depends on the geometric model chosen to build up the cell suspension. This model has to fit with the symmetry conditions of the suspension, i.e., whether the cells have a preferred or a random orientation in the cell column. As a first step, the determination of a diffusion coefficient from experiments with the capillary tube will be described. Then a geometric model for a column of homogeneously suspended particles will be introduced and applied to a suspension of packed cells. Finally, the calculation of a membrane permeability P_d is demonstrated to result from the effective diffusion coefficient D_{eff} of a solute through a column of packed cells and the cell packing density V_{rel}.

A. The Diffusion Coefficient

The capillary method using a diffusion tube open at only one side has frequently been used to determine diffusion coefficients of ions in aqueous solutions,[33-35] of proteins,[18] and of the self-diffusion of water.[14-17] For a capillary containing a test solute with a local concentration c at time t, the Fick equation

$$\frac{dc}{dt} = -D \cdot \frac{\partial^2 c}{\partial x^2} \tag{2}$$

reduces to the case of a linear finite system. The factor D herein is the diffusion coefficient of the test solute in the diffusion medium. The space coordinate x is identical with the axis of the capillary and marks the diffusion direction. Assuming a constant concentration of the test solute of zero outside of the capillary, Jost[36] has given the solution for the "diffusion out of a slab" of length ℓ both for the differential (Redwood et al.[13] and Garrick et al.[37-40]) and the integral (Osberghaus et al.[20] and Klösgen[21]) application.

In the integral technique,[14-17,20,21] only the space-averaged concentrations of the test solute are measured at the beginning (c_o) and at the end of an experiment ($\bar{c}(\Delta t)$) of duration time Δt. The diffusion coefficient D has then to be determined either by an iterative or by a graphic procedure from the solution of Equation 3

$$\bar{c}(\Delta t) = c_o \cdot \frac{8}{\pi^2} \sum_{\nu=o}^{\infty} \frac{1}{(2\nu + 1)^2} \cdot \exp\left[-(2\nu + 1)^2 \cdot \frac{D \cdot \Delta t}{\ell^2} \right] \tag{3}$$

the function converges fast, consideration of the first three to five terms proves to be sufficiently exact. The error for the determination of the diffusion coefficient is minimized[21] if, by choosing the appropriate experimental conditions, the value of $\bar{c}(\Delta t)/c_o$ comes to lie in the interval

$$0.25 \leq \frac{\bar{c}(\Delta t)}{c_o} \leq 0.65 \tag{4}$$

All experiments should be arranged to obey this relationship by adequate choice of either the capillary length ℓ, the measuring time Δt, or the effective diffusion coefficient D_{eff}.

In order to obtain a complete description of diffusion in a heterogeneous system such as a suspension of cells, the diffusion coefficients D_1 and D_2 for the extra- and intracellular pathways have to be measured in separate experiments. In the present work, this procedure will be demonstrated using as an example the diffusion of water through a suspension of red blood cells. In principle, this approach will work for any macroscopically homogeneous system that may be structured into small and identical elementary units. The theoretical investigation will result in a relationship like Equation 1, from which the membrane diffusion permeability P_d may be derived.

B. A Geometric Model for a Cell Suspension

A homogeneous suspension of randomly oriented red blood cells may be replaced by an isotropic model of a set of rectangular subsystems that constitute the suspension as shown

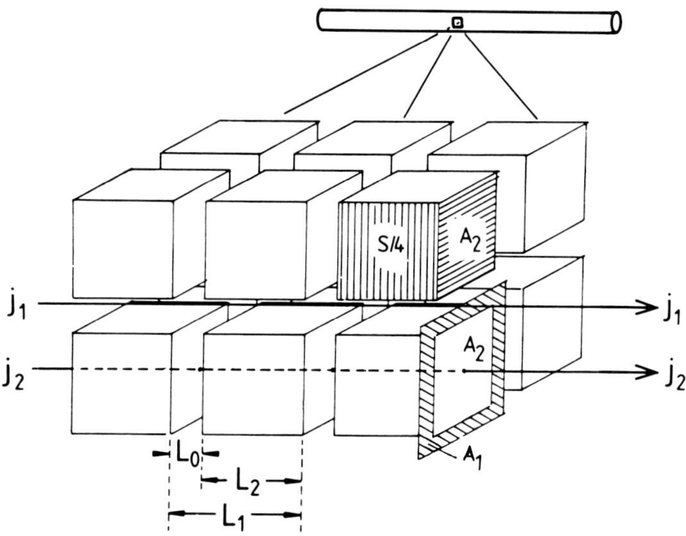

FIGURE 1. The cubic diffusion model. The cell suspension in the diffusion capillary is thought to be built from units of length L_1 each containing one model cell. These cells are imagined as cubes of length, L_2, with their front area, A_2, perpendicular to the direction of the flux, j_2, and in a distance, L_o, from each other. The extracellular flux, j_1, through the front surface, A_1, is coupled to the flux, j_2, by exchange of permeant at any site of the suspension volume including continuous exchange across the four lateral surfaces, S/4.

in Figure 1. As a convenient approximation each subsystem of total length L_1 is chosen to contain one cubic-shaped cell of length L_2. These model cells are thought to be equidistant from each other with a distance L_o in between them.

According to Crank,[32] diffusion through a complex system like this has to be described by a series-parallel-pathway model. There is an extracellular flux j_1 through the suspension medium with a diffusion coefficient D_1 perpendicular to the front surface A_1. In parallel, there is a second flux j_2 through the front surface A_2 of the cell. This flux j_2 describes the transport of the diffusing solute through an intracellular compartment of pathlength L_2 and through an extracellular one of length L_o. As the cell membrane is assumed to have identical properties all over the whole cell surface there is a coupling of the two fluxes j_1 and j_2 across the total side surface S of the cell.

There are no restrictions concerning the packing density except that it should be higher than that reached by natural sedimentation of the cell species in order to obtain macroscopic homogeneity. For symmetry reasons, a cube parallel to the x-axis chosen as model for a cell; other models have to account for the possible orientations of the cell.

When the mean cell volume $\overline{V}_{cell}$ is known, the value of the cell length L_2 which, in the case of a cube, is equal to the cell length in diffusion direction follows from

$$L_2 = \sqrt[3]{\overline{V}_{Cell}} \tag{5}$$

and the total lateral area of the cube is

$$\begin{aligned} S &= 4 \cdot A_2 \\ &= 4 \cdot L_2^2 \end{aligned} \tag{6}$$

In a homogeneous suspension the measured relative extracellular volume V_{rel} of the whole

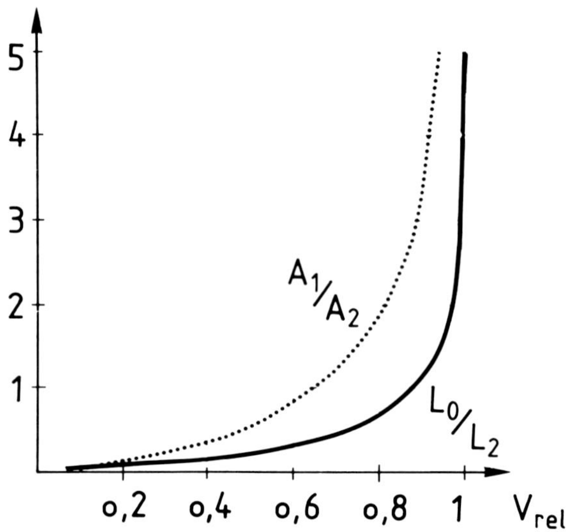

FIGURE 2. Macroscopic and microscopic geometric parameters. Relationship between the microscopic (A_1/A_2, L_o/L_2) and the macroscopic (V_{rel}) geometric parameters for the cubic cell model. The cell-to-cell distance L_o relative to the cell length L_2 and the change of the relative front surface A_1/A_2 are plotted as a function of the suspension's relative extracellular volume V_{rel}. A_1/A_2 is a measure of the relative portions of the pure extracellular flux, j_1, and the cell-permeating flux, j_2.

suspension applies to the total macroscopic system as well as to any microscopic subsystem with the total volume V and the extracellular portion V_1

$$V_{rel} = \frac{V_1}{V} \tag{7}$$

Therefore the geometric parameter L_o/L_2 that measures the extracellular pathway L_o in-between the cells relative to the intracellular pathway L_2 can be calculated from

$$\frac{L_o}{L_2} = (1 - V_{rel})^{-1/3} - 1 \tag{8}$$

This ratio is correlated to the tortuosity factor L_1/L_2 used by Redwood et al.[13]by

$$\frac{L_1}{L_2} = 1 + \frac{L_o}{L_2} \tag{9}$$

The geometric parameter A_1/A_2 follows from

$$\frac{A_1}{A_2} = (1 - V_{rel})^{-2/3} - 1 \tag{10}$$

Figure 2 is a graphic presentation of Equation 8 and Equation 10. Both functions $A_1/A_2 = f(V_{rel})$ and $L_o/L_2 = g(v_{rel})$ differ substantially from previous[13,20,37-40] conceptions. It should be noted, that even in the region of low extracellular volume V_1, the relative intercellular

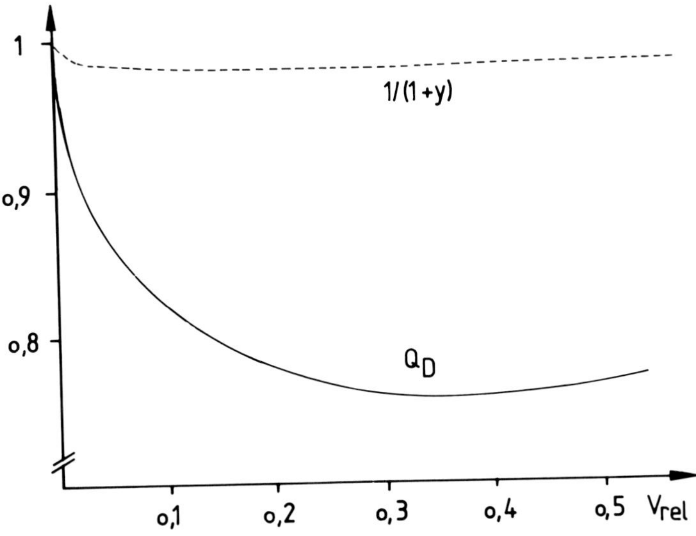

FIGURE 3. Graphic demonstration of the differences between the theoretical approach of Redwood et al.[13] and the one presented here. Introduction of correction factors allows to transfer both approaches[13,21] into each other. The factor $(1 + y)^{-1}$ (from Equations 11 and 16) formally distinguishes our relationship (Equation 11) between D_{eff}, V_{rel}, and P_d from that given by Redwood et al.[13] It arises from the exact solution of the diffusion equation in a cell suspension with explicit account of all extracellular diffusional flux components. The total factor Q_d from Equation 17 includes, besides the factor $(1 + y)^{-1}$, also the influence of Equations 11 through 16 on the solution that is inherent in the correct consideration of the geometric parameters A_1/A_2 and L_o/L_2 from Equations 8 and 10.

distance L_o/L_2 cannot be neglected. This becomes even more evident in the discussion of Figure 3.

C. The Membrane Permeability

A series-parallel-pathway model has to be used to derive an exact description of the diffusive solute transport in a suspension of biological cells. The theoretical analysis assumes a local steady-state model. This is permitted because of the small dimension $L_1 = L_o + L_2$ of the subsystem as compared to the capillary length ℓ. The details of the mathematical derivation are given in the Appendix.

The calculation results in the relationship

$$D_{eff} = \frac{(1 + r)}{(1 + z) \cdot (1 + y)} \cdot \frac{D_2 \cdot A_2}{A} \tag{11}$$

where

$$r = \frac{D_1 \cdot A_1}{D_2 \cdot A_2} \tag{12}$$

$$z = 2 \cdot [\pi_o \cdot (1 + r) + r \cdot \alpha \cdot \coth(\alpha/2)]^{-1} \tag{13}$$

$$\pi_o = \frac{P_d \cdot L_2}{D_2} \tag{14}$$

$$\alpha = [\frac{S}{A_2} \cdot \pi_o \cdot (1 + r^{-1})]^{1/2} \qquad (15)$$

$$y = \frac{\frac{L_o}{L_2}}{1 + \frac{L_o}{L_2}} \cdot \frac{z}{1 + z} \cdot \left[\frac{\alpha}{2} \cdot \coth(\alpha/2) - 1\right] \qquad (16)$$

The abbreviations are as defined before.

Equation 11 formally resembles a similar one given by Redwood et al.[13] Their approach was only valid for the approximation of an infinitely small extracellular volume V_1, whereas the new concept is theoretically valid for any extracellular volume.

There is a factor $(1 + y)^{-1}$ in Equation 11 which is not equal to unity in distinction from the corresponding formula given by Redwood et al.[13] This factor arises from the consideration of the extracellular diffusion in between the cells parallel to the diffusion direction x when the diffusion equation is solved. Moreoever, there is an additional difference that originates from the values of the geometric parameters A_1/A_2 and L_o/L_2 as functions of the relative extracellular volume as defined in Chapter 3. Because every term in Equation 11 depends on these geometric parameters, they largely affect the evaluation of a diffusion experiment. Therefore, the parameter Q_d has been introduced to represent the total influence of both corrections. This parameter Q_d is defined by the relationship

$$Q_d = \frac{D_{eff}}{D_{eff,o}} \qquad (17)$$

where D_{eff} is given by Equation 11 and $D_{eff,o}$ is calculated by the formula given by Redwood et al.[13] for negligible values of V_1. Q_d becomes equal to 1 when both approaches coincide for $V_1 \rightarrow 0$.

The influence of both corrections as dependent on $V_{rel} = V_1/V$ has been estimated for the case of water diffusion in a cell suspension and is shown in Figure 3. The curve of $(1 + y)^{-1}$ only slightly differs from unity, whereas the parameter Q_d deviates significantly. As expected, in the limit $V_1 \rightarrow 0$, both curves approach a value of 1 which demonstrates the importance of a precise consideration of the geometrical parameters of the system in the case of a nonnegligible extracellular volume. This is particularly true for cell populations that cannot be packed densely for mechanical reasons or because of their cell shape.

III. EXPERIMENTS

A. Materials and Methods

The self-diffusion of water is measured by using tritiated water (THO, 100 μCi/mℓ) as a tracer. Extracellular volume is determined by impermeable C^{14}-sucrose (10 μCi/mℓ). The suspension medium for the experiments is prepared from an isotonic solution of NaCl (9.5 gNaCl/ℓ H_2O) plus 10% (vol per vol) isotonic phosphate buffer (NaH_2PO_4/Na_2HPO_4, 125 mmol/ℓ, pH = 7.4). Osmolarity should be controlled and adjusted to about 330 mOsm/ℓ. High-precision microliter syringes like those used for gas-chromatographic injection are suitable as capillaries with variable length. This capillary length could be read to ±0.1 mm. The capillary diameter used in our studies was in the range of 0.802 ± .001 mm. The screw cap is cut away, and the face is carefully milled. The capillaries are filled with packed cells with a complete syringe.

A rack is used which will allow simultaneous measurements with 16 capillaries. Groups of four are positioned in the same incubation vessel, containing about 2.2ℓ suspension medium with addition of 5 m*M* NaN_3.

For studies on native human red cells, the cells obtained from freshly drawn human blood are isolated by centrifugation and washed three times in an excess of suspension medium before being suspended in this medium to a hematocrit of about 50%. They are then brought to high packing density by centrifugation for 30 min at 22000 × *g* at constant temperature of 0°C.

Heat fragmentation of washed blood cells is performed in a phosphate buffered saline solution containing 5 m*M* NaN_3 and 0.03% albumin. Fragmentation results from the denaturation of the membrane skeletal protein spectrin,[41-43] at about 49°C. As albumin was reported to stabilize the discoidal erythrocyte shape,[41,42] a very small amount is added to delay the fragmentation. The cell suspension is brought to a hematocrit of 10% and exposed for 15 min to a temperature of 51°C under continuous shaking to achieve spontaneous fragmentation into spheric particles.[41,43] Small vesicles that remain adherent to the large spheric fragments are sheared off by repetitive sucking up of the solution with a milliliter syringe and subsequent centrifugation.

B. Measurement of the Relative Extracellular Volume

The extracellular volume fraction of a packed erythrocyte suspension may be measured by means of either a dye[44] or a radioactive substance as an indicator.[38,45] The marker should distribute in the extracellular space comparable to Na^+. Macromolecules have been shown to be unsuitable[45] as indicators whereas smaller molecules, e.g., lactose, inulin, sucrose,[45] or serum albumin[38] provide for acceptable values. In a homogeneous suspension, the local relative extracellular volume, V_l/V, is identical to the packing density, V_{rel} of the whole system. This quantity is easy to measure in a dilution experiment using C^{14}-sucrose as a volume marker, as follows.

Packed red blood cells (1.25 mℓ) are mixed in a test tube with an equal amount of the suspension medium; 0.05 μCi C^{14}-sucrose are added and well distributed by careful stirring. The cells are brought to the desired high-packing density by centrifugation (30 min, 22000 × *g*). Under these conditions, the nonpermeable tracer remains in the aqueous supernatant; only a small portion, characteristic for the space in between the cells, is left in the sediment. A measured amount (0.5 mℓ, c_A) of the supernatant is removed. The residual supernatant is then sucked off completely. The remaining cell sediment is weighed (m_s) and is resuspended by the addition of a measured volume V_s (about 1 mℓ) of suspension medium and careful mixing. This suspension is again centrifuged and a defined amount (0.5 mℓ, c_S) of the supernatant is removed. The radioactivities c_A and c_S of both samples are counted in a liquid scintillation counter. The fractional extracellular volume V^o_{rel} of the packed cell suspension can then be calculated from

$$V_{rel}{}^{o} = \left[\frac{m_s}{\rho_{ery}}\right]^{-1} \cdot \frac{V_s \cdot c_s}{C_A - c_S} \tag{18}$$

using $\rho_{ery} = 1.096$ g/mℓ as the mean density of a human red blood cell population. From this value, V^o_{rel}, any desired fractional extracellular volume, V_{rel}, can be obtained by dilution of the packed cells with an additional volume ΔV of suspension medium.

The final value V_{rel} is then given by

$$V_{rel} = \frac{V_{tot} \cdot V_{rel}{}^{o} + \Delta V}{V_{tot} + \Delta V} \tag{19}$$

where $V_{tot} = m_s/\rho_{ery}$ is the total volume before dilution.

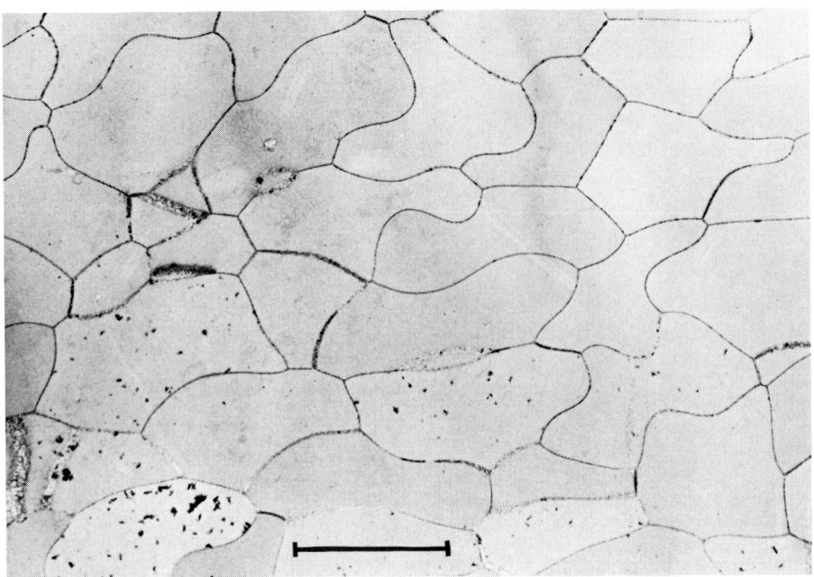

FIGURE 4. Electron micrograph of a section through a droplet of closely packed erythrocytes. Cells are deformable and fill almost the whole space. No preferred orientation observed. The bar marks a length of 4.5 μm.

C. Verification of the Geometric Assumptions

The calculation of P_d from V_{rel} and D_{eff} requires the choice of a geometric cell model in order to obtain the geometric parameters L_2, A_1/A_2, and L_o/L_2. Their values depend on the cell package density V_{rel} and on the cell model chosen to represent the real cell column (see Figure 2).

The choice of a cubic cell model is only justified if the cell column is macroscopically homogeneous. In order to test the validity of this assumption, electron micrographs (Figure 4) were prepared from a cut through a droplet of closely packed blood cells. The representative picture shows neither a preferred shape nor orientation of the cells so that the use of a cubic model is permitted.

The intracellular diffusion pathlength L_2 can be calculated from $(V_{cell})^{1/3}$ to yield

$$L_2 = 4.5 \ \mu m$$

The geometric parameters L_o/L_2 and A_1/A_2 are then determined from the relationships given in Equations 8 and 10.

D. Measurement of Diffusion Coefficients

Microliter syringes are set to the desired capillary length ℓ and mounted to the rack. Four different diffusion baths, each containing 2.2 ℓ of solution are used to provide for "infinitely large" diffusion surroundings. They are thermostated to 20°C.

Samples for diffusion measurements are prepared in a test tube by thoroughly mixing 1.25 mℓ packed erythrocytes with the same volume of the suspending solution and a small amount of THO (approximately 2 μCi). Thirty minutes waiting will provide for complete equilibration of the tracer and all rapidly equilibrating solutes. The samples for diffusion measurements and the samples for measuring the packing density are then packed by simultaneous centrifugation (30 min, 22,000 × *g*). Determination of the fractional extracellular volume is carried out as described in the previous chapter. The supernatants of the samples for diffusion measurements are carefully removed; the sediment is weighed (m_s) and brought from its tightly packed volume V_{rel}^{o} to the final value V_{rel} by an additional amount ΔV of suspension medium. This suspension should be homogenized very carefully. It is then filled into the capillaries. For that purpose, the needle of the filling syringe is inserted into the capillary syringe from above through the open capillary mouth down to the bottom. The needle is withdrawn from the capillary at the same rate as the capillary is filled with the suspension from below; in this way, the formation of air bubbles is avoided. Air bubbles would prevent any normalization of tracer activity c_o and $\bar{c}(\Delta t)$ to a defined unity volume, which is necessary for obtaining reproducible and correct results. An excess droplet of suspension reaching over the capillary mouth is removed with a razor blade and replaced by a droplet of suspension medium to prevent evaporation of water from the cell column and exsiccation of the outer surface during the subsequent manipulations. The rack with the filled capillaries (up to 16) is then slowly lowered into the incubation vessels by using a support in order to avoid vibrations. This is important because vibrations might cause a turbulence-induced washout of packed cells from the mouth of the capillary, the so-called $\Delta\ell$-effect.[33,35,46]

The initial radioactivity c_o of the capillary contents is determined by lysing a defined volume, usually 6 $\mu\ell$, in 1.5 $m\ell$ of distilled water. Proteins are precipitated by addition of 50 $\mu\ell$ 50% $HClO_4$ and separated by centrifugation. Samples to measure the initial concentration, c_o, are then taken from the supernatant.

At the end of the diffusion time, Δt (see Table 1 for appropriate diffusion times), the capillaries are removed from the incubation vessels. Their contents are ejected from the capillary syringe into 1.5 $m\ell$ distilled water. These samples are handled exactly as the c_o-sample to yield a $\bar{c}(\Delta t)$-value for each capillary. Both the c_o- and $\bar{c}(\Delta t)$-samples are counted for radioactivity and the count rates are corrected for equal suspension volumes.

The diffusion coefficient is calculated from the ratio $\bar{c}(\Delta t)/c_o$, the capillary length ℓ, and the diffusion time Δt by Equation 3. Iterative calculation of D_{eff} is terminated when the criterion

$$\left|1 - \frac{Q_{th}}{Q_{exp}}\right| \leq 10^{-4} \tag{20}$$

is fulfilled. Here, $Q_{exp} = (\bar{c}/c_o)exp$ is the measured radioactivity ratio and $Q_{exp} = (\bar{c}/c_o)_{exp}$ the theoretical one calculated from Equation 3.

The procedure described above is the same for all the diffusion measurements irrespective of the system investigated. However, in addition to the determination of the effective diffusion coefficient D_{eff}, the two independent diffusion coefficients, D_1 and D_2, of the intra- and extracellular medium have to be determined. For this purpose, the procedure reduces to the simple capillary technique.[14-18] All the resulting diffusion coefficients (D_1, D_2, and D_{eff}) are then combined to yield the membrane permeability P_d. Some details of the measuring procedures are specified in the following.

1. $\Delta\ell$-Effects

According to Equation 3, the lengtn, ℓ, of the column of suspended cells is of decisive importance for the determination of D_{eff}. The precision of the capillary length adjustment is increased if high capillary lengths are chosen. However, this may cause a loss of experimental precision as shown above Section II.A because of a too low tracer diffusion rate. This disadvantage can only be avoided if either the measuring time Δt or the diffusion coefficient D_{eff} are increased simultaneously so that the ratio $\bar{c}(\Delta t)/c_o$ falls into the desired interval (Equation 4). Nevertheless, the capillary length, ℓ, should be known exactly.

Except for errors in the adjustment of the correct length of the capillary, there may be a

Table 1
EXPERIMENTAL DIFFUSION PARAMETERS AS A FUNCTION OF PACKING DENSITY AND MEMBRANE PERMEABILITY

Pd/(cm/sec) =	$1 \cdot 10^{-2}$	$5 \cdot 10^{-3}$	$1 \cdot 10^{-3}$	$1 \cdot 10^{-4}$	$1 \cdot 10^{-5}$	
V_{rel}			D_{eff}/($10^{-6}cm^2/sec$)			α
0.05	3.228	2.182	1.059	0.752	0.726	0.7751
0.10	3.934	2.885	1.776	1.479	1.448	0.6319
0.15	4.600	3.577	2.505	2.220	2.190	0.5239
0.20	5.264	4.276	3.247	2.976	2.947	0.4402
0.25	5.934	4.988	4.005	3.747	3.720	0.3731
0.30	6.615	5.714	4.781	4.536	4.510	0.3182
0.95	18.70	18.57	18.45	18.41	18.41	0.0155

$$\alpha = 1 - \frac{D_{eff}(P_d = 10^{-5} \text{ cm/sec})}{D_{eff}(P_d = 10^{-2} \text{ cm/sec}}$$

P_d/cm/sec	V_{rel} = 0.05, ℓ = 0.48 cm	V_{rel} = 0.05, ℓ = 0.72 cm	V_{rel} = 0.15, ℓ = 0.48 cm	V_{rel} = 0.15, ℓ = 0.72 cm	V_{rel} = 0.30, ℓ = 0.48 cm	V_{rel} = 0.30, ℓ = 0.72 cm
	Δt/hr	Δt/hr	Δt/hr	Δt/hr	Δt/hr	Δt/hr
$1 \cdot 10^{-5}$	21.06	47.39	6.99	15.73	3.39	7.63
$1 \cdot 10^{-4}$	20.33	45.74	6.89	15.49	3.38	7.61
$1 \cdot 10^{-3}$	14.44	32.49	6.11	13.74	3.20	7.19
$5 \cdot 10^{-3}$	7.74	15.76	4.28	9.62	2.67	6.02
$1 \cdot 10^{-2}$	7.00	10.66	3.33	7.48	2.32	5.21

Note: Lowering of the packing density may be used to bring $\bar{c}(\Delta t)/c_o$ into the optimal range as defined by Equation 4 without increasing the measuring time Δt or reducing the capillary length ℓ. D_1 and D_2 for water have been used for calculation. The parameter α describes the decreasing discrimination of P_d values upon diminishing the packing density.

time-dependent loss of cell column material. This corresponds to a decrease of the effective column length and thus to an overestimation of the diffusion rate. There are two possible reasons for such an undetected loss of cell column material. One reason may be a perturbation due to turbulences when the capillaries are transferred into the incubation vessel. This perturbation can be diminished by depositing an excess droplet of incubation medium onto the capillary top after filling, in order to keep it wet. The second reason is a slow loss of material due to convection caused by stirring.[35,46] For theoretical reasons,[36] the stirring is indispensible because it has to keep the tracer concentration nominally at zero in the incubation vessel outside the capillary. Thus, the stirring frequency has to be kept at a level which does not cause a washout of cells from the column. This can be checked by measuring the hemoglobin contents of the cell column.

2. The Self-Diffusion Coefficient of THO in Aqueous Salt Solution

The self-diffusion coefficient, D_1, of water may be assumed to be nearly the same in isotonic salt medium and in pure water.[34] For its determination, an amount of 5 μℓ THO (specific activity: 100 μCi/mℓ) is given into 2.5 mℓ incubation medium. Capillaries of varying diffusion length ℓ are then filled with medium. The diffusion time Δt should be selected on the basis of literature data[47] to provide for $\bar{c}(\Delta t)/c_o$ ratios in the optimal range (Equation 4).

$\Delta\ell$-effects will probably limit the experimental precision in the case of short capillary lengths. Except for this limitation, the results obtained should not depend on capillary length. In fact, we obtain a mean value of

$$D_1 = 2.13 \cdot 10^{-5} cm^2/sec$$

for the self-diffusion of THO in the suspension medium which is in good agreement with data reported by others.[13,20,17,34,38,47] However, there remains a difference of up to 10% among the different authors. Mills[47] has reported very sophisticated work on the determination of the water self-diffusion coefficient and his value $D_1 = 2.00 \cdot 10^{-5}$ cm²/sec seems to be the most reliable one.

3. The Diffusion Coefficient of THO in Aqueous Hemoglobin Solution

The intracellular diffusion resistance of erythrocytes is likely to be mainly dependent on the hemoglobin concentration. Therefore, the diffusion characteristics of the intracellular phase have to be investigated in the range of physiological hemoglobin concentrations (330 g/ℓ).

a. Preparation of Hemoglobin Solutions

Hemoglobin (Hb) solutions are prepared by osmotic lysis of washed red blood cells in distilled water. For that purpose, the cells are mixed with a ninefold volume of double-distilled water and thoroughly shaken; then the ghost membranes are removed by centrifugation (30 min, 26,000 × *g*). The lysate is dialyzed against phosphate buffer (pH = 7.4) to restore isotonic conditions (exclusion limit: M = 10,000 g/mol). This hemoglobin solution is then reconcentrated by ultrafiltration using a membrane with an exclusion limit of *M* = 10,000 g/mol. The exact concentration of hemoglobin is determined as cyanomethemoglobin by routine techniques.

b. Diffusion Experiment in Hemoglobin Solutions

Samples of approximately 1 mℓ of buffered hemoglobin solution with well-defined concentration are mixed with 1.5 μℓ THO by thorough stirring. After a sufficient time of equilibration, the capillaries are filled with the labeled solution as described above and lowered into the incubation vessel containing phosphate-buffered saline instead of hemoglobin solution. The error arising from a possible loss of hemoglobin into the incubation vessel is supposed to be very small because of the short measuring time Δt of approximately 6 hr and the low aqueous diffusion coefficient of hemoglobin of $D_{Hb} = 7.2 \cdot 10^{-7}$ cm²/sec.[48]

The results are presented in Figure 5 and demonstrate a good linear relationship (correlation coefficient: r = −0.987) between the value of D_2 and the hemoglobin concentration c_{Hb}. The regression line is given by

$$D_2 = 2.18 \cdot 10^{-5} cm^2/sec - 2.63 \cdot 10^{-8} (cm^2 \cdot \ell)/(sec \cdot g) \cdot c_{Hb}$$

from which the diffusion coefficients of THO in the intracellular compartment is read for physiological conditions to be

$$D_2(330 g/\ell) = 1.31 \cdot 10^{-5} cm^2/sec$$

which is in sufficient agreement with previous data.[13,20] The extrapolation of the regression

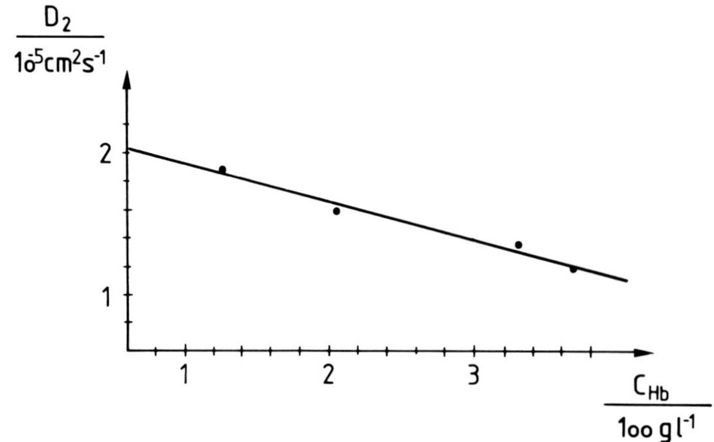

FIGURE 5. Diffusion coefficient D_2 for water in hemoglobin solution as a function of the hemoglobin concentration. Data determined by the capillary technique (T = 20°C).

to zero hemoglobin concentration gives another value for the extracellular diffusion coefficient in bulk water:

$$D_1 = 2.18 \cdot 10^{-5} cm^2/sec$$

This value agrees with that found in the experiments described in the chapter before.

4. The Diffusion Coefficient D_{eff} of THO in a Packed Cell Column

The values of the effective diffusion coefficient D_{eff} for suspensions of packed cells vary, if all other parameters such as D_1, D_2, cell species, and cell model are kept constant, with the suspension packing density. Therefore, each diffusion coefficient, D_{eff}, is correlated with the value of the corresponding extracellular volume, V_{rel}, by a function as Equation 1. The relationship between D_{eff} and V_{rel} depends on the membrane permeability P_d. To estimate this relationship, let us suppose that only a few cells are present in the suspension. Then the diffusion will be dominated by the extracellular pathway (D_1) and D_{eff} will nearly equal D_1 and will be independent of P_d. On the other hand, in a highly packed suspension, where V_{rel} is negligibly small, the value of D_{eff} approaches D_2 in the case of a very low membrane diffusion resistance. In an intermediate range, which is the general case, it will be a function of both D_2 and P_d. This consideration shows that upon varying the fractional extracellular volume, V_{rel}, the value of D_{eff} varies over a broad range. So we obtain an additional variable besides the measuring time, Δt, and the capillary length, ℓ. This may be used to shift the quantity $\bar{c}(\Delta t)/c_o$ into the optimal range (Equation 4).

The limited survival time of cells in vitro makes it impossible to extend the measuring time up to values which would allow measurements over the whole range of permeability coefficients P_d in closely packed suspensions. An alternative possibility for increasing the ratio $(D \cdot \Delta t/\ell^2)$ consists in decreasing the capillary length ℓ or increasing the effective diffusion coefficient D_{eff}. Shortening the capillary will enlarge the systematic errors and the influence of the $\Delta\ell$-effect. Thus, in order to include slow permeants, the effective diffusion coefficient turns out to become an important experimental parameter to shift the system investigated into a suitable range. The lowering of the suspension packing density leads to higher effective diffusion coefficients D_{eff} and thus to a decrease of the incubation time that is required to reach optimal experimental precision for the determination of diffusion coefficients (Equation 4). Since former methods[13,20,37-40] were restricted to the condition of nearly vanishing extracellular volume, the only available variables were the measuring time, Δt,

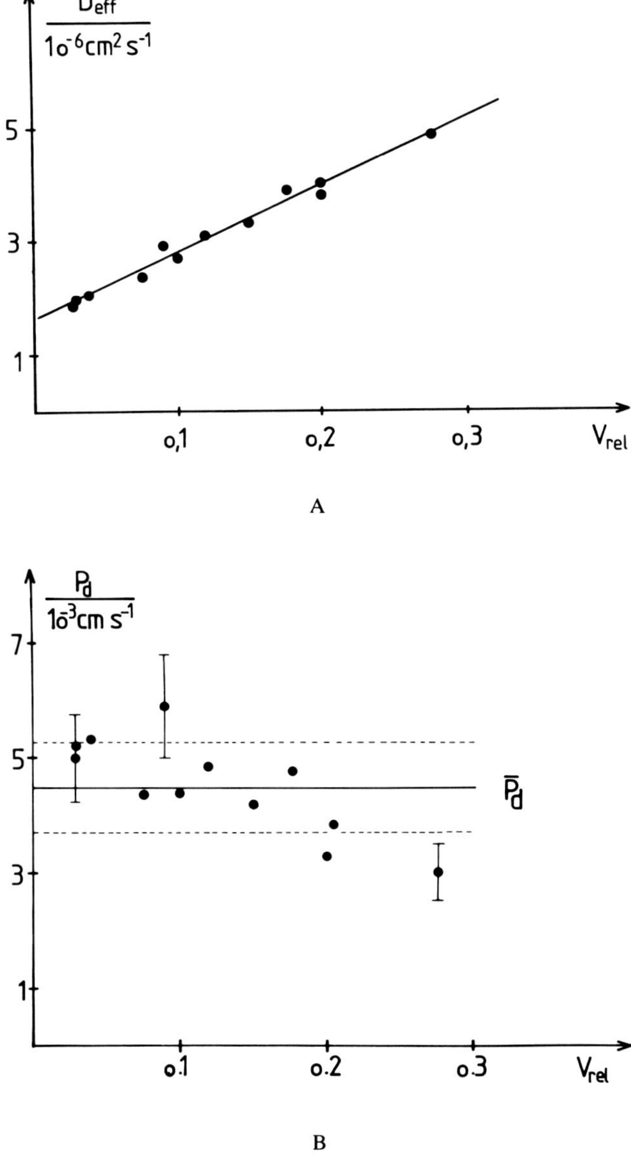

FIGURE 6. Diffusion of tritiated water through suspensions of erythrocytes. (A) Bulk diffusion coefficient D_{eff} as a function of the packing density, V_{rel}. (B) Diffusional membrane permeability P_d as a function of the fractional extracellular, V_{rel} calculated from the data in Figure 6A.

and the capillary length, ℓ. Now that all effects have been considered (cf. Section II) the controlled variation of D_{eff} enlarges the range of accessible permeability coefficients. The range of possible dilutions is limited only by the requirement that the suspension must stay homogeneous.

To demonstrate the validity of this new approach, measurements of D_{eff} have been made under systematic variation of V_{rel}. The technical procedure is identical to that in the previous description, with an amount of approximately 2 μCi of tritiated water as a diffusion marker. The results shown in Figure 6A demonstrate a linear correlation (correlation coefficient: $r = 0.993$) between D_{eff} and V_{rel} over the range of packing densities from about 2 to 30%

fractional extracellular volume. The membrane permeability P_d for water, which may be calculated by use of Equations 8, 10, and 11 to 16 as discussed below, should be independent of V_{rel}. To illustrate this fact, the corresponding values of P_d are compiled in Figure 6B. Linear regression leads to

$$P_d(V_{rel}) = 5.57 \cdot 10^{-3}\text{cm/sec} - 8.64 \cdot 10^{-3}\text{cm/sec} \cdot V_{rel}$$

with a regression coefficient $b = -8.64 \cdot 10^{-3}$ cm/sec. Statistical evaluation* of the experimental data demonstrates that this slope is not significantly different from zero. Thus, P_d is independent of V_{rel} within the limits of our experimental accuracy.

In view of these results, the experimental parameters V_{rel} and D_{eff} are free to be chosen for the determination of the membrane permeability P_d within the theoretical limitations. In the following section, the results of such determinations are tested in order to prove the consistency between the new theoretical approach and the experimentally obtained dependence of D_{eff} on V_{rel}.

E. Evaluation of the Water Permeability Coefficient P_d of Human Erythrocytes

From each set of experimentally obtained values for V_{rel} and D_{eff}, a permeability coefficient P_d can be calculated by Equations 8, 10, and 11 to 16. The final water permeability coefficient $\bar{P}_d$ is obtained as the statistical mean of these individual values because it is independent of the cell column packing density. The water diffusion experiments result in a mean permeability coefficient

$$\bar{P}_d = 4.47 \cdot 10^{-3} \text{ cm/sec}$$

with a relative standard error of about 18%.

A covariance analysis is applied to the measured values of D_{eff} and V_{rel} to obtain the statistical error limits. The results of this consistency test are shown in Figure 7 where both experimental and theoretical curves of the effective diffusion coefficient as a function of the relative extracellular volume are shown.

The 95% confidence interval obtained by covariance analysis is represented by a dark region symmetrical to the experimental regression line $D_{exp} = f(V_{rel})$. The theoretical course of $D_{th} = f(V_{rel})$ may be calculated from the permeability coefficient $P_d = 4.47 \cdot 10^{-3}$ cm/sec by use of Equations 8, 10, and 11 through 16.

It can be seen that the line calculated from the theoretical approach falls into the experimentally determined error interval. This result demonstrates the good consistency between the function $D_{th} = f(V_{rel})$ that follows from the theoretical description and the experimentally determined function $D_{exp} = f(V_{rel})$.

F. Measurements on the Water Permeability of Spheric Erythrocyte Membrane Vesicles

Bulk diffusion techniques for the determination of high permeabilities might be of particular relevance for measurements in small vesicles (with a diameter of ≤ 1 μm) because in this size range the conventional tracer flux techniques cannot be used. Spheric, undeformable vesicles of equal size can theoretically be packed to an extracellular space of 26%. Because of this large extracellular space, suspensions of such vesicles cannot be studied by

* As a simple test, the confidence interval of the regression coefficient[82] should be calculated. In the case discussed here, it is found to be $\Delta b = \pm 9.908\ 10^{-3}$cm/sec. This includes the case of a horizontal line. The slope of the regression line is, thus, not well determined. The independence of P_d from the packing density is further substantiated by calculating the rank coefficient, r_S, of Spearman.[83] It is found to be $r_S = 0.6993$. This is less than the corresponding critical value $r^*_S = 0.7273$ for an error level of $\alpha = 1\%$.

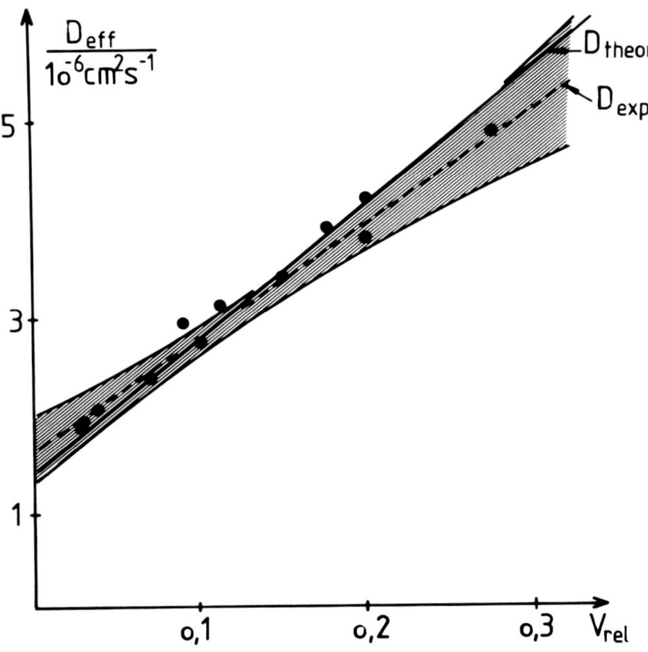

FIGURE 7. Consistency between experimental and theoretical results. Analysis of covariance yields the shaded 95% confidence interval containing the experimentally obtained data. The theoretically determined function $D_{eff} = f(V_{rel})$ (for $P_d = 4.47 \cdot 10^{-3}$cm/sec) completely develops within this confidence region.

the original cell column method.[13] Suspensions of spheric particles are, however, very appropriate for the new bulk diffusion approach presented here.

Essentials of the procedure are outlined in the following for vesicles prepared from red blood cells by heat fragmentation as described above. The size distribution of the vesicles has to be measured in order to determine the mean spherical volume V_{sph} and the intracellular diffusion length $L_2 = (V_{sph})^{1/3}$.

Osmotic fragility curves indicate that the vesicle populations are not homogeneous.[87] This is also substantiated by the finding that the fragments can be packed to an extracellular volume of 17.5%. This is about 8% denser than the densest possible packing of homogeneous spheres. Thus, there have to be several size classes or the vesicles are not completely spherical.

The details of the vesicle population should be investigated by means of electron microscopy. A droplet of packed vesicles is given into a buffered solution of formaldehyde, glutaraldehyde, and OsO_4 for fixation. The fixed droplet is dehydrated and embedded in a matrix of epoxyacrylate to be cut with a microtome for stereologic evaluation.

After contrasting with uranylacetate, we obtain a typical electron micrograph like that in Figure 8 showing vesicle sections from a population of heat-fragmented cells. The cut is taken near the surface of the droplet where the vesicle suspension is diluted from the fixation suspension. The population of heat-fragmented cells can be seen to consist of a minor fraction of very small vesicles and a dominant fraction of large vesicles some of which are still deformable to a slightly nonspherical shape. The measurement of the two semiaxes from electron micrograph intersections proves that the vesicles may be approximated as spheres. The large vesicles have small spherules associated to the surface. Even intense shearing could not remove these spherules except under appreciable hemolysis.

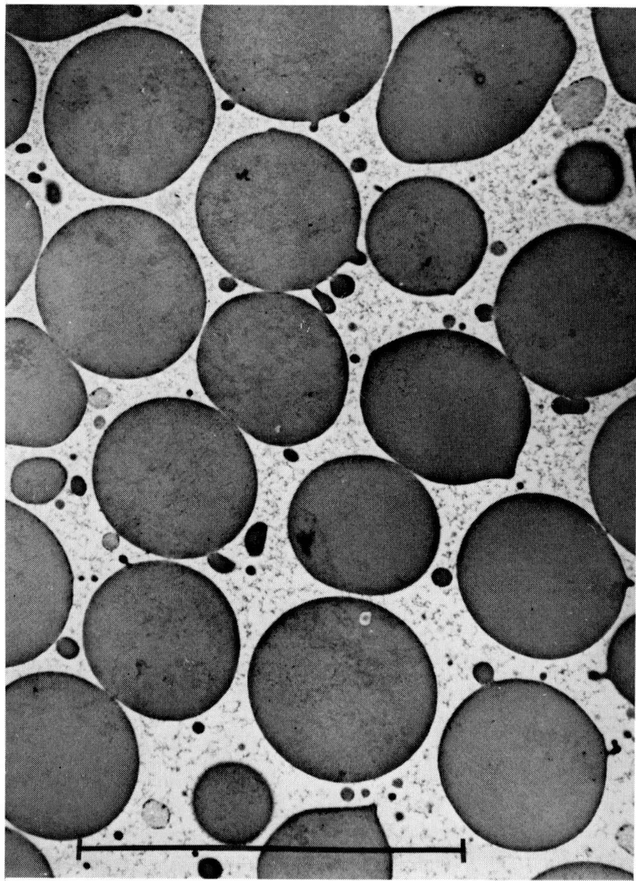

FIGURE 8. Electron micrograph of a population of spheric vesicles. Population of vesicles (mean spheric radius R_m = 1.86 μm) prepared by heat fragmentation of erythrocytes. The bar indicates a length of 10 μm.

The vesicle sections may be classified and counted to obtain information about the three-dimensional size distribution of the vesicles in the suspension. This stereologic investigation yields a circle intersection histogram. The size distribution of the vesicle population is calculated by evaluation of the circle intersection histogram according to the method of Lenz.[49,50] This procedure yields a representative vesicle radius,

$$R_m = 1.86\ \mu m$$

corresponding to a representative volume,

$$V_{sph} = 27\ \mu m^3$$

which well supports the concept of heat fragmentation of the red cell into about three large fragments.[41] According to Lenz,[50] a standard error, σ, in the determination of R_m, can be calculated. This standard error is found to be about 21%, which seems to be rather high at first sight. Nevertheless, this value may not be interpreted to suggest an inhomogeneous population of spheres. This has been proven by Anton and Voit,[51] who made computer simulations of extensive series of thin cuts through equal spheres. Even in this case, they did not obtain low values for the standard error σ, which thus turns out to be irrelevant for the interpretation of a histogramm.

Again the cubic cell model is applied. A cube with the same mean volume as our spherical vesicles has an intracellular diffusion pathlength of

$$L_2 = 3\ \mu m$$

The spherical volume still amounts to one third of the native cell volume. Thus, there is no considerable change in curvature during fragmentation which might affect the permeability properties. A change in the hemoglobin concentration that might have come along with the fragmentation process would not change to the experimental results as the method is hardly sensitive[13,21] to variations in the value of D_2. Thus, diffusion experiments with packed vesicles may be carried out analogous to those in erythrocyte suspensions. Using the particular values $D_1 = 2.13 \cdot 10^{-5}$ cm²/sec and $D_2 = 1.31 \cdot 10^{-5}$cm²/sec for the extra- and intracellular diffusion coefficients, our experiments resulted in the determination of a water permeability coefficient,

$$P_{d,sph}\ 5.24 \cdot 10^{-3} \text{cm/sec}$$

This value agrees well with that measured in suspensions of unmodified erythrocytes within the limits of experimental precision.

IV. DISCUSSION

The more generalized formulation of tracer diffusion through a suspension of single cells presented here avoids the restrictions inherent in the approach of Redwood et al.[13] The new theoretical description considers exactly all the diffusion pathways, and the cells need no longer be packed up to a negligible extracellular volume. As a prerequisite, the cell suspension under investigation has to be homogeneous and the cell dimensions have to be small compared to the diffusion column. In that case, the contributions of the different diffusion pathways may be estimated from the measurement of the fractional extracellular volume, V_{rel}, under the presupposition of a known distribution of cell sizes and by application of an appropriate geometric cell model. Under these circumstances, the diffusion of a solute through a homogeneous system of cells may be described by a measurable bulk diffusion coefficient, D_{eff}. From that quantity, the membrane permeability P_d may be derived.

The capillary technique used[14,15-18,33] is experimentally simple; its application to cell suspensions is presumably very adequate for the investigation of very rapid transmembrane exchange processes since it provides for appropriate time resolution. In spite of high permeabilities, there are no errors due to unstirred layers,[52] because all diffusion steps are taken into account. As another important advantage, the cells need no longer be closely packed. Thus, the method can be applied to nondeformable cells and vesicles.

In order to obtain sufficient experimental precision, the term $(D_{eff} \cdot \Delta t/\ell^2)$ in Equation 3 should be large. This can be achieved either by a short capillary length, ℓ, or by large values of Δt or of D_{eff}. This is particularly important in the case of relatively low permeabilities. Short capillary lengths will be difficult to adjust properly; furthermore, the influence of $\Delta\ell$ effects will be increased. Long measuring times, Δt, will conflict with the limited lifetime of cells in vitro. So the variation of the value of D_{eff} via the appropriate setting of a fractional extracellular volume, V_{rel}, seems to be promising. This is illustrated in Table 1; the effect of lowering the packing density of the suspension on the value of the effective diffusion coefficient, D_{eff}, has been calculated for different values of the membrane permeability, P_d. For the extra- and intracellular pathways the diffusion coefficient of water has been used. It can be seen from Table 1 that the higher the fractional extracellular volume V_{rel}, the less pronounced is the dependence of D_{eff} on P_d. As an extreme, the values of D_{eff} for 95% extracellular volume are shown; they are very close to the value of D_1, the extracellular diffusion coefficient. Obviously, under these conditions, D_{eff} cannot be used to obtain precise information on the permeability, P_d. This general tendency is shown in the last column of Table 1 where the total change of D_{eff} over the chosen permeability range relative to the maximal value has been expressed by introducing the parameter α. This parameter is defined by

$$\alpha = 1 - \frac{D_{eff}(P_d = 1 \cdot 10^{-5} \text{cm/sec})}{D_{eff}(P_d = 1 \cdot 10^{-2} \text{cm/sec})}$$

If this relationship is high, the permeability is well determinable; if it is low, there is a lack of resolution. At the same time, when the fractional extracellular volume increases, the change of the effective diffusion coefficient, D_{eff}, with membrane permeability, is diminished. Combinations of the experimental parameters V_{rel}, ℓ, and Δt are shown in Table 1. The following features of the system become evident:

1. The method is not suitable for the measurement of low permeability values
2. Very long exposure times, Δt, can be avoided by lowering the packing density. This, however, results in decreasing the resolution for different values of P_d

Thus, there are two conficting tendencies whose influence becomes more important at low permeability values. They have to be optimized for the system that is going to be investigated. To what extent the extracellular volume can be increased has to be decided for each system individually. In many cases, the requirement that the suspension should stay homogeneous will establish a limit to further dilution. In the case of water diffusion, we obtained good results up to 30% extracellular volume for suspensions of erythrocytes.

In the following, the new, improved capillary method described will be considered within the framework of the other available strategies for measuring water permeabilities in single cells. Generally, the experimental techniques for the investigation of water membrane permeabilities may be divided into nonequilibrium methods and quasi-equilibrium methods. A list of measured water permeability values for mammalian cells is given in Table 2; all results have been converted into units of the permeability P_d (cm/sec), which may be interpreted as a solute velocity to traverse a membrane barrier of thickness δ. The experimental techniques used are also indicated so that values resulting from methods measuring the hydraulic permeability can be distinguished from values obtained by measuring the diffusive permeability coefficient. Obviously, the choice of the method strongly influences the results. A detailed introduction into the different experimental techniques is given by Stein[9,10] and, especially for the water transport in erythrocytes, by Forster.[12]

The nonequilibrium methods make use of osmotic phenomena which result in cell shrinking or cell swelling up to final hemolysis. The related quantity is the hydraulic permeability coefficient, L_p[9-11,53] that can be transformed into the more convenient filtration coefficient, $P_f = R \cdot T \cdot L_p/\bar{V}_w$ (cm/sec), where T (°C) is the temperature and $\bar{V}_w$ is the partial molar volume of water. L_p may be determined by monitoring cell volume change[11,54,55] either by a direct method, like the evaluation of microcinephotography[55] or the measurement of time-dependent 90° light scattering in a flow apparatus.[5,25,54,56] The nonequilibrium condition provides for a net flux of water which is, because of the additional driving forces,[57] different from the pure diffusive exchange. Therefore, measurements of L_p yield P_f values larger than the permeability values, P_d, obtained by equilibrium techniques.[9,22,53,57] The difference between the osmotic permeability, P_f, and the diffusional permeability, P_d, has been interpreted in terms of the presence of water-filled (protein) channels. Inhibitor studies (PCMBS[22,23,26,58])

Table 2
WATER TRANSPORT ACROSS NATIVE CELL MEMBRANES

P cm / 10^{-3}sec	T / °C	Parameter determined	System	Ref.
9.15	20	L_p, stop flow	Human Red Blood Cell (RBC), saline	78
12.0	20	L_p, stop flow	RBC, saline	5
20.0	23	L_p, stop flow	RBC, saline	56
18.3	20			
17.9	23	L_p, stop flow	RBC, saline	79
*16.3	20			
*4.76	20	L_p, stop flow	RBC	25
12.0	20	L_p, swelling	RBC	55
2.8	37	L_p, stop flow	Rabbit, corneal	
2.1	23	L_p, stop flow	endothelium	80
			Sarcoplasm. reticulum	69
*3.8	20	ω, THO, rapid flow	RBC, saline	4
5.6	23	ω, THO, rapid flow	RBC, saline	3
4.5	20			
5.3	23	ω, THO, stop flow	RBC, saline	2
*4.2	20			
*3.3	20	ω, THO, stop flow	RBC, saline	54
*3.0	20	ω, NMR, T_1, T_2, T_{12}	RBC, saline	61, 62
*2.1	20		RBC, plasma	
4.0	20	ω, NMR T_1	RBC, saline	30
*3.3	20	ω, NMR, T_1	RBC, saline	81
*5.0		ω, NMR, T_2		
*5.6	20	ω, NMR, T_2	RBC, saline	64
*5.1	20	ω, NMR, T_2	RBC, saline	65
*9.2	37	ω, NMR, T_2	RBC, saline	85
*5.1	20			
*6.6	37		RBC ghosts	
*3.5	20			
*2.4	20	ω, rapid flow	RBC ghosts	26
*2.9	20			
1.2	20	ω, D_2O, scattering, ghosts	RBC ghosts	28
1.2	20	ω, D_2O, scattering, ghosts	RBC ghosts	27
3.7	20	P_d, THO, capillary	RBC, saline	20
4.5	20	P_d, THO, capillary	RBC, saline	21, 87
7.6	37	P_d, THO, capillary	Dog lung alveolar	37
2.9	20			
8.0	37	P_d, THO, capillary	Dog RBC	39
5.5	20	P_d, THO, capillary	Dog RBC	
3.0	37	P_d, THO, capillary	Calf, pulmonary artery endothelium	86

Note: Water permeability coefficients P (cm/sec) measured in mammalian cells by different experimental techniques. Data for human red blood cells have been converted to the same temperature (T = 20°C) using an activation energy E_a = 22.2 · kJ/mol[62] and a volume-to-surface ratio V/A = 0.684 · 10^{-4}cm[84] in order to provide for better comparison. Converted data are marked by *. Values for temperatures other than 20°C are given to allow comparison with data obtained for other species or ghosts from human red blood cells.

have been performed in order to distinguish between different parallel transfer pathways. For human erythrocytes, an inhibitor-induced diminution of P_f/P_d has been demonstrated,[22,23,59,60] which indicates the possible contribution of porous structures to the water transport.

The second group of methods uses quasi-equilibrium techniques and investigates the purely diffusive self-exchange of water to determine the coefficient, ω, that is correlated to P_d by the relation, $P_d = \omega \cdot R \cdot T$. The diffusive flux is either derived from tracer fluxes[4,13,20,21,26] or is determined by NMR techniques[61-67] by refractive index measurements,[27,29,68-70] or by neutron diffraction.[71] These methods make use of one of the following conditions (1) either the state of water is different on both sides of the membrane, or (2) there is a water tracer, THO and D_2O. Except for the capillary methods[13,20,21] that involve packed cell systems, all these quasi-equilibrium techniques are based on the experimental investigation of diffusional exchange over individual cell membranes. The theoretical analysis[72] is based on first-order exchange kinetics in a closed two-compartment system with the ratio of cell volume to cell surface area as a parameter. It is based on the assumption that membrane penetration is the rate-limiting step and the two compartments that are separated from each other by the membrane, which might contribute to the total diffusion resistance, too, are well stirred. This presumption is surely true for slowly permeating test solutes[22,52,72-76] and in the case of small cells with a diameter < 10 μm.[22] However, problems due to the solute diffusion in adjacent compartments (the unstirred layers) become increasingly important in the case of fast permeants such as water or certain small nonelectrolytes. This results in an underestimation of the permeability values obtained from the evaluation of exchange kinetics data.

As a consequence, many investigators[1,6-8,22,73-75] have tried to correct for this unwanted effect of the extra- and intracellular diffusion. For example, the experimental values obtained from the osmotic methods mentioned above have been evaluated using the measurable time delay of cell shrinkage that is caused by an additional diffusion resistance to correct for the unstirred layers.[5] For the quasi-equilibrium measurements, another attempt consists in dividing the measured apparent permeability[1,7,8,52,76] in one component that is responsible for the solute transfer through the unstirred layer and in a second one that is taken as the pure membrane permeability. This latter value can be derived from the measured apparent permeability coefficient by use of a series diffusion model. By doing so, the total diffusion resistance is postulated to be made up from a series of diffusion barriers that are constituted by the membrane itself and by an adherent membrane-bound layer of solvent. This layer is thought to have a defined and constant thickness and a constant concentration gradient. Both the diffusion coefficient of the solute in this layer and the thickness of the layer have to be measured in a separate experiment.[1,52,76] For example, Brahm[1] has determined the operational thickness of the outer unstirred layer of human red blood cells in a very dilute suspension (hematocrit $< 0.6\%$) to be about 2 μm.

But all these estimations are questionable since the postulate of a uniform concentration gradient over a defined outer boundary[52,76] is not realistic; it is only an appropriate approximation for flat cell surfaces. For small cells having an appreciable curvature, a constant concentration gradient is not an appropriate presumption and the unstirred layer is "smeared out".[8] Thus, the diffusion resistance of a thin unstirred layer will depend on the assumed cell shape and size.[8] Approximations based on unstirred layers with the properties of plane sheets will give results different from approximations based on the assumption of spherical sheets.

All these corrections do not consider the diffusion within the intracellular compartment. Therefore, Hansson Mild[73,74] has performed a theoretical investigation on the dependence of apparent permeability values on intracellular diffusion effects for both a spherical and a rectangular cell model. The measured apparent permeability coefficient was calculated as a function of the intracellular diffusion coefficient, D_2, while the extracellular region was assumed to be well stirred. The extrapolation of D_2 to infinitely large values is equivalent to the assumption that the intracellular medium is well stirred and thus allows to estimate the "true" permeability. For any given pair of the true membrane permeability, P_d, and the intracellular diffusion coefficient, D_2, of a solute, the intracellular space may be regarded to be well stirred up to a critical spherical cell radius.[22] In the case of water exchange rate, this radius is about 10 μm.[22] Since the human red blood cell and most other animal cells are smaller than this value, the influence of the intracellular diffusion is not preponderant. However, for the investigation of the exchange of other solutes such as alcohols, for example, these aspects have to be considered in small cells, too, because their intracellular diffusion coefficient D_2 is usually lower than the water diffusion coefficient. These calculations still neglect the influence of exofacial unstirred layers.

Summing up, the experimental techniques discussed above are confined to the equilibration of solutes between single cells and their surroundings in very dilute suspensions. They suffer from time-resolution limitations because of sampling problems and are thus appropriate only for slow transfer processes but can hardly be applied to the investigation of rapid solute exchange. In contrast, the range of application of the capillary method is not limited by high permeabilities. The newly introduced approach may possibly close the experimental gap between low and high permeabilities. Furthermore, there are no unstirred layer effects because the capillary setup[13,21] strictly avoids all convective interferences. The diffusion characteristics of a composite system such as a cell column are determined by the diffusion coefficients and by the volumetric fraction of each compartment, respectively. It is modeled by a series-parallel-pathway approach. The experimentally measured quantity is the bulk diffusion coefficient D_{eff} for the whole cell suspension instead of a rate constant for the solute exchange between the extra- and intracellular compartments. By this means, the total diffusion resistance for the diffusing substance is enlarged. This results in an increase of the equilibration time and thus expands the spectrum of permeants that can be measured towards more permeable species.

The theory of Redwood et al.[13] requires a system of very closely packed cells with vanishing intercellular space L_o. That is why the method is restricted in application to very high permeability values. Naturally, only populations of cells that fulfill the condition to be packed up to nearly 100%ww can be investigated. In order to satisfy this theoretically founded requirement only cell species can be used for investigation that are very deformable. Erythrocytes, for example, can be closely packed up to ≈98% without substantial cell damage, but other species that are spherically shaped cannot reach values more than 74% at most. Cells with these properties can now be studied on the basis of the more general approach described here. At vanishing extracellular space the solution that has been found coincides with that given by Redwood et al.[13] The new approach is not confined to tightly packed cells. It can be used as long as the cell suspension remains homogeneous on a macroscopic scale. The cells need not be deformable, which has been tested using spheric vesicles from heat-fragmented human red blood cells. The variability of the packing density also allows measurements of permeability values in the intermediate range.

Geometrically, the suspension is assumed to be made up from identical units each containing one single cell and its extracellular surroundings (Figure 1). Since we used well-homogenized suspensions, a cube could be chosen as a cell model for ease of calculation. However, other geometric bodies will do if all the possible orientations relative to the diffusion axis are considered. Evaluation of water-diffusion measurements on human red blood cells using our generalized approach result in $P_{d,C} = 4.47 \times 10^{-3}$cm/sec for the cubic cell model and, as a test, for a quader model in $P_{d,Q} = 4.21 \cdot 10^{-3}$ cm/sec.[21] For spheric vesicles from heat-fragmented human erythrocytes, we get $P_{d,sph} = 5.24 \cdot 10^{-3}$ cm/sec. Obviously, the choice of the geometry does not much influence the data obtained. Summing up, an improved and generalized variation of the capillary technique,[14-18] combined with a bulk-diffusion method,[13,20] has been presented.[21] As compared to previous methods using cells in dilute suspensions, this method is of great advantage as there are no problems of time resolution which is usually limited by the sampling procedure. The technique should be applicable to all kinds of single cell populations no matter how deformable they are and what shape they have. The experimental results are rather insensitive of the geometric cell model chosen for the calculation, but the cell size has to be well defined in order to account for the relative portions of the different diffusion compartments. The method does not suffer from unstirred layer problems and is well suited for highly permeant solutes such as water or alcohols.[87] As a restriction, the cell suspension is required to be homogeneous and the cells have to be sufficiently small with regard to the capillary used. Permeability values $<10^{-3}$ cm/sec are difficult to be measured because of the increasing measuring time. This can be partially compensated for by the variation of the value of the effective diffusion coefficient as a function of the cell packing density. Reproducible permeability values for water permeation have been obtained both for suspensions of human erythrocytes and for suspensions of spheric vesicles from heat-fragmented red blood cells. Many difficulties discussed above have been overcome and the capillary method has turned out to be a very applicable technique for the investigation of rapid exchange processes in composite systems.

V. APPENDIX

A. Solution of the Diffusion Equation

The cell suspension is assumed to consist of identical quaders. The microscopic diffusion properties of such a subsystem yield the same effective diffusion coefficient, D_{eff}, that is also measured as the macroscopic bulk diffusion coefficient of the whole suspension. This implies that the real concentration profile which exists inside the capillary of length ℓ may be approximated by a polygone line whose linear segments are made up by the individual subsystems with their total length, L_1. So each subsystem has to be very small as compared to the capillary's dimension in order to achieve very short local relaxation times, L_2^2/D, and thus a situation of local quasi-stationarity that accounts for an approximately linear local concentration gradient. This is surely granted for cell suspensions with a typical subsystem length of about 10^{-6} m and corresponding relaxation times of about 1 sec. In contrast, bulk relaxation time is defined by the capillary length, ℓ, and is approximately 10^4 to 10^5 sec. Provided that no further transport mechanism has to be accounted for, the transport problem reduces to the solution of a quasi-stationary diffusion equation within a subsystem. The geometric setup of such a typical subsystem is outlined in Figure 9 and suggests further structuring according to the volume elements α, β, and γ. In each of these elements, the diffusion may be calculated individually and then connected by continuity conditions to the diffusion within neighboring elements.

The most complicated section is the region denoted by index β because there are two fluxes, an extracellular flux, j_1, and an intracellular flux, j_2, that have to be coupled across the lateral surface S of the model cell. Concerning this region, calculation is performed according to the procedure of Redwood et al.[13] In the following, the extracellular properties are denoted by the index "1" and the intracellular ones by the index "2". There are further abbreviations introduced that have not been used in the previous sections; a summary of them is given at the end of this chapter. The total subsystem flux j is given by

$$j = j_1 + j_2 = J_1 \cdot A_1 + J_2 \cdot A_2 \tag{21}$$

and J_1 and J_2 are the diffusion currents with their related face areas, A_1 and A_2.

Using Fick's law, $dc/dt = -D \cdot \partial^2 c/\partial x^2$ the total number, $d\hat{n}_1$, of particles that are passing a face surface, A_1, perpendicular to diffusion direction x in the extracellular region on a distance between x and x + dx is given by

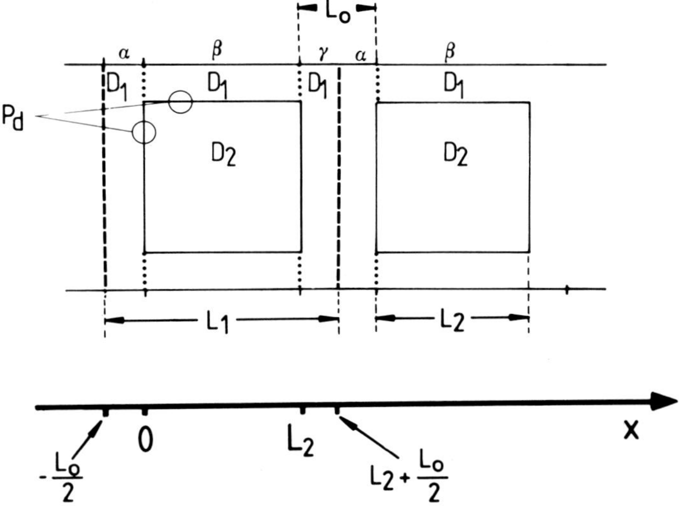

FIGURE 9. Schematic view of the cell suspension model. Cells are imagined as cubes of length L_2 and to be homogeneously distributed within their extracellular medium with a constant distance L_o between each other. To facilitate the calculation, a unit of the suspension with length L_1 consisting of three regions α, β, and γ is introduced. Regions α and γ are simple compartments filled with the extracellular suspension medium contributing equal diffusion resistances $1/D_1$, respectively. Region β is a complex one containing both the model cell (P_d, D_2) itself and also a portion of extracellular suspension medium along the lateral cell surface (D_1).

$$\begin{aligned} d\hat{n}_1 &= n_1(x) - n_1(x + dx) \\ &= [J_1(x) - J_1(x + dx)] \cdot A_1 \cdot dt \end{aligned} \tag{22}$$

There is an additional transverse current, $J_{1,p}$ across the lateral surface, S, of the cell governed by the membrane permeability, P, which results in a transverse particle exchange, $dn_{1,p}$:

$$\begin{aligned} dn_{1,p} &= J_{1,p} \cdot dS \cdot dt \\ &= -P \cdot dS \cdot dt \cdot (c_{1,\beta} - c_{2,\beta}) \end{aligned} \tag{23}$$

where the lateral surface dS is given by

$$dS = \frac{S}{L_2} \cdot dx \tag{24}$$

Thus, the total transfer dn_1 is

$$\begin{aligned} dn_1 &= dn_1 + dn_{1,p} \\ &= A_1 \cdot D_1 \cdot \frac{\partial^2 c_{1,\beta}}{\partial x^2} \cdot dx \cdot dt - \frac{S}{L_2} \cdot P \cdot (c_{1,\beta} - c_{2,\beta}) \cdot dx \cdot dt \end{aligned} \tag{25}$$

The quasi-stationarity condition, $dn_1 = 0$, leads to

$$A_1 \cdot D_1 \cdot \frac{\partial^2 c_{1,\beta}}{\partial x^2} - \frac{S}{L_2} \cdot P \cdot (c_{1,\beta} - c_{2,\beta}) = 0 \tag{26}$$

The intracellular particle exchange, dn_2, leads to a similar equation,

$$A_2 \cdot D_2 \cdot \frac{\partial^2 c_{2,\beta}}{\partial x^2} + \frac{S}{L_2} \cdot P \cdot (c_{1,\beta} - c_{2,\beta}) = 0 \quad (27)$$

because the transverse diffusion current $J_{2,P}$ from inside the cell to the outside is equal to $-J_{1,P}$.

This set of two differential equations has to be true everywhere in the region β; especially at the boundaries, x = 0 and x = L_2, quasi-stationarity conditions demand the amount of solute that is passing the membrane from region α into β and then from region β into γ to be equal. So, there are the boundary conditions for the fluxes into and out of the region β:

$$c_{1,\beta}\Big|_{x=0} = c_\alpha\Big|_{x=0} = c_o$$

$$c_{1,\beta}\Big|_{x=L2} = c_\gamma\Big|_{x=L2} = c_L \quad (28)$$

where c_α and c_γ denote the solute concentrations in region α and in region γ, respectively.

The continuity conditions at the boundaries of region β are

$$-A_2 \cdot D_1 \cdot \frac{\partial c_\alpha}{\partial x}\Big|_{x=o} = P \cdot A_2 \cdot c_o - c_{2,\beta}(x = 0)) = -A_2 \cdot D_2 \cdot \frac{\partial C_{2,\beta}}{\partial x}\Big|_{x=o} \quad (29)$$

and

$$-A_2 \cdot D_2 \cdot \frac{\partial c_{2,\beta}}{\partial x}\Big|_{x=L2} = P \cdot A_2 \cdot (c_{c,\beta}(x = L_2) - c_L) = -A_2 \cdot D_1 \cdot \frac{\partial c_\gamma}{\partial x}\Big|_{x=L2} \quad (30)$$

According to the proposals of Redwood et al.[13] a set of abbreviations is introduced

$$a_1^2 = \frac{P \cdot S \cdot L_2}{D_1 \cdot A_1} \qquad \alpha_2^2 = \frac{P \cdot S \cdot L_2}{D_2 \cdot A_2}$$

$$\alpha^2 = \alpha_1^2 + \alpha_2^2$$

$$r = \frac{\alpha_2^2}{\alpha_1^2} = \frac{D_1 \cdot A_1}{D_2 \cdot A_2} \quad (31)$$

and Equations 26 and 27 are adapted to the microscopic diffusion problem by transforming the coordinate x into a cell-normalized coordinate which is defined by:

$$\chi = x/L_2 \quad (32)$$

So, Equations 26 and 27 convert into

$$D_1 \cdot A_1 \cdot \frac{\partial^2 c_{1,\beta}}{\partial \chi^2} - P \cdot S \cdot L_2 \cdot (c_{1,\beta} - c_{2,\beta}) = 0 \quad (33)$$

and

$$D_2 \cdot A_2 \cdot \frac{\partial^2 c_{2,\beta}}{\partial \chi^2} - P \cdot S \cdot L_2 \cdot (c_{2,\beta} - c_{1,\beta}) = 0 \tag{34}$$

These two equations can be added to give

$$D_1 \cdot A_1 \cdot \frac{\partial^2 c_{1,\beta}}{\partial \chi^2} + D_2 \cdot A_2 \cdot \frac{\partial^2 c_{2,\beta}}{\partial \chi^2} = 0 \tag{35}$$

$$D_1 \cdot A_1 \cdot \frac{\partial c_{1,\beta}}{\partial \chi} + D_2 \cdot A_2 \cdot \frac{\partial c_{2,\beta}}{\partial \chi} = \kappa \tag{36}$$

Here, κ is an integration constant.

It can be interpreted according to Equation 21 and Fick, as:

$$-\left(D_1 \cdot A_1 \cdot \frac{\partial c_{1,\beta}}{\partial \chi} + D_2 \cdot A_2 \cdot \frac{\partial c_{2,\beta}}{\partial \chi}\right) = J_1 \cdot A_1 \cdot L_2 + J_2 \cdot A_2 \cdot L_2 = j \cdot L_2 \tag{37}$$

The second integration gives

$$D_1 \cdot A_1 \cdot c_{1,\beta}(\chi) + D_2 \cdot A_2 \cdot c_{2,\beta}(\chi) = -j \cdot L_2 \cdot \chi + k \tag{38}$$

where k is another integration constant. So,

$$c_{2,\beta}(\chi) = \frac{1}{D_2 \cdot A_2} \cdot [-j \cdot L_2 \cdot \chi - D_1 \cdot A_1 \cdot c_{1,\beta}(\chi) + k] \tag{39}$$

This relation is introduced into Equation 33 which then results in

$$\frac{\partial^2 c_{1,\beta}}{\partial \chi^2} - \alpha^2 \cdot c_{1,\beta}(\chi) = = \alpha_1^2 \cdot \frac{j \cdot L_2}{D_2 \cdot A_2} \cdot \chi - \alpha_1^2 \cdot \frac{k}{D_2 \cdot A_2} \tag{40}$$

The solution of this differential equation is

$$c_{1,\beta}(\chi) = E \cdot e^{\alpha x} + F \cdot e^{-\alpha x} + B \cdot \chi + G \tag{41}$$

where the first two terms solve the homogeneous part of the differential Equation 40 and the last two terms are a special solution of the inhomogeneous part. Equation 41 is inserted into Equation 40, which leads to

$$-\alpha^2 \cdot (G + B \cdot \chi) - \alpha_1^2 \cdot \frac{j \cdot L_2}{D_2 \cdot A_2} + \alpha_1^2 \cdot \frac{k}{D_2 \cdot A_2} = 0 \tag{42}$$

The coefficients G, B, E, and F are now left to be determined. As a first step, separation of variables in Equation 42 results in

$$G = \frac{k}{D_1 \cdot A_1 + D_2 \cdot A_2} \tag{43}$$

$$B = -\frac{j \cdot L_2}{D_1 \cdot A_1 + D_2 \cdot A_2} \tag{44}$$

because Equation 42 is valid for all χ. Using c_1 (χ) as given by Equation 41 and the related continuity and boundary conditions (Equations 28 through 30) for $\chi = 0$ and for $\chi = 1$ results in:

$$c_o = G + E + G \tag{45}$$

$$c_L = G + B + E \cdot e^{\alpha} + F \cdot e^{-\alpha} \tag{46}$$

Furthermore, a combination of Equation 37 and the differentiated form of Equation 41 leads to

$$-D_2 \cdot A_2 \cdot \frac{\partial c_{2,\beta}}{\partial \chi} = j \cdot L_2 + D_1 \cdot A_1 \cdot (\alpha \cdot E \cdot e^{\alpha}\chi - \alpha \cdot F \cdot e{-}^{\alpha}\chi + B) \tag{47}$$

the left-hand side of which is known from the continuity conditions. So we have, using Equation 43,

$$D_1 \cdot A_1 \cdot C_{1,\beta} + D_2 \cdot A_2 \cdot c_{2,\beta} = G \cdot (1 + r) \cdot D_2 \cdot A_2, \text{ for } \chi = 0, \tag{48}$$

and

$$D_1 \cdot A_1 \cdot c_{1,\beta} + D_2 \cdot A_2 \cdot c_{2,\beta} = -j \cdot L_2 + k, \qquad \text{for } \chi = 1. \tag{49}$$

The introduction of three further abbreviations,

$$\pi_o = \frac{P{\cdot}L_2}{D_2} \tag{14}$$

and

$$a_{-} = r \cdot \left(1 - \frac{\alpha}{\pi_o}\right) \qquad a_{+} = r \cdot \left(1 + \frac{\alpha}{\pi_o}\right) \tag{50}$$

finally gives

$$c_o = G - \frac{B}{\pi_o} - a_{-} \cdot E - a_{+} \cdot F \tag{51}$$

and

$$c_L = G + B \cdot \left(1 + \frac{1}{\pi_o}\right) - a + \cdot E \cdot e^{\alpha} - a_{-} \cdot F \cdot e^{-\alpha} \tag{52}$$

Now, there are four equations (Equations 45, 46, 51, and 52) for the determination of the four coefficients G, B, E, and F. G and B have already been defined with Equations 43 and 44. Combination of Equations 45 and 51 and insertion of Equations 43 and 44 finally yields:

$$\frac{1}{\pi_o} = \frac{E}{B} \cdot (1 + a_-) + \frac{F}{B} \cdot (1 + a_+) \tag{53}$$

and, the same procedure being applied to Equations 46 and 52,

$$\frac{1}{\pi_o} = \frac{E}{B} \cdot e^{\alpha} \cdot (1 + a_+) + \frac{F}{B} \cdot e^{-\alpha} \cdot (1 + a_-) \tag{54}$$

which leads to

$$E = \frac{B}{\pi_o} \cdot \frac{(1 + a_-) \cdot e^{-\alpha} + (1 + a_+)}{(1 + a_-)^2 \cdot e^{-\alpha} - (1 + a_+)^2 \cdot e^{\alpha}} \tag{55}$$

and to

$$F = e^{\alpha} \cdot E \tag{56}$$

Subtraction of Equations 45 and 52 results in

$$\begin{aligned} \frac{c_o - C_L}{-B} &= 1 + 2 \cdot \frac{E}{B} \cdot (e^{\alpha} - 1) \\ &= 1 + \frac{2}{\pi_o \cdot (1 + r) + r \cdot \alpha \cdot \coth(\pi/2)} \end{aligned} \tag{57}$$

wherein the second term is abbreviated with z, so that Equation 57 turns into

$$\frac{c_o - c_L}{-B} = 1 + z \tag{58}$$

With the definition of coefficient B as specified in Equation 44, the total flux j in region β is given by

$$j = \frac{1 + r}{1 + z} \cdot \frac{D_2 \cdot A_2}{L_2} \cdot (c_o - c_L) \tag{59}$$

This flux j is the same within all three regions, α, β, and γ. The differential equation for quasi-stationary diffusion in region α,

$$A \cdot D_1 \cdot \frac{\partial^2 c_\alpha}{\partial \chi^2} = 0 \tag{60}$$

can be integrated to yield

$$A \cdot D_1 \cdot c_\alpha = R \cdot L_2 \cdot \chi + S \tag{61}$$

where the integration constants R and S follow from the boundary equations and from the continuity conditions. From the boundary conditions $c_\alpha = c_{1,\beta} = c_O$ for $\chi = 0$ and from the continuity condition, we get

$$\frac{\partial c_\alpha}{\partial_\chi}\bigg|_{\chi=o} = \frac{\partial c_{1,\beta}}{\partial_\chi}\bigg|_{\chi=o}$$

So we obtain

$$A \cdot D_1 \cdot c_o = S$$

and

$$\frac{R \cdot L_2}{A \cdot D_1} = B + \alpha \cdot E - \alpha \cdot F \tag{62}$$

We find another form of Equation 62 that is completely expressed in terms of coefficients that are already known from the solution of the diffusion equation in region β:

$$c_\alpha(\chi) = [B + \alpha \cdot E - \alpha \cdot F] \cdot \chi + c_o \tag{63}$$

The same procedure concerning region γ leads from

$$A \cdot D_1 \cdot \frac{\partial^2 c_\gamma}{\partial \chi^2} = 0 \tag{64}$$

to

$$A \cdot D_1 \cdot c_\gamma = U \cdot L_2 \cdot \chi + V \tag{65}$$

The boundary condition is c_γ $(\chi = 1) = c_{1,\beta} = c_L$ and the continuity condition is

$$\frac{\partial c_\gamma}{\partial_\chi}\bigg|_{\chi=1} = \frac{\partial c_{1,\beta}}{\partial_\chi}\bigg|_{\chi=1}$$

Application to Equation 64 for $\chi = 1$ yields

$$V = A \cdot D_1 \cdot c_L - U \cdot L_2 \tag{66}$$

and

$$U = \frac{A \cdot D_1}{L_2} \cdot [B + \alpha \cdot E \cdot e^{\alpha} - \alpha \cdot F \cdot e^{-\alpha}] \tag{67}$$

This may be used in Equation 65, which then results in

$$c_\gamma(\chi) = [B + \alpha \cdot E \cdot e^{\alpha} - \alpha \cdot F \cdot e^{-\alpha}] \cdot (\chi - 1) + c_L \tag{68}$$

analogous to Equation 63. The unit cell of the cell suspension extends between

$$-L_o/2 \quad \text{and} \quad L_2 + L_o/2$$

as depicted in Figure 9 over a distance of

$$L_2 + L_0$$

There is a total concentration decrease Δc over this distance, given by

$$\Delta c = \left| c_\gamma\left(x = L_2 + \frac{L_0}{2}\right) - c_\alpha\left(x = -\frac{L_0}{2}\right) \right| \tag{69}$$

The flux j that is caused from this concentration gradient is equal to

$$j = D_{eff} \cdot A \cdot \frac{c_\gamma\left(x = L_2 + \frac{L_0}{2}\right) - c_\alpha\left(x = -\frac{L_0}{2}\right)}{L_2 + L_0} \tag{70}$$

where D_{eff} is the measured overall diffusion coefficient. Here, c_α and c_β have to be inserted from Equations 63 and 68. It appears advisable to introduce

$$y = \frac{\frac{L_0}{L_2}}{1 + \frac{L_0}{L_2}} \cdot \frac{z}{1 + z} \cdot \left[\frac{\alpha}{2} \cdot \coth\left(\frac{\alpha}{2}\right) - 1\right) \tag{71}$$

as an abbreviation. So, after some arithmetic effort, a final expression for the effective diffusion coefficient D_{eff} is found as

$$D_{eff} = \frac{(1 + r)}{(1 + z) \cdot (1 + y)} \cdot \frac{D_2 \cdot A_2}{A} \tag{72}$$

which provides for the characteristic properties of diffusion in a suspension of permeable compartments.

B. List of Abbreviations

A	Face surface of elementary unit of the suspension
A_1	Face surface of extracellular portion of elementary unit
A_2	Face surface of model cell
c_0	Starting concentration of tracer (cpm)
$\bar{c}(\Delta T)$	Final tracer concentration after measuring time Δt
c_1	Extracellular solute concentration in region β
c_2	Intracellular solute concentration in region β
c_α	Solute concentration in region α
c_γ	Solute concentration in region γ
c_A	Tracer concentration in the supernatant
c_{Hb}	Hemoglobin concentration
c_s	Tracer concentration in the (packed) cell sediment
ϵ	Extinction coefficient
J	Total diffusion current according to Fick's law
j_1	Extracellular diffusion flux density in region β
j_2	Intracellular diffusion flux density in region β
L_0	Extracellular distance in between two cells
L_1	Total length of an elementary unit
L_2	Total length of the model cell in diffusion direction

L_p	Hydraulic permeability coefficient
$\Delta\ell$	Variation of capillary length
ℓ	Capillary length
M_r	Molecular weight
m_s	Mass of cell sediment
P_f	Osmotic permeability (cm/sec)
P_d	Diffusive permeability (cm/sec)
π_o	Reduced permeability
ρ_{ery}	Density of erythrocytes (g/cm^3)
R_m	Spherical radius, from stereologic evaluation of cell sections
S	Sum of side surfaces of a model cell
Δt	Measuring time
V_1	Extracellular portion of the volume of an elementary unit
V_2	Volume of a model cell
V	Total elementary unit volume: $V = V_1 + V_2$
$\bar{V}_{cell}$	Mean cell volume
V_{tot}	Cell sediment volume
ΔV	Additional solution volume
V_{rel}^o	Relative extracellular volume of the suspension before diluting
V_{rel}	Relative extracellular volume of the suspension
$\bar{V}_{sph}$	Mean spherical volume
$\bar{V}_w$	Molar volume of water
ω	Diffusive permeability coefficient

ACKNOWLEDGMENTS

The authors wish to thank Dr. W. H. Schröder and the staff from the Institut für Neurobiologie, Kernforschungsanlage Jülich, for preparing the electron micrographs. This work was supported in part by a grant from the Deutsche Forschungsgemeinschaft, SFB 160.

REFERENCES

1. **Brahm, J.,** Permeability of human red cells to a homologous series of aliphatic alcohols, *J. Gen. Physiol.*, 81, 283, 1983.
2. **Paganelli, C. V. and Solomon, A. K.,** The rate of exchange of tritiated water across the human red cell membrane, *J. Gen. Physiol.*, 41(2), 259, 1957.
3. **Barton, T. C. and Brown, D. A. J.,** Water permeability of the fetal erythrocyte, *J. Gen. Physiol.*, 47, 839, 1964.
4. **Vieira, F. L., Sha'afi, R. I., and Solomon, A. K.,** The state of water in human and dog red cell membranes, *J. Gen. Physiol.*, 55, 451, 1970.
5. **Sha'afi, R. I., Rich, G. T., Sidel, V. W., Bossert, W., and Solomon, A. K.,** The effect of the unstirred layer on human red cell water permeability, *J. Gen. Physiol.*, 50, 1377, 1967.
6. **Fettiplace, R. and Haydon, D. A.,** Water permeability of lipid membranes, *Am. Physiol. Soc.*, 60(2), 510, 1980.
7. **Mierle, G.,** A method for estimating the diffusion resistance of the unstirred layer of microorganisms, *Biochim. Biophys. Acta*, 812, 827, 1985.
8. **Mierle, G.,** The effect of cell size and shape on the resistance of unstirred layers to solute diffusion, *Biochim. Biophys. Acta*, 812, 835, 1985.
9. **Stein, W. D.,** *The Movement of Molecules Across Cell Membranes*, Academic Press, New York, 1967.
10. **Stein, W. D.,** *Transport and Diffusion Across Cell Membranes*, Academic Press, Orlando, Fla., 1986.
11. **Sha'afi, R. I. and Gary-Bobo, C. M.,** Water and nonelectrolytes permeability in mammalian red cell membranes, *Prog. Biophys. Mol. Biol.*, 26, 105, 1973.

12. **Forster, P. E.,** The transport of water in erythrocytes, in *Current Topics in Membranes and Transport,* Vol. 2, Bronner, F. and Kleinzeller, K., Eds., Academic Press, New York, 1971, 42.
13. **Redwood, W. R., Rall, E., and Perl, W.,** Red cell membrane permeability deduced from bulk diffusion coefficients, *J. Chem. Physiol.,* 64, 706, 1974.
14. **Anderson, J. S. and Saddington, K.,** The use of radioactive isotopes in the study of the diffusion of ions in solution, *J. Chem. Soc.,* Suppl. 2, 381, 1949.
15. **Wang, J. H.,** Self-diffusion and structure of liquid water. I. Measurement of self-diffusion of liquid water with deuterium as tracer, *J. Am. Chem. Soc.,* 73, 510, 1951.
16. **Wang, J. H.,** Self diffusion and structure of liquid water. II. Measurement of self-diffusion of liquid water with O^{18} as a tracer, *J. Am. Chem. Soc.,* 73, 4181, 1951.
17. **Wang, J. H., Robinson, C. V., and Edelman, I. S.,** Self-diffusion and structure of liquid water. III. Measurement of the self-diffusion of liquid water with H^2, H^3, and O^{18} as tracers, *J. Am. Chem. Soc.,* 75, 466, 1953.
18. **Wang, J. H., Anfinsen, C. B., and Polestra, F. M.,** The self-diffusion coefficients of water and ovalbumin in aqueous ovalbumin solutions at 10°C, *J. Am. Chem. Soc.,* 76, 4763, 1954.
19. **Passiniemi, P., Liukkonen, S., and Rastas, J.,** Open-ended plastic capillary method for diffusion experiments with β-active tracers, *Z. Naturforsch.,* 32a, 513, 1977.
20. **Osberghaus, U., Schönert, H., and Deuticke, B.,** A simple technique of measuring high membrane permeabilities of human erythrocytes, *J. Membr. Biol.,* 68, 29, 1982.
21. **Klösgen, B.,** Schnelle Permeation durch Erythrozytenmembranen, Diffusionsmessung in Zellsuspensionen mit der Kapillarmethode und Erweiterung des Messverfahrens auf Suspensionen mittlerer Zellkonzentration, Ph.D. thesis, RWTH, Aachen, Aachen, 1985.
22. **Hansson Mild, K. and Løvtrup, S.,** Movement and structure of water in animal cells. Ideas and experiments, *Biochim. Biophys. Acta,* 822, 155, 1985.
23. **Macey, R. I.,** Inhibition of water and solute permeability in human red blood cells, *Biochim. Biophys. Acta,* 211, 104, 1970.
24. **Macey, R. I.,** Transport of water and urea in red blood cells, *Am. J. Physiol.,* 15, C195, 1984.
25. **Karan, D. M. and Macey, R. I.,** The permeability of the human red cell to deuterium oxide (heavy water), *J. Cell. Physiol.,* 104, 209, 1980.
26. **Brahm, J.,** Diffusional water permeability of human erythrocytes and their ghosts, *J. Gen. Physiol.,* 79, 791, 1982.
27. **Lawaczeck, R.,** Water permeability through biological membranes by isotopic effects of fluorescence and light scattering, *Biophys. J.,* 45, 491, 1984.
28. **Pitterich, H. and Lawaczek, R.** On the water and proton permeabilities across membranes from the erythrocyte ghosts, *Biochim. Biophys. Acta,* 821, 233, 1985.
29. **Engelbert, H.-P. and Lawaczek, R.,** The H_2O/D_2O exchange across vesicular lipid bilayers. Lecithin and mixtures of lecithins, *Ber. Bunsenges. Phys. Chem.,* 89, 754, 1985.
30. **Dix, J. A. and Solomon, A. K.,** Role of membrane proteins and lipids in water diffusion across red cell membranes, *Biochim. Biophys. Acta,* 773, 219, 1984.
31. **Colton, C. K., Smith, K. A., Merrill, E. W., and Reece, J. M.,** Diffusion of organic solutes in stagnant plasma and red cell suspensions, *Chem. Eng. Prog. Symp. Ser.,* 66, 85, 1968.
32. **Crank, J.,** *The Mathematics of Diffusion,* 2nd ed., Clarendon Press, Oxford, 1975, chap. 2 and chap. 12.
33. **Wang, J. H.,** Tracer-diffusion in liquids. I. Diffusion of tracer amount of sodium ion in aqeuous potassium chloride solutions, *J. Am. Chem. Soc.,* 74, 1182, 1952.
34. **Fortes, J.-M., Mercier, M., and Molenat, J.,** Phénomènes de transport dans les solutions aqueuses concentrées d'halogénures alcalins et alcalino-terreux, *J. Chim. Phys. Phys. Chim. Biol.,* 71, 164, 1974.
35. **Spedding, P. L. and Mills, R.,** Tracer diffusion measurements in mixtures of molten alkali carbonates, *J. Electrochem. Soc.,* 113(6), 599, 1966.
36. **Jost, W.,** Diffusion in solids, liquids, gases, in *Physical Chemistry, a Series of Monographs,* Huchinson, E. and van Rysselberghe, P., Eds., Academic Press, New York, 1960.
37. **Garrick, R. A. and Redwood, W. R.,** Membrane permeability of isolated lung cells to nonelectrolytes, *Am. J. Physiol.,* 233, 104, 1977.
38. **Garrick, R. A., Patel, B. C., and Chinard, F. P.,** Permeability of dog erythrocytes to lipophilic molecules: solubility and volume effects, *Am. J. Physiol.,* 238, C107, 1980.
39. **Garrick, R. A., Patel, B. C., and Chinard, F. P.,** Erythrocyte permeability to lipophilic solutes changes with temperature, *Am. J. Physiol.,* 242, C74, 1982.
40. **Garrick, R. A., Patel, B. C., and Chinard, F. P.,** Effects of sulfhydryl and other reagents on the diffusional permeability of dog erythrocytes to small solutes, *Biochim. Biophys. Acta,* 734, 105, 1983.
41. **Ponder, E.,** Red cell structure and its breakdown, in *Protoplasmatologia, Handbuch der Protoplasmaforschung,* Vol. X, No. 2, Heilbrunn, L. V. and Weber, F., Eds., Springer-Verlag, Berlin, 1955.
42. **Deeley, G. O. Th. and Coakley, W. T.,** Interfacial instability and membrane internalization in human erythrocytes heated in the presence of serum albumin, *Biochim. Biophys. Acta,* 727, 293, 1983.

43. **Herrmann, A., Lentzsch, P., Lassmann, G., Ladhoff, A.-M., and Donath, E.,** Spectroscopic characterization of vesicle formation on heated human erythrocytes and the influence of the antiviral agent amantadine, *Biochim. Biophys. Acta,* 812, 277, 1985.
44. **Vettore, L., Falezza, G., De Matteis, M. C., Cetto, G., and Zandegiacomo, M.,** A new method for the determination of sodium and potassium in human red blood cells, using indocyanine green as a marker for trapped plasma, *Clin. Chim. Acta,* 55, 345, 1974.
45. **Maizels, M. and Remington, M.,** Percentage of intercellular medium in human erythrocytes centrifuged from albumin and other media, *J. Physiol.,* 145, 658, 1959.
46. **Nanis, L., Richards, S. R., and Bockris, J. O'M.,** The $\Delta\ell$-effect in capillary-reservoir diffusion measurements, *Rev. Sci. Instrum.,* 36(5), 673, 1965.
47. **Mills, R.,** Self-diffusion in normal and heavy water in the range 1—45°C, *J. Phys. Chem.,* 77, 685, 1973.
48. **Lamm, O. and Polson, A.,** The determination of diffusion constants of proteins by a refractometric method, *Biochemistry,* 30, 528, 1936.
49. **Lenz, F.,** Zur Größenverteilung von Kugelschnitten, *Z. Wiss. Mikrosk.,* 63, 50, 1956.
50. **Lenz, F.,** Die Bestimmung der Größenverteilung von in einem Festkörper eingebetteten kugelförmigen Teilchen mit Hilfe der durch einen ebenen Schnitt erhaltenen Schnittkreise, *Optik* XI(11), 524, 1954.
51. **Anton, H. J. and Voit, E.,** Die Darstellung der Größenverteilung kugeliger Kerne durch Schnittflächenhistogramme, *Microsk. Acta,* 84(1), 17, 1981.
52. **Barry, D. H. and Diamond, J. M.,** Effects of unstirred layers and membrane phenomena, *Physiol. Rev.,* 64, 3, 1984.
53. **Kedem, O. and Katchalsky, A.,** Thermodynamic analysis of the permeability of biological membranes to nonelectrolytes, *Biochim. Biophys. Acta,* 27, 229, 1958.
54. **Sha'afi, R. I., Gary-Bobo, C. M. and Solomon, A. K.,** Permeability of red cell membranes to small hydrophilic and lipophilic solutes, *J. Gen. Physiol.,* 58, 238, 1971.
55. **Jay, A. W. L.,** Hydraulic permeability coefficients of individual human erythrocytes, *Can. J. Physiol. Pharmacol.,* 56, 458, 1978.
56. **Mlekoday, H. J., Moore, R., and Levitt, D. G.,** Osmotic water permeability of the human red cell, *J. Gen. Physiol.,* 81, 213, 1983.
57. **Katchalsky, A. and Curran, P. F.,** *Nonequilibrium Thermodynamics in Biophysics,* Harvard University Press, Cambridge, Mass., 1965.
58. **Toon, M. R., Dorogi, P. L., Lukacovic, M. F., and Solomon, A. K.,** Binding of DTNB to band 3 in the human red cell membrane, *Biochim. Biophys. Acta,* 818, 158, 1985.
59. **Moura, T. F., Macey, R. I., Chien, D. Y., Karan, D., and Santos, H.,** Thermodynamics of all-or-none water channel closure in red cells, *J. Membr. Biol.,* 81, 105, 1984.
60. **Solomon, A. K., Chasan, B., Dix, J. A., Lukacovic, M. F., Toon, M. R., and Verkman, A. S.,** The aqueous pore in the red cell membrane: band 3 as a channel for anions, cations, nonelectrolytes, and water, *Ann. N. Y. Acad. Sci.,* 414, 97, 1983.
61. **Fabry, M. E. and Eisenstadt, M.,** Water exchange between red cells and plasma, measurement by nuclear magnetic relaxation, *Biophys. J.,* 15, 1101, 1975.
62. **Fabry, M. E. and Eisenstadt, M.,** Water exchange across red cell membranes. II. Measurement by nuclear magnetic resonance T_1, T_2, and T_{12} hybrid relaxation. The effects of osmolarity, cell volume and medium, *J. Membr. Biol.,* 42, 375, 1978.
63. **Morariu, V. V. and Benga, G.,** Evaluation of a nuclear magnetic resonance technique for the study of water exchange through erythrocyte membranes in normal and pathological subjects, *Biochim. Biophys. Acta,* 469, 301, 1977.
64. **Conlon, T. and Outhred, R.,** Water diffusion permeability of erythrocytes using an NMR technique, *Biochim. Biophys. Acta,* 288, 354, 1972.
65. **Conlon, T. and Outhred, R.,** The temperature dependence of erythrocyte water diffusion permeability, *Biochim. Biophys. Acta,* 511, 408, 1978.
66. **Small, W. C. and Goldstein, J. H.,** The effect of changing extracellular osmolality on water transport in the human red blood cell as measured by the cell water residence time and the activation energy of water transport, *Biochim. Biophys. Acta,* 640, 430, 1981.
67. **Benga, Gh., Pop, V. I., Popescu, O., Ionescu, M., and Mihele, V.,** Water exchange through erythrocyte membranes: nuclear magnetic resonance studies on the effects of inhibitors and of chemical modification of human membranes, *J. Membr. Biol.,* 76, 129, 1983.
68. **Lawaczek, R.,** On the permeability of water molecules across vesicular lipid bilayers, *J. Membr. Biol.,* 51, 229, 1979.
69. **Kasai, M., Kanemasa, T., and Fukumoto, Sh.,** Determination of reflection coefficients for various ions and neutral molecules in sarcoplasmatic reticulum vesicles through osmotic volume change studied by stopped flow technique, *J. Membr. Biol.,* 51, 311, 1979.
70. **Levitt, D. G. and Mlekoday, H. J.,** Reflection coefficient and permeability of urea and ethylene glycol in the human red cell membrane, *J. Gen. Physiol.,* 81, 239, 1983.

71. **Franks, N. P. and Lieb, W. R.,** Rapid movement of molecules across membranes, *J. Mol. Biol.*, 141, 43, 1980.
72. **Gardos, G., Hoffmann, J. F., and Passow, H.,** Flux measurements in erythrocytes, in *Laboratory Techniques in Membrane Biophysics,* Passow, H. and Stämpfli, R., Eds., Springer-Verlag, Berlin, 1969.
73. **Hansson Mild, K.,** Diffusion exchange between a membrane bounded sphere and its surroundings, *Bull. Math. Biophys.*, 34, 93, 1972.
74. **Hansson Mild, K.,** On the theoretical aspects of tracer exchange in red blood cells, *J. Theor. Biol.*, 39, 183, 1973.
75. **Miller, D. M.,** The effect of unstirred layers on the measurement of transport rates in individual cells, *Biochim. Biophys. Acta,* 266, 85, 1972.
76. **Dainty, J.,** Water relations of plant cells, *Adv. Bot. Res.*, 1, 279, 1963.
77. **Walter, A., Hastings, D., and Gutknecht, J.,** Weak acid permeability through lipid bilayer membranes, role of chemical reactions in the unstirred layer, *J. Gen. Physiol.*, 79, 917, 1982.
78. **Naccahe, P. and Sha'afi, R.,** Effect of PCMBS on water transport across biological membranes, *J. Cell. Physiol.*, 83, 449, 1974.
79. **Craescu, C. T., Cassoly, R., Galacteros, F., and Prehu, C.,** Kinetics of water transport in sickle cells, *Biochim. Biophys. Acta,* 812, 811, 1985.
80. **Liebovitch, L. S., Fischbarg, J., and Koatz, R.,** Osmotic water permeability of rabbit corneal endothelium and its dependence on ambient concentration, *Biochim. Biophys. Acta,* 646, 71, 1981.
81. **Chien, D. Y. and Macey, R. I.,** Diffusional water permeability of red cells independence on osmolality, *Biochim. Biophys. Acta,* 464, 45, 1977.
82. **Sachs, L.,** *Angewandte Statistik,* Springer-Verlag, Berlin, 1978, chap. 2 and chap. 5.
83. **Spearman, C.,** The proof and measurement of association between two things, *Am. J. Psychol.*, 15, 72, 1904.
84. **Richieri, G. V. and Mel, H. C.,** Temperature effects on osmotic fragility, and the erythrocyte membrane, *Biochim. Biophys. Acta,* 813, 41, 1985.
85. **Benga, Gh., Borza, V., Popescu, O., Pop, V. I., and Muresan, A.,** Water exchange through erythrocyte membranes: nuclear magnetic resonance studies on resealed ghosts compared to human erythrocytes, *J. Memb. Biol.*, 89, 127, 1986.
86. **Garrick, R. A., Ryan, U. S., and Chinard, F. P.,** Endothelial cell permeability to water, *Biochim. Biophys. Acta,* 862, 227, 1986.
87. **Klösgen, B., Schönert, H., and Deuticke, B.,** Measurement of rapid membrane permeation in cell suspensions by application of a generalized capillary method, *Biochim. Biophys. Acta,* 939, 29, 1988.

Chapter 5

THE LINEAR DIFFUSION METHOD AND ITS APPLICATION IN STUDIES OF THE PERMEABILITY OF LUNG CELLS TO WATER

Rita Anne Garrick

TABLE OF CONTENTS

I. INTRODUCTION

The linear diffusion technique was reported first for use with erythrocytes and subsequently developed for use with other cell types. In this chapter, I will describe first the linear diffusion technique emphasizing the practical aspects required to use the method. Secondly, I will discuss the permeability of cells from the pulmonary system to water. Much of this work has been done with the linear diffusion technique.

II. LINEAR DIFFUSION TECHNIQUE

The linear diffusion technique is based on the observation that the movement of a solute through a tissue is slow compared to the rate of diffusion through individual cells.[1] The resistance to movement is assumed to be due to, among other things, the presence of many plasma membranes in series. If the resistance due to the plasma membranes can be separated from the other resistances present, then the permeability of the plasma membrane can be measured in a fairly simple experimental system. The technique developed is similar to that used by Gary-Bobo et al.[2] for measuring water diffusion through lipid-water phases and incorporates a modification of a gel-slicing technique for measuring diffusion coefficients in agar gel.[3] Basically, cells are separated, mixed with an extracellular marker, packed into tubing to form columns or sticks of cells, and a radiolabeled solute (including tracer water) is applied to one end of the tubing and allowed to diffuse through the stick. Subsequently, the stick is sliced, each slice is counted, and the diffusion coefficient calculated.

The mean plasma membrane permeability coefficient is calculated from these data with the series-parallel pathway model developed by Perl in Redwood et al.[1] Calculation of the permeability coefficient with this model employs physical measurements of the factors that can contribute resistance to solute movement through the cells, and it separates these resistances from that provided by the plasma membrane. This system provides an experimentally simple, highly reproducible procedure for measuring diffusional movement through a variety of cell types. The application of this technique requires a comprehensive understanding of the experimental procedures. The experimental procedures are described in the following section. This description emphasizes those points of procedure which are not clearly described in previous publications, which are required to use the system with cells other than erythrocytes, or which I have found are confusing to others using this system.[1,4-6]

A. Cell Preparation

The cells are prepared for use by isolation, washing, and packing by centrifugation. The specific procedures will differ with different cell types.[4,6,7] The cells should be packed tightly enough so that the subsequent packing in the polyethylene tubing does not reduce the cells to $<60\%$ of the total column length. The cell count should be approximately 10^7 cells or greater. The actual number of cells in a column will differ with the packing geometry and cell size. A measured amount of the cell pellet is removed with a pipette, placed into a small tube, and a radiolabeled extracellular marker is added. A small portion of the pellet must be left for a background stick of cells. We found that 6 $\mu\ell$ of marker per 0.1 $m\ell$ of cell was a satisfactory concentration with ^{14}C-labeled sucrose, inulin, or raffinose and ^{125}I-human serum albumin at 1 $\mu Ci/\mu\ell$ (with appropriate carrier solutes).[3-7] The cells with the extracellular marker are mixed on a vortex mixer for 2 min at a setting of 2 to 4. A different mixing procedure may be used if distribution of the marker is uniform.

B. Preparation of Diffusion Tubes

The cell suspension with the extracellular marker is drawn into polyethylene tubing (e.g., internal diameter 0.86 mm, Intramedic PE 90®, Clay Adams Inc., Division of Becton,

Dickson and Co., Parsippany, N.J.) by means of a syringe. (An 18-gauge needle fits the PE 90 tubing.) A length of tubing, approximately 8 cm long, is cut with a razor blade and one end clayed securely with Seal-ease (Clay Adams). This is repeated for all of the tubing which contains the cells with the radiolabeled marker. The cells which have been left for a background stick are drawn into tubing and sealed as is done for the other cells. Each length of tubing is then placed, clay end first, into a glass support tube for centrifugation. For PE 90 tubing, 100 $\mu\ell$ micropipettes which are heat sealed at one end and broken approximately 7.5 cm from the sealed end are the correct size. Not all 100-$\mu\ell$ pipettes have the same internal diameter; those from Dade Diagnostics are satisfactory at present. The sticks of cells in the glass supports are placed in a microhematocrit centrifuge and centrifuged for between 10 and 45 min.

A microhematocrit centrifuge has a set speed (13,000 *g* for IEC model MB); this may not be satisfactory for all cells.[8] In that case, the speed may be stepped down with a rheostat or a different centrifuge may be used. In the latter case, a different combination of tubing size and support tubes may be necessary. (See Alpini et al.[8] for an example of this with hepatocytes.) In either case, it is necessary to determine: (1) that the cells are viable at the end of the centrifugation and (2) that they are packed tightly enough for subsequent analysis (see Sections II.B and I.1).

After centrifugation, the stick of cells in the polyethylene tubing is removed from the glass support and the supports are discarded. The tube is sliced with a razor blade just above the cell-supernatant fluid interface and then, below this interface. This yields Section A (Figure 1), which is saved for the extracellular volume determination, Section B, which is discarded, and Section C, which is used for the diffusion experiment. The distance from the interface to the point at which section B is cut is determined by the packing of the cells in the tubing. The tube must be cut below the point at which the extracellular volume becomes constant. For erythrocytes, Section B is cut about 0.5 cm below the interface.

The packed columns of cells are then taped to plastic trays for the diffusion experiments. The sticks must be taped to the trays so that they do not lie against the tray surface during the pulsing procedure and subsequent incubation. In addition, the open end of the tubing should not be covered with the tape. We have found that cellophane (Scotch®) tape is most satisfactory. A length of tape is cut, which is shorter than the tubing, the tubing is placed lengthwise, approximately one third of the way from the bottom of the tape, and the tape is folded up, over the tubing. This allows the tape to be attached to the tray and the column of cells to stand away from the tray during pulsing and diffusion. If the diffusion experiments are to be conducted at other than room temperature, the trays and test solute are incubated for approximately 15 min before continuing.

C. Diffusion Experiment

Each tube is pulsed at its open end with the radiolabeled test solute. Approximately 1 μCi in 1 $\mu\ell$ is added with a microliter syringe. The pulsing is most easily accomplished by standing the trays vertically with the open end of the tube upright. The solute must be dissolved in an appropriate physiological buffer since it is applied directly on to the cells and to prevent "running" of the pulse. When applied, the pulse should "sit on top of the tube" as a distinct drop. If it creeps down the side of the tube, it cannot be used. This is most often a problem when first using the technique or with very lipid-soluble substances. This is probably the source of the problems described by Osberghaus et al.,[9] although sufficient details are not given in their publication to confirm this. In addition, they demonstrate the problems that will be encountered if the solutes are not dissolved in buffer first in their injection of pure hexanol into the polyethylene tubing. If the solute is very lipophilic and can be used at low enough concentration that it can be dissolved in physiological buffers, this limitation can be overcome.[5] After 2 min of contact, the excess tracer is removed by

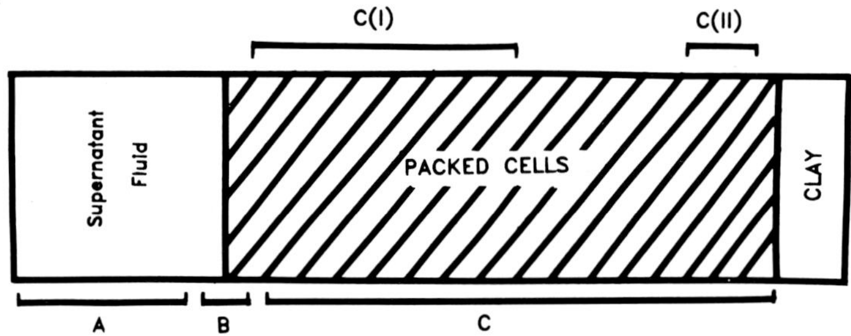

FIGURE 1. Diagrammatic representation of the diffusion tube after centrifugation. A is the supernatant fluid, B the interface, and C the packed cells. C(I) is the section used for the diffusion experiment and C(II) is used for the extracellular volume determination.

touching blotting paper to the side of the drop. The tube is then sealed with a small amount of petroleum jelly applied gently to the open end of the tube.

The trays are placed in an incubator or other temperature-controlled chamber (control should be ±0.1°C) at the desired temperature and diffusion allowed to proceed for a set period of time. The start of the diffusion time is recorded at the time of pulsing and is halfway between the time the pulse is applied and when the excess is removed (i.e., with a 2-min pulse, 1 min after the time the pulse is actually applied). The length of the diffusion time is determined by the diffusion coefficient for the test solute and the slice width desired as is explained below in Section II.D).

D. Slicing Procedure

The tubes are removed from the incubator and frozen immediately with solid CO_2, which covers the entire diffusion tube. This is recorded as the end of the diffusion time. The frozen tubes on the trays are placed on the platform of a gel slicer (Mickle). The razor blades used for slicing and the slicing arm are also cooled. The tube is sliced at a predetermined width which is calculated from the equation:

$$\delta = \sqrt{D \times t_d}/1.5 \qquad (1)$$

where δ is the slice width in mm, D is the estimated diffusion coefficient in units of 10^{-5} $cm^2\ sec^{-1}$ and t_d is the diffusion time in hours. This assumes that 15 slices will be cut to determine the diffusion coefficient. This is designed to provide 15 slices with sufficient counts and an adequate decrease in radioactivity to allow reliable estimation of the half-distance squared in the semilogarithmic plot of the activity vs. the distance from the pulsing point as described below (Figure 2). On a practical level, it is difficult to work with slices smaller than 0.5 mm. Thus, the experimenter can vary the slice width or the diffusion time in order to obtain a reliable data plot.

The slicing procedure that is employed reduces the possibility of radioactive contamination.[1] A new blade is used for each diffusion tube. The tube of cells is positioned under the blade at Section C(I) (Figure 1) approximately 1 cm from the clay-sealed end and three to five slices taken for counting. These will contain the extracellular marker only and are used to calculate the relative extracellular volume. The tube of cells is repositioned with the pulsing point at the open end of Section C(II) under the blade. The tube is sliced sequentially, from the pulsed end, for 15 slices. Each slice is removed as it is cut and put into a scintillation vial which contains 1 mℓ of water. Section A of the tube is also frozen and sliced at the same width as the diffusion tube. In addition, the background stick of cells is sliced and provides the background to be subtracted in the subsequent calculations. The cell pellet in

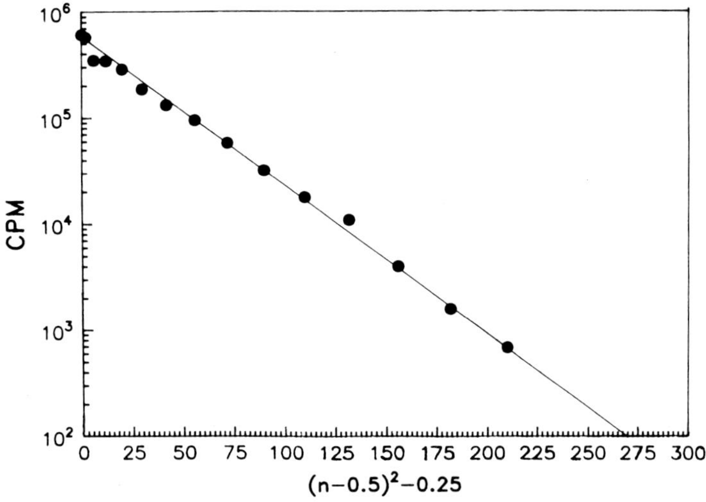

FIGURE 2. Typical semilogarithmic plot of the radioactivity (cpm) as a function of the distance from the pulsing point plotted as $(n - 0.5)^2 - 0.025$, where n is the ordinal number of the slice. From this plot the half-distance squared is obtained and used in Equation 3 to calculate the diffusion coefficient.

the scintillation vial is mixed vigorously with a vortex mixer until the pellet is dispersed in the water. Scintillation fluid is then added for a homogeneous suspension and mixed again. Double-labeled slices (i.e., a ^{3}H-labeled solute and ^{14}C-labeled extracellular marker) are counted on a two-channel liquid scintillation counter and corrected for crossover and background. If an extracellular marker such as ^{125}I-labeled serum albumin is used, the slices from C(I) and A which are used to measure the relative extracellular volume are added to 2 mℓ of water, mixed, and counted in a gamma counter.

Diffusion coefficients through the intracellular medium and through the extracellular medium must also be measured for the subsequent calculation of the permeability coefficient. The extracellular medium is the fluid in which the cells are suspended when drawn into the tubing.[1,4] The procedure for preparing the intracellular material differs with the cell type.[1,4] The diffusion coefficients are measured as for the packed cells but extracellular markers and centrifugation of the tubing are excluded from acellular preparations. We usually use a smaller diameter tubing (PE 50) for these studies.[1,4]

E. Calculation of the Diffusion Coefficient

The diffusion coefficient is calculated as described in Redwood et al.[1] The experimental conditions are assumed to correspond to one-dimensional semi-infinite diffusion through a homogeneous medium with an impulse deposition of tracer of amount m_o (cpm/cm^2) at $x = 0$ and $t = 0$, the solution of which has been defined by Crank[1] as:

$$c(x, t) = m_o (\pi Dt)^{-1/2} - x^2/4 Dt \qquad (2)$$

The development of this equation for this experimental situation is described in Redwood et al.[1] The practical aspects of using it will be described here.

The data are plotted as the logarithm of the cpm vs. the distance from the pulsing point. The distance from the pulsing point is plotted as $(n - 0.5)^2 - 0.25$ where n is the ordinal number of the slice. An example is given in Figure 2. From this plot and using the logarithm of Equation 2, the diffusion coefficient can be calculated. A straight line may be drawn by eye through the 15 data points and the half-distance squared, $X_{1/2}$, calculated. We use a computer program for this step. The half-distance squared, the slice width, δ, in mm, and the diffusion time, t_d, in hours, are used in the following equation to calculate the diffusion coefficient:

$$D = 0.10019(\delta^2 X_{1/2}/t_d) \text{ in units of } 10^{-5}\text{cm}^2 \text{ sec}^{-1} \qquad (3)$$

It was found that the diffusion columns expanded slightly in the longitudinal direction upon freezing. This means that the experimental value $\delta(n - 0.5)$ must be corrected for this expansion in order to calculate the correct diffusion distance. The degree of expansion is measured in the tubing for each of the preparations (i.e., cells, intracellular, and extracellular material) used. The "freezing factors (ff)" indicate a change in column length of 3 to 7% in different preparations so that the ffs range from 0.93 to 0.97. The diffusion coefficient calculated from Equation 3 is corrected for the expansion by the equation:[1]

$$D_c = ff^2 D \qquad (4)$$

Each of the diffusion coefficients is abbreviated as D, for the diffusion coefficient through the packed column of cells, D_1, the diffusion coefficient through the column of extracellular fluid, and D_2, the diffusion coefficient through the column of intracellular material.

F. Measurement of Other Parameters

1. Relative Extracellular Volume

The relative extracellular volume for each column of cells is calculated as described previously.[1] Briefly, the counts in the slices from the supernatant portion (Section A) are corrected for background and crossover and averaged. The same procedure is followed for the slices from Section C(I). The ratio of the counts in the pellet to the counts in the supernatant fluid equals the relative extracellular volume, V_1/V. This should also be corrected for the freezing factor.

2. Tortuosity and Intercellular Diffusion Distance

Additional measurements must be made in order to obtain parameters required for the subsequent calculation of the permeability coefficient. These include measurements to calculate: L_1/L_2, the tortuosity of the extracellular pathway in the tube of packed cells; L, the cell length in the diffusion direction, usually the cell diameter; L_o, the mean intercellular distance in the diffusion direction derived from L and V_1/V; and A/A_2, the cross-sectional area, which is calculated within the permeability program using V_1/V and L_1/L_2.[1,4] The tortuosity is calculated from measurements of diffusion of an extracellular marker, e.g., insulin, through packed cells (D) and through the extracellular material (D_1) and used in the following equation:

$$L_1/L_2 = -1 \pm \sqrt{\frac{1 + 4 \times (D_1/D)}{2x}} \qquad (5)$$

where $x = (1 - V_1/V)(V_1/V)^{-1}$. Usually the tortuosity range is from 1, as for erythrocytes, to 1.2, as for lung cells.[1,4] L_o for a cell assumed to approximate a sphere is calculated as:

$$L_o = \frac{L}{3} \times \frac{(V_1/V)}{(1 - V_1/V)} \qquad (6)$$

3. Relative Exchange Area

The final quantity required before calculating the permeability coefficient is S/A_2, which

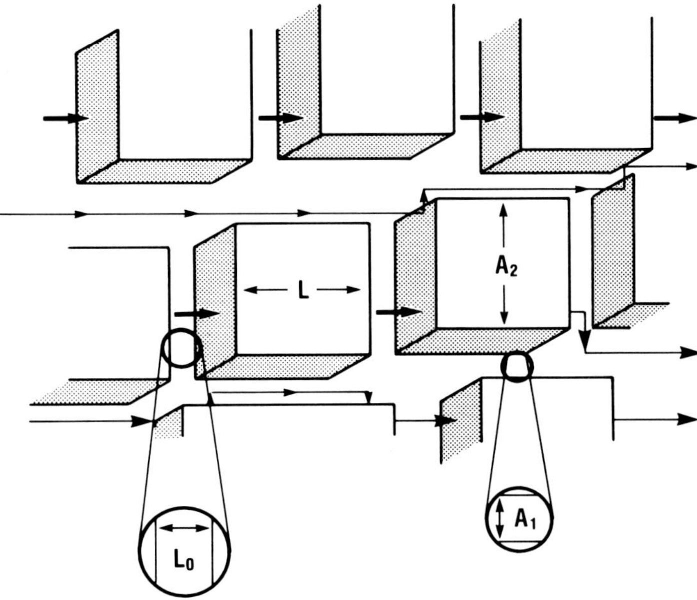

FIGURE 3. Schematic representation of a longitudinal section through the packed cells. The diffusion flux is shown demonstrating some of the possible pathways the diffusing solute might follow. L is the cell length in the diffusion direction, L_o is the intercellular diffusion distance, A_1 is the extracellular area, and A_2 is the intracellular area.

is basically the ratio of the side-face exchange area to the end-face exchange area. In a spherical cell which can be represented by a square, S/A_2 is 4, for erythrocytes, S/A_2 is 8.[1,4] It is also assumed that the permeability characteristics of the end-face and side-face exchange areas do not differ.

G. Permeability Coefficient — Series-Parallel Pathway Model

The permeability coefficient, P_o, is calculated by the series-parallel pathway model as developed by Perl in Redwood et al.[1] The cells are assumed to be packed in a regular manner (Figure 3). The local steady-state diffusional flux is idealized as a one-dimensional intracellular pathway in parallel with a one-dimensional extracellular pathway with solute exchange occurring within the series pathway and between pathways.

1. Basis of the Model

Development of the model is based on the concept that the mean membrane resistance to diffusional movement can be separated from the other resistances present in the packed cell preparation. The resistance provided by other components is determined by measuring the diffusion coefficients through intracellular and extracellular material, the volume of the different pathways, and the tortuosity of the extracellular pathway. The resistance provided by the membrane is deduced then from the diffusion coefficient for the packed cells after accounting for the other resistances present. If the resistance provided by the plasma membrane is close to that provided by another component of the system (i.e., if D_1 or D_2 approaches D), then it becomes more difficult to separate the membrane component. This is indicated by the ease or difficulty in solving Equation 8 below. The area for permeation is calculated from the total cross-sectional area, A, based on the measured values for V_1/V and L_1/L_2 which does not require an estimate of the cell surface area, a definite advantage. A can be calculated as:

$$A/A_2 = 1 + A_1/A_2 = 1 + (V_1/V)(1 - V_1/V)^{-1}(L_1/L_2) \quad (7)$$

2. Solution for the Permeability Coefficient

A program which can be used to solve Equation 5 in Redwood et al.[1] so that:

$$y = (D_2/D) \times (1/AA_2) \times (1 + R)/(1 + Z) = 1 \quad (8)$$

is included as an appendix. The quantities that must be entered to solve the equation are D, D_1, D_2 (as $\times$ 10^{-5} cm^2 sec^{-1}), V_1/V, L_1/L_2 and S/A_2. The quantity that is obtained is defined as π_o. The permeability coefficient is calculated from the relation:

$$\pi_o = P_oL/D_2 \quad (9)$$

In cellular preparations other than erythrocytes, the P_o calculated is redefined as P_o', which must then be corrected for the series resistance of the extracellular pathway in the diffusion direction. This is accomplished by the equation:[4]

$$1/P_o = 1/P_o' + L_o/D_1 \quad (10)$$

This yields the mean membrane permeability coefficient P_o or P_d.

It is important to recall that the value of the permeability coefficient by itself does not indicate anything about the manner in which water moves through the plasma membrane. It is necessary to measure the permeability coefficient under different conditions and to compare it to that measured for other cells, artificial membranes, or other systems.

H. Advantages of the Linear Diffusion Technique

The linear diffusion technique has advantages over other techniques which are commonly used for determining diffusional permeability. These advantages include: (1) relatively simple procedures and commonly available equipment, (2) simple procedures for control of experimental temperature over a wide temperature range, (3) ease of use of many different cell types, (4) use of solutes which penetrate the cells more rapidly than water, and (5) ease of evaluation of changes in the calculated P_d with variation in experimental parameters.

1. Equipment Requirements

The diffusional permeability of erythrocytes has been measured with the rapid-flow technique,[10] rapid-reaction stop-flow apparatus for influx[11] and for efflux,[12] and with a nuclear magnetic resonance (NMR) technique.[13] All of these require complex equipment and operation. Some require high solute concentrations[12] or, as for NMR, are limited by the solutes that can be tested. By contrast, the linear diffusion technique requires centrifuges, gel slicers, incubators, or water baths, if appropriately designed chambers can be made, and counters for gamma or beta emitters. All of these are usually available in the laboratory or are inexpensive to obtain. The linear diffusion technique has been taught to many students and used effectively by them for studies confined to 2 to 3 months in the summer. The procedures, if considered step by step, are simple to follow with a minimum of dexterity required. Calculation of the diffusion coefficients can be done by hand and is readily performed with a microcomputer. It is not reasonable to calculate the permeability coefficient by hand because an iterative process is used, but it is readily calculated with a microcomputer using the program supplied in the appendix.

2. Temperature Studies

With the rapid reaction stop-flow technique, permeation studies at other than room temperature require elaborate procedures and are limited by increasing instrument noise levels as the temperature is increased.[14] Brahm[12] found it necessary to use two different techniques to cover the temperature range from 5 to 37°C in studying the permeation of alcohols into erythrocytes. In the linear diffusion technique, the diffusion columns are placed into a temperature-controlled chamber which can be as simple as an incubator with good thermostatic control. For temperatures lower than room temperature, we use either a refrigerated water bath or a water bath in a 4°C room, which is then heated to the desired temperature.[15] The same can be done with an incubator. None of the other equipment has to be maintained at other than room temperature. The temperature range covered is limited only by the diffusion times necessary for particular solutes (see Section II.E).

From these studies, we can calculate the temperature dependence of permeation. If this is compared to other defined systems, some inferences may be drawn about the way in which water moves through the membrane. Usually, the Arrhenius activation energy, E_a, is calculated. It is important to recognize that some of the assumptions inherent in the Arrhenius analysis are not satisfied when applied to cellular or artificial membranes. Specifically, the assumption is that a single molecular process is rate limiting when an Arrhenius plot is obtained in permeability studies or that a nonlinear plot indicates a change from one rate-limiting process to another. When we compare P_o or E_a for different solutes or in different membranes, we recognize that similarity of these values does not necessarily mean the processes are the same in the systems we compare. The comparison of these coefficients does allow us to proceed with a more systematic approach to formulating the questions that could allow eventual description of these processes.

3. Possible Cell Types

All of the techniques described above were developed to measure diffusional permeation into mammalian erythrocytes. With the development of techniques for isolating viable cells from tissues and the culture of many cell types, the possibility of making similar measurements with other cells presented itself. Many of the rapid-flow or stop-flow instruments have tube diameters which are too small to accommodate other cells, most of which are larger than erythrocytes. While NMR techniques can be used with different cell types, the limitation on available solutes makes this reasonable only for measuring water permeability. The linear diffusion technique has been used successfully to measure the permeation characteristics of a variety of cell types.[4,6-8] At the present time, it is the only method available for measuring diffusional permeability of isolated cells to solutes with a permeability coefficient larger than 10^{-5} cm^2 sec^{-1}.

An additional requirement for calculating the permeability coefficient with these techniques is an estimate of the cell surface area. This has been established with reasonable certainty for many mammalian erythrocytes although differences in the value used by different investigators are still cited as a source of differences in calculated permeability coefficients.[12] With preparations of other cell types, the calculation of surface area is fraught with many more problems.[16] The calculation of the permeability coefficient with the series-parallel pathway model does not require surface area estimation. Because the calculation considers the total cross-sectional area available for diffusion, it can accommodate the presence of microvilli or other surface features of the cells, provided these are distributed on the side-face and end-face cell surfaces. The total cross-sectional area is estimated from experimentally determined values for the relative extracellular volume and the tortuosity (See Section II.F.3). The final calculations require the cell length in the diffusion direction which for most cells is the cell diameter, a value which can be measured with reasonable accuracy.

4. Rapidly Permeating Solutes

Among the solutes that enter cells by diffusional permeation, rapidly permeating substances are of particular interest. It is precisely these substances which can be most difficult or impossible to measure in stop-flow systems.[11,17] In systems that require water movement to measure permeation, only substances that move more slowly than water can be measured.[11] Rapidly permeating solutes are those best suited for measurement in the linear diffusion technique.

I. Limitation of the Linear Diffusion Technique

As with all techniques, there are some limitations to the use of the linear diffusion technique. The limitations can be due to: (1) experimental variables, (2) the test solutes, and (3) the cells used in the experiments.

1. Experimental Variables

The experimental variables include sufficiently tight packing of the cells and adequate cell disruption to determine D_2. The relative extracellular volume that can be tolerated and still allow separation of the membrane resistance will depend upon the percent of the total resistance that is provided by the cellular and extracellular pathways. With erythrocytes we found that with $V_1/V > 0.02$, we could not solve for π_o (i.e., $y \neq 1$) and hence, could not calculate a permeability coefficient.[1,5,15] With cells that can be described as a sphere, we have found that we could tolerate V_1/V of 0.15 to 0.20. Part of the difference in these preparations is the length of the extracellular pathway as represented by L_1/L_2. This is one with erythrocytes and greater than one with spherical cells.

The extracellular volume tolerated depends upon the rate of permeation of the test solute. For more slowly penetrating solutes, a lower V_1/V is necessary to be able to solve for π_o. As mentioned previously, the packing must be consistent throughout the cell column. It is assumed that the extracellular marker chosen will, in fact, be extracellular. If there is any question of this, it must be confirmed. Equilibrium studies over time are the usual method used. In addition, inconsistency in the V_1/V measurements in the diffusion experiments is an indication of cellular uptake.

Determination of D_2 is idealized as measurement of diffusion through the cytoplasmic contents with the plasma membrane stripped away. This is, in fact, hard to attain. In erythrocytes, a hemoglobin preparation does approximate this condition.[5] In other cells, we developed a method of freezing and thawing the cells which are already packed in the tubing to obtain our ''intracellular'' material. We reasoned that, although the plasma membrane was still present, it was disruptured and did not provide a significant barrier to diffusion. This can be evaluated by determining that D_2 is higher than D. This method was chosen in order to have the least dilution of the intracellular material.[4] We did determine that dilution of the intracellular material did not alter the D_2 up to about 15% extracellular volume.[7] In some cells it is necessary to use other methods to prepare the intracellular material.[6] It is important to confirm that 90% or more of the cells are ruptured.

2. Solute-Cell Interaction

Characteristics of the solute interaction with the cell preparation which can limit its use in the technique include metabolism of the solute, uptake by processes other than diffusion, or a low permeation rate which requires a long diffusion time. The first two characteristics must be determined by other techniques if they are thought to be possible. This limitation would apply to all techniques for measuring diffusional permeability. The linear diffusion technique was developed primarily for rapidly permeating solutes. An indication of the permeation rate for a particular solute is obtained by the distance the solute travels in the diffusion tube. If this is very short in initial studies (indicated by a plot of ln c vs. x which falls to zero considerably before 15 slices) then the diffusion time must be lengthened or smaller slices taken. If diffusion times are lengthened, it is especially important to evaluate cell viability at the end of the diffusion time in a companion control tube. A very flat plot indicates a much faster diffusion rate than estimated.

3. Cell Types

For cells to be used in the linear diffusion technique, they must remain viable for the length of the diffusion time, pack sufficiently well for a reasonable V_1/V, and preferably be homogeneous, although this will depend upon the object of the experiment. At the same time, the linear diffusion technique is the only method, of which we are aware, that can be used for cells other than erythrocytes for determining diffusional permeation to water or to solutes that permeate very rapidly. For all of the work of which we are aware, no cell preparation has proved unusable in this technique.

J. Applications of the Linear Diffusion Technique

The applications of the linear diffusion technique include calculation of P_d, assessment of the changes in P_d with variation in temperature or incubation conditions or alteration in membrane properties of the cells, determination of diffusion coefficients through various types of media or cells, and evaluation of the intracellular diffusion and the contribution of this to the total resistance to permeation.[4-8,15,18] Interpretation of these data enables the investigator to evaluate the factors regulating permeation, the permeation processes for solutes, how the permeation could be changed, and the contribution of intracellular or extracellular resistances to permeation.[4-8,15,18] Perhaps the most significant point is that the linear diffusion technique allows measurement of water and solute permeation in many cell types.

III. WATER PERMEABILITY OF LUNG CELLS

The first cells, other than erythrocytes, used in the linear diffusion technique were a preparation of cells from rabbit lungs.[4] Consideration of the permeability of lung cells to water provides both an overview of the approaches used to study these cells and an example of the applications possible with the linear diffusion technique.

A. Background

The movement of water across or between the cells of the intact lung has been an object of study for many years. In the mammalian lung, the epithelial cells on the alveolar side and the endothelial cells on the capillary side provide the main barrier to exchange between these two areas. Under normal physiological conditions, exchanges between these areas are closely controlled. In pathological conditions, such as the development of edema, the normal barrier fails to function.

In order to address the question of the processes responsible for the failure in pathological conditions, it is necessary to first define those factors which regulate movement under normal conditions. These problems have been studied for many years in in vivo and structured organs, usually by means of the multiple-indicator dilution technique.[19] The results of these studies have been interpreted to show that the alveolar side is a much tighter barrier to movement of water or solutes than the capillary side.[20] However, the question of the role of the endothelial and epithelial cells in defining this differential permeability has not been answered. Analyses of these experiments do not provide separation of movement across the cells, by the transcellular pathway, or between the cells, by the paracellular pathway.

Models of greater or lesser complexity have been developed in recent years to achieve this separation using data from indicator dilution experiments.[21,22] In these models, the permeability of the cells is assumed to be similar to that for erythrocytes since these had been the only cells for which diffusional permeability coefficients had been available. In recent years it has become increasingly evident that for rapidly permeating substances, such as water, a value for the permeability coefficient cannot be calculated from indicator dilution studies.[22]

Table 1
DIFFUSION COEFFICIENTS IN PACKED CELLS, D, IN EXTRACELLULAR FLUID, D_1, AND INTRACELLULAR MATERIAL, D_2, FOR TRITIATED WATER ^{3}HHO

	°C	D	D_1	D_2
		× 10^{-5} cm^2 sec^{-1}		
Lung cells	10	0.172	1.05	0.388
	15	0.253	1.22	0.506
	20	0.381	1.66	0.826
	37	0.602	2.05	0.933
Alveolar macrophages	15	0.111	0.892	0.371
	20	0.139	1.07	0.471
	37	0.262	1.51	0.781
Endothelial cells	37	0.682	2.45	0.932

With the development of the linear diffusion technique and of methods for obtaining viable cells from the lung,[23,24] it became possible to measure the diffusional permeability of the lung cells directly. We have conducted studies of this type with mixtures of cell types isolated enzymatically from the lung,[4,7] alveolar macrophages obtained by lavage,[7,25] and endothelial cells from the pulmonary artery.[26] Additional work is available which measures osmotic permeability with alveolar macrophages,[25] lung fibroblasts,[27] and bullfrog lung[28] I will consider first those studies which measure diffusional permeability and then those which measure osmotic permeability.

B. Enzymatically Isolated Lung Cells

Viable cells are isolated enzymatically from rabbit lungs as has been described.[4] The cell population consists of 55 to 65% Type II cells, 20% alveolar macrophages, and the remainder a mixture of endothelial cells, monocytes, or Type I cells. The bulk of the population, therefore, are epithelial cells. The cells are used in the linear diffusion technique as described previously. The extracellular markers used include ^{14}C-inulin, ^{14}C-sucrose, and ^{125}I-albumin. With each of these, a relative extracellular volume of 0.12 to 0.17 was measured.[4,7]

Diffusion coefficients measured from 10 to 37°C for tracer water (^{3}HHO) in packed cells, intracellular material, and extracellular fluid (supernatant from the cellular separation) are listed in Table 1. From these data, it is obvious that D is lower than D_2 and D_2 is lower than D_1 as is required in this technique (Section II.G.1). The relative resistance provided by the plasma membrane and the cytoplasm can be evaluated with the ratio $(D_2\text{-}D)/D_2$.[4] The closer this ratio is to zero, the less of the resistance to movement is provided by the plasma membrane relative to the resistance provided by the cytoplasm. The ratios calculated for the mixed lung cells at temperatures from 10 to 37°C are listed in Table 2. I interpret these data to indicate that the cytoplasm can be an important component of the resistance to water diffusion and that the relative membrane resistance is lower at higher temperatures. This indicates that possible cytoplasmic resistances should be considered when evaluating movement across a cellular barrier and provides some values which can be used in these calculations. In addition, comparison of the ratio $(D_2\text{-}D)/D_2$ for different solutes or for water provides information about the plasma membrane that is useful in analyzing data concerning diffusional permeability.[4]

The permeability coefficients are calculated using V_1/V of 0.15, L_1/L_2 of 1.2, L_o of 0.7 × 10^{-4} cm, L of 10 × 10^{-4} cm, and S/A_2 of 4 for the mixed lung cells. The permeability coefficients calculated between 10 and 37°C for the mixed population of cells from the lung

Table 2
THE RELATIVE INTRACELLULAR AND MEMBRANE CONTRIBUTION TO RESISTANCE TO DIFFUSIONAL MOVEMENT OF ^{3}HHO REPRESENTED BY THE RATIO $(D_2\text{-}D)/D_2$

	$(D_2\text{-}D)/D_2$ °C			
	10	**15**	**20**	**37**
Lung cells	0.49	0.50	0.54	0.35
Alveolar macrophages	—	0.70	0.71	0.66
Endothelial cells	—	—	—	0.27

Table 3
PERMEABILITY COEFFICIENTS CALCULATED WITH THE SERIES-PARALLEL PATHWAY MODEL FOR ^{3}HHO AT 37°C. THE ACTIVATION ENERGY IS LISTED FOR THE LUNG CELLS AND THE MACROPHAGES BETWEEN 20 AND 37°C

	P_o, x 10^{-5} cm sec^{-1} °C				E_a
	10	**15**	**20**	**37**	**kcal mol^{-1}**
Lung cells	68	156	264	755	12
Alveolar macrophages	—	59	70	110	5
Endothelial cells	—	—	—	304	—

are listed in Table 3.[7] There is an increase in P as the temperature is increased from 64 to 755 $\times$ 10^{-5} cm sec^{-1}. The Arrhenius equation can be used to estimate the relation between the change in P and the change in temperature, which is expressed as E_a, the activation energy. Use of this equation does not assume that these data can be analyzed on a molecular level but it does allow comparisons to be made to other data. The activation energy calculated is 12 kcal mol^{-1} between 15 and 37°C and higher than this below 15°C.

C. Alveolar Macrophages

Alveolar macrophages, which are obtained by lavage of the lungs, have been used in the linear diffusion method.[7] These cells, which are a 90% or greater pure population of macrophages, are obtained without enzymatic treatment. The diffusion coefficients obtained for extracellular, intracellular, and cellular preparations are listed in Table 1 and the relative cytoplasmic and membrane contributions to diffusional resistance evaluated in Table 2. The diffusion coefficients D_1 and D_2 are similar to those in the lung cells but D for the cellular preparation is much lower than in the mixed lung cells. If we compare the relative membrane resistance and cytoplasmic resistance to water movement in the two populations of cells, we can see that the membrane provides significantly more of the total resistance to diffusion in the macrophages compared to the lung cells.

The permeability coefficients calculated for the macrophages at the different temperatures are listed in Table 3. The Ps are calculated with V_1/V of 0.12, L_1/L_2 of 1.2, L_o of 0.7 $\times$ 10^{-4} cm, L of 15 $\times$ 10^{-4} cm, and S/A_2 of 4. It is obvious that the permeability coefficients for water in the alveolar macrophages, 59 to 110 $\times$ 10^{-5}cm sec^{-1}, are much lower than

Table 4
HYDRAULIC CONDUCTIVITY, L_p, AND OSMOTIC PERMEABILITY COEFFICIENT, P_f, AT 20 TO 25°C

	°C	$L_p \times 10^{-10}$ cm $(cm\ H_2O)^{-1}$ sec^{-1}	$P_f \times 10^{-5}$ cm sec^{-1}
Alveolar macrophages	20	15.7	217
Fibroblasts	20	27	375
	22	11	153
Bullfrog epithelium	25	34	472

those in the mixed population from the alveolar surface of the lung. The E_a calculated with the data for the macrophages is also much lower, 5 kcal mol^{-1} over the temperature range of 15 to 37°C, than that calculated for the isolated lung cells.

D. Endothelial Cells

I have used one other population of cells from the pulmonary system in the linear diffusion technique; these are endothelial cells isolated from calf pulmonary artery.[26] The cells are isolated without the use of enzymes and cultured as described by Garrick et al.[26,29] These have been studied at 37°C for water permeation. The diffusion coefficients are listed in Table 1, the value of $(D_2\text{-}D)/D_2$ in Table 2, and the permeability coefficient in Table 3. The permeability coefficient for the endothelial cells is calculated with V_1/V of 0.15, L_1/L_2 of 1.14, L_o of 1.07×10^{-4} cm, L of 20.25×10^{-4} cm, and S/A_2 of 4. The diffusion coefficient D and the relative membrane resistance to diffusion are more similar in the lung cells and the endothelial cells than in the macrophages. The permeability coefficient calculated for the endothelial cells, 304×10^{-5} cm sec^{-1} at 37°C, is lower than that for the lung cells and higher than that for the macrophages.

E. Osmotic Permeability of Alveolar Macrophages, Fibroblasts, and Epithelial Cells

The permeability of the alveolar macrophages and lung fibroblasts to water under an osmotic gradient has been measured also.[25,27] Calculation of the osmotic permeability (P_f or P_{os}) is based on the volume changes which occur in the cells when they are exposed to a hyper- or hypoosmotic solution. This change can be monitored in different ways as is described in other parts of this volume. The results are expressed as the hydraulic conductivity, L_p, or can be converted to the permeability coefficient, P_{os}. The values for L_p and P_{os} at 20°C for the alveolar macrophages are listed in Table 4. These values can be compared to those for the lung fibroblasts at 20 and 22°C. The unusual pattern for the fibroblasts has been explained by the authors as due to a change in the plasma membrane lipids which must then provide the area through which the water is moving under the osmotic gradient.[27] A report of the permeability to water of alveolar epithelial cells from intact bullfrog lung lists an L_p equivalent to 34×10^{-10} cm $(cm\ H_2O)^{-1}$ sec^{-1} and a P_{os} of 472×10^{-5} cm sec^{-1} as listed in Table 4.[28] These values are within the same range as for the macrophages.

F. Interpretation of the Results

From interpretation of the data that has been presented, the pathways or processes for water movement into the cells have been suggested. The mixed lung cells have a fairly rapid rate of diffusional water permeation over the temperature range studied. The rate is similar to that reported for lipid bilayers although this can vary with the membrane composition.[30] The temperature dependence for diffusional water permeability, 12 kcal mol^{-1} from 15 to 37°C, is also similar to that for some lipid bilayers. The E_a increases substantially below 15°C. The value of the ratio of the membrane and cytoplasmic resistance is taken to indicate that the cytoplasm can provide a significant resistance to diffusion and that the membrane contribution decreases with increasing temperature. All of these factors are compatible with water moving primarily through the lipid areas of the membranes. Additional evidence comes from work with other solutes.[7]

The alveolar macrophages have a much lower diffusional permeability to water than the lung cells over the same temperature range. In addition, the membrane contributes much more of the total diffusional resistance in the macrophages. The temperature dependence of diffusional water permeability in the macrophages is also much lower, $E_a = 5$ kcal mol^{-1}. This value of E_a is similar to that reported for the diffusion of water in water and water permeation in erythrocytes.[5] The osmotic permeability of water in the alveolar macrophages is considerably higher than the diffusional permeation.

The relation between the osmotic permeability and the diffusional permeability of a cell population is often expressed as the ratio P_f/P_d (where P_f is assumed to be equivalent to P_{os}).[11,25] The interpretation of this ratio is based on work which was performed with artificial membranes where the composition and organization of the membrane was known. This ratio is 3.1 for the macrophages, similar to that in erythrocytes. In the erythrocytes, where there is a high permeation for water under both diffusional and osmotic gradients, the combination of the temperature dependence and the P_f/P_d ratio led to the interpretation that water moves through distinct protein areas of the membrane.[11,15] In the macrophages, a similar interpretation is possible but the low permeation rate would mean a very restricted area of protein for permeation. At this point, the correct interpretation is that additional work with membrane-active agents must be done to determine if water moves primarily through protein areas of the membrane in the alveolar macrophage.

The data available for the isolated endothelial cells do not yet allow any interpretation as to the processes by which water moves through the membrane. However, the patterns that are seen are more similar to those in the isolated lung cells than in the macrophages. An important difference between the lung cells and the endothelial cells may be the lack of enzymatic treatment of the latter. This leaves intact the glycocalyx and any resistance it may provide to water permeation. For the fibroblasts, the authors conclude that water moves through lipid areas while for the bullfrog epithelial cells, a porous pathway is assumed; both are studies of osmotic permeability.

The pattern that emerges is one which demonstrates the variability of possible processes for water permeation into cells from the pulmonary system. Additional studies with pure cell populations, as the endothelial cells which can be isolated without enzymes, will allow a clearer definition of the role of the intracellular compartment, the protein and lipid areas of the membrane, and the glycocalyx in providing resistance to water movement into or across cells. Our ability to do this work is greatly enhanced by the linear diffusion technique.

APPENDIX
BASIC PROGRAM FOR THE CALCULATION OF P (PERMEABILITY COEFFICIENT)

```
5   CLS
20  PRINT "ENTER D,D1,D2,V1V,L1L2,PS":INPUT D,D1,D2,V1V,L1L2,PS
30  D = D*.00001
40  D1 = D1*.00001
50  D2 = D2*.00001
60  A1A2 = V1V/((1 - V1V)*(L1L2))
70  R = (D1/D2)*(A1A2)
75  AA2 = 1 + A1A2
80  PRINT "ENTER GUESSED VALUE OF PI AND THE INCREMENT OF PI FOR
    THE ITERATION."
```

```
90 INPUT PI,DEX
100 LPRINT" ITERATION NO. Y DIFFERENCE PI"
110 N = 1
120 ALPHA = SQR(PS*PI*(1 + 1/R))
130 COTH = EXP(ALPHA) + 1
140 COTH = COTH/(EXP(ALPHA) - 1)
150 Z = 2/ (PI* (1 + R) +R*ALPHA*COTH)
160 Y = (D2/D)*(1/AA2)*(1+R)/(1 + Z)
170 DIFF = ABS (1 - Y)
180 LPRINTTAB(9); N; TAB(20); Y; TAB(32); DIFF; TAB(55); PI
190 IF DIFF < .0001 THEN 260
200 IF N > 9 THEN 220
210 PI = PI + DEX:N = N + 1:GOTO 120
220 PRINT "10" ITERATION STEPS HAVE BEEN DONE, DO YOU WANT TO ENTER
    NEW VALUES FOR PI AND INCREMENT?"
230 INPUT ANS$
240 IF ANS$ = "YES" THEN 80
245 IF ANS$ = "NO" THEN 500
250 PRINT "YES OR NO PLEASE":GOTO 230
260 LPRINT"PI = "; PI
270 LPRINT"NUMBER OF ITERATIONS="; N
280 LPRINT"DIFFERENCE" = ; DIFF
500 DAN = 16
0
```

ACKNOWLEDGMENTS

The work reported from our laboratory was supported by: NIH Heart, Lung and Blood Institute grants HL 12879 and HL 12974 and the Fordham University Faculty Research Council.

REFERENCES

1. **Redwood, W. R., Rall, E., and Perl, W.,** Red cell membrane permeability deduced from bulk diffusion coefficients, *J. Gen. Physiol.,* 64, 706, 1974.
2. **Gary-Bobo, C. M., Lange, Y., and Rigaud, J. L.,** Water diffusion in lecithin-water and lecithin-cholesterol-water lamellar phases at 22°C, *Biochim. Biophys. Acta,* 233, 243, 1971.
3. **Schantz, E. J. and Lauffer, M. A.,** Diffusion measurements in agar gel, *Biochemistry,* 4, 658, 1962.
4. **Garrick, R. A. and Redwood, W. R.,** Membrane permeability of isolated lung cells to nonelectrolytes, *Am. J. Physiol.,* 233, C104, 1977.
5. **Garrick, R. A., Patel, B. C., and Chinard, F. P.,** Permeability of dog erythrocytes to lipophilic molecules: solubility and volume effects, *Am. J. Physiol.,* 238, C107, 1980.
6. **Polefka, T. G., Redwood, W. R., Garrick, R. A., and Chinard, F. P.,** Permeability of Novikoff hepatoma cells to water and monohydric alcohols, *Biochim. Biophys. Acta,* 642, 67, 1981.
7. **Garrick, R. A. and Chinard, F. P.,** Membrane permeability of isolated lung cells to nonelectrolytes at different temperatures, *Am. J. Physiol.,* 243, C285, 1982.
8. **Alpini, G., Garrick, R. A., Jones, M. J. T., Nunes, R., and Tavoloni, N.,** Nonelectrolyte permeability of isolated rat hepatocytes, *Am. J. Physiol.,* 251, C872, 1986.
9. **Osberghaus, U., Schonert, H., and Deuticke, B.,** A simple technique of measuring high membrane permeabilities of human erythrocytes, *J. Membr. Biol.,* 68, 29, 1982.
10. **Paganelli, C. V. and Solomon, A. K.,** The rate of exchange of tritiated water across the human red cell membrane, *J. Gen. Physiol.,* 41, 259, 1957.

11. **Sha'afi, R. I., Gary-Bobo, C. M., and Solomon, A. K.,** Permeability of red blood cell membrane to small hydrophilic and lipophilic solutes, *J. Gen. Physiol.,* 58, 238, 1971.
12. **Brahm, J.,** Permeability of human red cells to a homologous series of aliphatic alcohols, *J. Gen. Physiol.,* 81, 283, 1983.
13. **Conlon, T. and Outhred, R.,** Water diffusion permeability of erythrocytes using an NMR technique, *Biochim. Biophys. Acta,* 288, 354, 1972.
14. **Galey, W. R., Owen, J. D., and Solomon, A. K.,** Temperature dependence of nonelectrolyte permeation across red cell membranes, *J. Gen. Physiol.,* 61, 727, 1973.
15. **Garrick, R. A., Patel, B. C., and Chinard, F. P.,** Erythrocyte permeability to lipophilic solutes changes with temperature, *Am. J. Physiol.,* 242, C74, 1982.
16. **Grinstein, S., Rothstein, A., Sarkadi, B., and Gelfand, E. W.,** Responses of lymphocytes to anisotonic media: volume-regulating behavior, *Am. J. Physiol.,* 246, C204, 1984.
17. **Brahm, J.,** Diffusional water permeability of human erythrocytes and their ghosts, *J. Gen. Physiol.,* 79, 791, 1982.
18. **Garrick, R. A., Patel, B. C., and Chinard, F. P.,** Effects of sulfhydryl and other reagents on the diffusional permeability of dog erythrocytes, *Biochim. Biophys. Acta,* 734, 105, 1983.
19. **Chinard, F. P., Vosburgh, G. J., and Enns, T.,** Transcapillary exchange of water and other substances in certain organs of the dog, *Am. J. Physiol.,* 183, 221, 1955.
20. **Taylor, A. E. and Garr, K. A.,** Estimation of equivalent pore radii of pulmonary capillary and alveolar membranes, *Am. J. Physiol.,* 218, 1133, 1970.
21. **Bassingthwaighte, J. B. and Goresky, C. A.,** Modeling in the analysis of solute and water exchange in the microvasculature, in *Handbook of Physiology Section 2: The Cardiovascular System,* Vol. 4(Part 1), Renkin, E. M. and Michel, C. C., Eds., American Physiological Society, Bethesda, Md., 1984, 549.
22. **Chinard, F. P. and Cua, W. O.,** Endothelial extraction of tracer water is independent of temperature in dog lungs, *Am. J. Physiol.,* 250, H1017, 1986.
23. **Ayuso, M. S., Fisher, A. B., Parilla, R., and Williamson, J. R.,** Glucose metabolism by isolated rat lung cells, *Am. J. Physiol.,* 225, 1153, 1973.
24. **Kikkawa, Y. and Yoneda, K.,** The type II epithelial cell of the lung. I. Method of isolation, *Lab. Invest.,* 30, 76, 1974.
25. **Garrick, R. A., Polefka, T. G., Cua, W. O., and Chinard, F. P.,** Water permeability of alveolar macrophages, *Am. J. Physiol.,* 251, C524, 1986.
26. **Garrick, R. A., Ryan, U. S., and Chinard, F. P.,** Endothelial cell permeability to water, *Biochim. Biophys. Acta,* 862, 227, 1986.
27. **Rule, G. S., Law, P., Kruuv, J., and Lepock, J. R.,** Water permeability of mammalian cells as a function of temperature in the presence of dimethylsulfoxide: correlation with the state of the membrane lipids, *J. Cell Physiol.,* 103, 407, 1980.
28. **Schaeffer, J. D., Kim, K.-J., and Crandall, E. D.,** Effects of cell swelling on fluid flow across alveolar epithelium, *J. Appl. Physiol.,* 56, 72, 1984.
29. **Ryan, U. S., Mortara, M., and Whitaker, C. T.,** Methods for microcarrier culture of bovine pulmonary artery endothelial cells avoiding the use of enzymes, *Tissue Cell,* 12, 619, 1980.
30. **Fettiplace, R. and Haydon, D. A.,** Water permeability of lipid membranes, *Physiol. Rev.,* 60, 510, 1980.

Chapter 6

WATER PERMEABILITY OF NOVIKOFF HEPATOMA CELLS

Thomas G. Polefka

TABLE OF CONTENTS

I. INTRODUCTION

The idea that a discrete boundary surrounds a cell was established in the late 19th century as a result of permeability studies. These studies showed that cells were normally impermeable to dyes and that some barrier must be responsible for the osmotic properties of the cells.[1,2] In 1899, Overton suggested that this invisible barrier was lipoidal and behaved as both a selective solvent, permitting rapid penetration of lipid soluble solutes, and as a set of molecular sieves, allowing selective permeation of small polar solutes.[3] In 1922, Chambers demonstrated that eosin would not penetrate an ameoba; however, when the dye was injected into the cell, the dye quickly distributed itself throughout the cytoplasm, only to be arrested at the cell boundary.[4,5] Direct visualization of the plasma membrane by electron microscopy confirmed the existence of this cell barrier which had been proposed by Nageli and Cramer almost a century earlier.[1]

Permeability studies continue to provide a wealth of information on cell membrane structure and function. Solomon and co-workers have used permeability studies to probe the red cell membrane.[6-9] The morphological and biochemical simplicity of this cell yields information which may be interpreted unequivocally in terms of the molecular architecture of this membrane. To date, there are only limited data on the permeability properties of other isolated mammalian cell types.[10-14]

In this review, water transport across the Novikoff hepatoma cell membrane will be examined.[15,16] Two phenomenological coefficients, namely, the diffusional permeability coefficient for water, (Pd), and osmotic permeability coefficient, (Lp), have been measured and will be used to show how permeability studies can yield information on membrane structure and organization.

II. THE HYDRAULIC PERMEABILITY COEFFICIENT

The hydraulic permeability coefficient, Lp, is a measure of the resistance water experiences as it crosses a barrier under the influence of a hydrostatic or osmotic pressure gradient.[17] Lp is measured conveniently by exposing a cell suspension to nonisoosmolar solutions of a nonpermeating solute which initiates water movement into or out of the cells. The rate of swelling or shrinking is a measure of the hydraulic permeability coefficient according to Equation 1:

$$Lp = \frac{d(Vol)/dt}{S \cdot (\Delta\pi) \cdot R \cdot T} \qquad (1)$$

where d (Vol)/dt is the cell volume change with time; S, the cell membrane area; $(\Delta\pi)$, the difference between initial and final osmotic pressures; R, the universal gas constant; and T the absolute temperature.

To apply Equation 1 to the calculation of Lp, the cells must behave as ideal osmometers, i.e., cell volume change must be inversely proportional to changes in osmotic pressure. Figure 1 illustrates this relationship for the Novikoff hepatoma cell. The figure shows that for osmolalities between 290 (isosmolal) and 450 mOsm/kg, cell volume decreases as a linear function of the reciprocal osmolality (1/Osm).

Although there are several methods available to monitor cell volume changes, the spectrophotometric techniques based on measurements of the intensity of scattered light are the simplest and most widely used techniques. The technique used to measure Lp for the Novikoff cell was modified from Blok et. al.[18] Briefly, a 3-mℓ aliquot of Novikoff cells (4.0×10^6 cells/mℓ) in 290 mOsm/kg Na^+/K^+ Ringers buffer was transferred to a cuvette maintained at constant temperature. After temperature equilibration and the establishment of a zero-

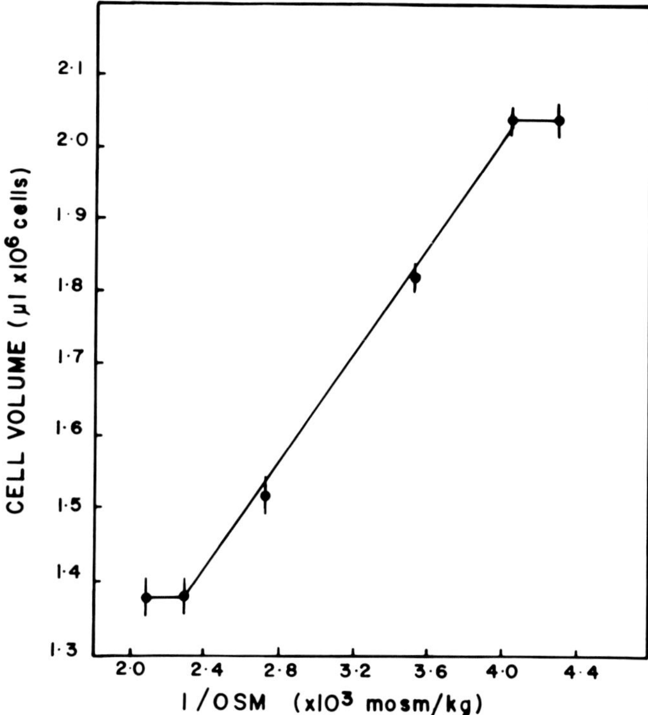

FIGURE 1. Novikoff cell volume as a function of the reciprocal of buffer osmolality. Cell volume measured from cytocrit and corrected for extracellular space. (From Polefka, T. et al., *Biochim. Biophys. Acta*, 642, 79, 1981. With permission.)

time baseline, a solute concentration gradient was induced by rapidly injecting 25 to 100 $\mu\ell$ of either 1.5 *M* NaCl or 2.7 *M* sucrose into the cell suspension. Cell volume change was followed as a function of time in a spectrophotometer at a wavelength of 600 nm. Data from a representative shrinking experiment and isotonic control run are shown in Figure 2. The figure shows that after an initial decrease in absorbance due to dilution of the cell suspension, there is an increase in absorbance as water moves out of the cells and the cells shrink.

Since volume changes are not directly measured in the spectrophotometric technique, ideal osmotic behavior (cf. Figure 1) is assumed to exist if the reciprocal absorbance (1/A) is proportional to 1/osmolality. The parameter 1/A is given by the following equation:

$$(1/A) = (1/A_o - 1/A_\infty)\ Ao \tag{2}$$

where A_o and A_∞ are the absorbance at $t = 0$ and $t = \infty$, respectively. Figure 3 shows that with respect to scattered light, the Novikoff cell may be considered an ideal osmometer. This result permits the change in reciprocal absorbance with time to be related to cell volume change with time according to the following equation:

$$\frac{d(Vol)}{dt} = K\,\frac{d(1/A)}{dt} \tag{3}$$

where d(Vol)/dt has the same meaning as Equation 1, (1/A) is from Equation 2, and K is a constant of proportionality.

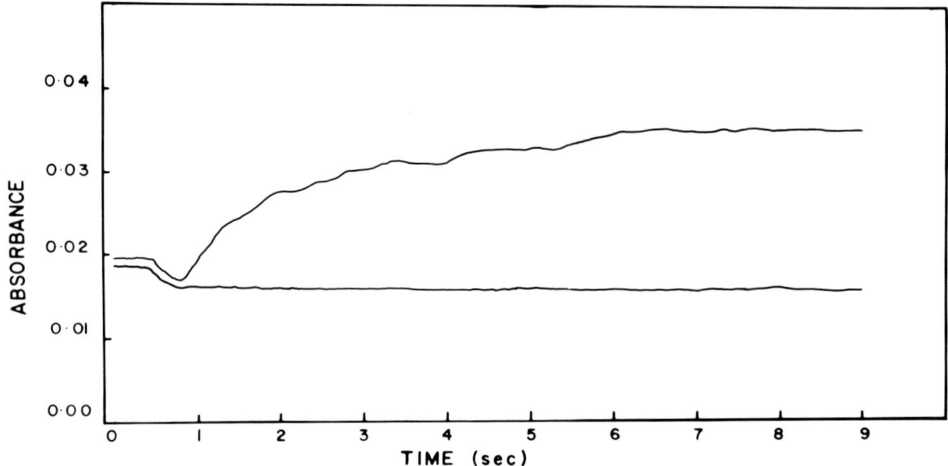

FIGURE 2. Shrinking of Novikoff cell in hypertonic buffer with time. Recorder tracing shows change in absorbance (600 nm) of Novikoff cell suspension following injection of 100 μℓ of 1.5 *M* saline (upper curve) or 100 μℓ of isotonic buffer (lower curve). (From Polefka, T. et al., *Biochim. Biophys. Acta*, 642, 79, 1981. With permission.)

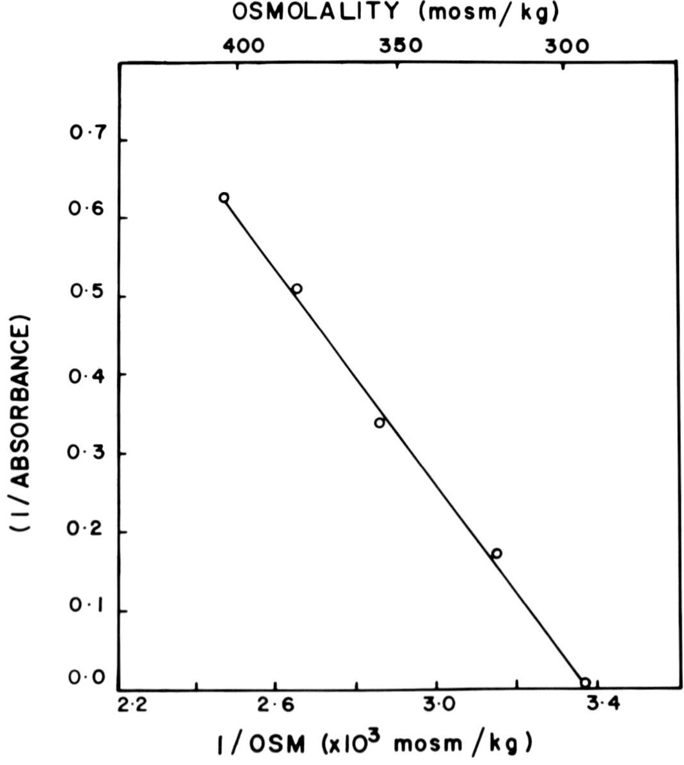

FIGURE 3. Relationship between the relative absorbance change of Novikoff cell suspension (at equilibrium) with the reciprocal of buffer osmolality. (From Polefka, T., et al., *Biochim. Biophys. Acta*, 642, 79, 1981. With permission.)

The rate of cell volume change due to water flux can be calculated in two ways. In the differential method, the initial ascending portion of the shrinkage curve (of Figure 2) is used to calculate d(Vol)/dt from Equations 2 and 3. The alternate approach uses the entire shrinkage curve by integrating the equation described by Lucké et al.[19] as done by Hempling.[12]

However, it should be noted that both methods do not always yield the same value for Lp. Indeed, Garrick et al.[20] have found that Lp calculated for alveolar macrophages by the differential method was one third the value determined by the integral method. For the Novikoff cells, Lp calculated by the differential method was one half the value calculated by the integral method. The difference in Lps is attributed to the different cell membrane areas used by each method in the calculation of Lp. In the integral method, cell membrane area is extrapolated from cell volume measurements (assumes spherical geometry) and therefore represents a minimum value for cell area. For the differential method as applied to the Novikoff cell, membrane area was calculated by a morphometric analysis which included the area contribution provided by the microvilli. This area was 247% greater than the area calculated from spherical geometry. The effect of this additional area, which is unaccounted for in the integral method, is to reduce Lp by approximately 50%. Rule et al.[21] have found that similar values of Lp are obtained for lung fibroblasts by either rate methods. This may reflect the relatively smooth surface of these cells which yield similar cell membrane areas.

Lp values reported for the Novikoff cell membrane in this report were calculated by the differential method. This method was used because it reflected the additional membrane area provided by the microvilli and to avoid the influence of ''unstirred layers'' due to limited mixing in the cuvette.

III. DIFFUSIONAL PERMEABILITY COEFFICIENT OF WATER

The diffusional permeability coefficient is a measure of the resistance a solute experiences as it traverses a barrier under the influence of a concentration gradient.[17] Three techniques have been employed to measure the diffusional permeability coefficient for water (Pd). In the rapid-flow technique used by Paganelli and Solomon,[7] a cell suspension is mixed in a rapid-flow mixing chamber with an isotonic buffer solution containing tritiated water. The mixture is forced down a tube which has filter-covered ports to permit sampling at various intervals along the tube. Distance along the tube is converted into units of time from the known velocity of the flow. The second method for measuring Pd is known as the linear diffusion technique.[22] In this technique, cells are drawn into polyethylene tubing and packed by centrifugation. The resulting packed-cell column is pulsed with tritiated water and diffusion is permitted to proceed for several hours. From the distribution of radioactivity along the cell column, a diffusion coefficient is calculated. This diffusion coefficient is used along with diffusion coefficients measured through buffer (reflects diffusion through extracellular fluid) and cell homogenate (reflects diffusion through intracellular compartment) to calculate a permeability coefficient by means of a series-parallel pathway model. A complete description of this model is given in Redwood et al.[22] The last and most recently developed technique for measuring Pd is nuclear magnetic resonance. This technique measures proton exchange between cell water and extracellular water in terms of spin-lattice relaxation time, T_1. The reader is referred to the studies of Marariu and Benga[23,24] and the chapter by Benga (Chapter 3) in this volume for a detailed discussion of this technique.

The water permeability coefficient for the Novikoff cell membrane was measured with the linear diffusion technique described by Redwood et al.[22] The general usefulness and adaptability of this technique to cells other than erythrocytes have been shown previously by Garrick and Redwood for isolated lung cells.[10] A detailed description of the experimental procedures used for lung cells is described by Dr. Garrick (Chapter 5 in this volume) and the procedure for the Novikoff cells can be found in Polefka et al.[15]

Table 1
OSMOTIC PERMEABILITY COEFFICIENTS MEASURED AT DIFFERENT OSMOTIC PRESSURE GRADIENTS

Osmotic gradient (mOsm/kg)	Lp × 10^{-10} cm (cm H_2O)$^{-1}$ sec^{-1}	P_{os} × 10^{-5} cm sec^{-1}
60	6.29 ± 0.33 (12)	87 ± 5 (12)
88	5.54 ± 0.23 (9)	77 ± 3 (9)
117	5.70 ± 0.30 (9)	79 ± 5 (9)

Note: Values are means ± S.E. with number of determinations in parentheses. P_f was calculated from Lp data by the following relationship: P_{os} = Lp(RT/Vw), where Vw is the partial molar volume of water.
From Polefka, T. et al., *Biochim. Biophys. Acta,* 642, 79, 1981. With permission.

Table 2
HYDRAULIC PERMEABILITY COEFFICIENT FOR SEVERAL MEMBRANE SYSTEMS

Cell type/model system	Lp × 10^{-9} cm (cm H_2O)$^{-1}$ sec^{-1}	Ref.
Human erythrocyte	12.44	26
Ehrlich ascites cell	2.43	12
Human leukocyte	1.33	13
Human lymphocyte	0.75	14
Human leukemic leukocyte	0.82	13
Dog alveolar macrophage	1.57	20
Novikoff ascites cell	0.59	15
Lipid bilayer	1.21	27
Lipid bilayer (treated with amphotericin B)	29.10	27

IV. WATER TRANSPORT ACROSS THE NOVIKOFF CELL MEMBRANE

The paucity of published data on the permeability properties of mammalian cells other than erythrocytes provided the impetus for the current investigation. The red cell, until recently, was the only mammalian cell membrane whose passive permeability properties were characterized in sufficient detail to permit a description of its barrier properties to water.[25] Three lines of evidence will be presented here to support the hypothesis that water movement across the Novikoff cell membrane does not appear to occur through pores.

Hydraulic permeability coefficients (Lp) for the Novikoff cell were calculated according to Equation 1 using the initial ascending portion of the shrinkage curve (Figure 2) to estimate the rate of cell volume change. The hydraulic permeability coefficients (Lp) for the Novikoff cell as a function of several osmotic gradients are presented in Table 1. The permeability coefficients did not vary significantly ($p > 0.05$) with the osmotic gradient inducing the flux and a group mean hydraulic permeability coefficient of 6.0 ± 0.9 (SE) × 10^{-10} cm (cm H_2O)$^{-1}$ sec^{-1} was calculated. This value is compared with Lp measured in other cell types and for model membranes (Table 2). The data indicate that water flux across most membranes falls within a narrow range (this close agreement is interesting when one considers the sensitivity of Lp to the estimate of cell membrane area, cf. Equation 1) and only in the case of human red cells and bilayers treated with a pore-forming antibiotic is water transport substantially more rapid.

Table 3
WATER PERMEABILITY COEFFICIENTS AT VARIOUS TEMPERATURES

Temperature (°C)	Pd × 10^{-5} cm sec^{-1}
20	97 ± 10 (18)
30	125 ± 9 (6)
37	163 ± 26 (10)

Note: The values reported are means ± S.E. with the number of determinations given in parentheses. From Polefka, T. et al., *Biochim. Biophys. Acta*, 642, 67, 1981. With permission.

The diffusional permeability coefficients for tritiated water for the Novikoff cell membrane are listed in Table 3. These values were calculated from the measured linear diffusion coefficients by means of the series-parallel pathway model.[22] The value reported at 20°C is not different from the value measured for planar lipid bilayer membranes (138 × 10^{-5} cm sec^{-1}) and dog alveolar macrophages (70 × 10^{-5} cm sec^{-1}).[27,28] However, Pd for the Novikoff cell membrane is one fourth the value measured in human erythrocytes (approximately 400 × 10^{-5} cm sec^{-1}) and one third the permeability observed for a mixed population of lung cells (264 × 10^{-5} cm sec^{-1}).[7,24,28]

V. THE USE OF Lp AND Pd TO DISTINGUISH WATER PERMEATION PATHWAYS

Estimates of membrane porosity have been obtained by comparing the diffusional permeability coefficient for water with the osmotic permeability coefficient. The rationale for this comparison is that differences between these coefficients reflect quantitative differences in water movement. Classically, a ratio of the osmotic permeability coefficient and the diffusional permeability coefficient (Pos/Pd) greater than unity may be taken as evidence for viscous flow through channels. Sha'afi and Gary-Bobo have reported Pos/Pd values >3.3 for mammalian erythrocytes.[25] A similar value has been reported for dog alveolar macrophages.[20] For the Novikoff cell, a Pos/Pd value equivalent to 0.9 is calculated at 20°C. In the absence of pore-forming antibiotics, bilayers have been shown to yield Pos/Pd values of unity and a value >3.8 in the presence of these antibiotics. Taken together, the Pos/Pd data suggest that water movement across the Novikoff cell membrane is utilizing the same route regardless of the driving force.

Another parameter employed to differentiate permeation pathways is based on the temperature dependence of water transport across membranes. According to Stein,[29] permeation of water through unrestricted pores would be characterized by an activation energy (E_a) similar to the 4.5 kcal mol^{-1} value measured for the self-diffusion of water. The apparent activation energies for osmotic and diffusional water flow across the Novikoff cell membrane are presented in Table 4 along with the values for human red cells and lipid bilayers. These examples represent the only systems in which E_a has been determined for both Pos and Pd. With respect to osmotic water flow, E_a reported for the Novikoff cell is significantly greater than the values reported for the red cells which indicates that the Novikoff cell membrane is more restrictive to water flow than the red cell membrane. The E_a reported for the Novikoff cell membrane is within the range (10 to 15 kcal/mol) measured for lipid model membranes

Table 4
APPARENT ACTIVATION ENERGIES FOR OSMOTIC AND DIFFUSIONAL WATER MOVEMENT ACROSS SEVERAL MEMBRANE SYSTEMS

Cell type/model system	Osmotic (kcal mole^{-1})	Diffusional (kcal mole^{-1})	Ref.
Self-diffusion of water		4.5 ± 0.3	31
Human erythrocyte	3.3 ± 0.4	6.0 ± 0.2	25
Novikoff cell	10.4 ± 0.4	6.7 ± 1.9	15, 16
Lipid bilayers	14.6	12—14	31, 32

Note: The values reported are means ± S.E.

which lack aqueous pores. From these data, water movement across the Novikoff cell is believed to occur via a lipid pathway.

The apparent activation energies for the diffusional movement of water across the red cell and the Novikoff cell are similar. Both values are not very different from the E_a for the self-diffusion of water and suggest that these membranes are rather unrestrictive. For the red cell, the E_a for both osmotic and diffusional water flow are consistent with permeation through pores. The data for the Novikoff cell is much more difficult to interpret. Although E_a for the diffusional movement of water across the Novikoff cell membrane is 30% lower than the value measured for osmosis, the values are not statistically different ($p > 0.05$).[16] The large variance associated with the E_a for diffusion across the Novikoff cell membrane does not permit differentiation between permeation through aqueous pores or across a lipid matrix. However, other data presented in this report suggest strongly that water permeation across the Novikoff cell membrane is via a lipid pathway.

Vieira et al.[33] have shown for the red cell that the product of Lp and the bulk viscosity of water (η_w) remain virtually independent of temperature (i.e., $Lp \cdot \eta_w$ = constant). These authors have taken this as evidence for viscous flow through cylindrical pores according to Poiseuille's law.[25] An equivalent analysis for the Novikoff cell is presented in Table 5. The data indicate that $Lp \cdot \eta_w$ is not independent of temperature. This result is interpreted to indicate that water flow is not occurring through an aqueous phase as described for the red cell. However, if water movement is occurring across a lipid phase, Lp may be related to membrane lipid viscosity (e.g., membrane fluidity). Esko et al.[34] have measured the viscosity (η_M) of LM cell membranes by fluorescence polarization. If the assumption is made that the viscosity of the Novikoff cell is not very different from the LM cell membrane, then an estimate of the dependence of Lp on membrane fluidity can be made. The data in Table 5 indicate that the product of the hydraulic permeability coefficient and membrane viscosity ($Lp \cdot \eta_M$) is constant over the temperature range 20 to 40°C. This result suggests that filtration across the Novikoff cell membrane may be a function of membrane lipid viscosity (fluidity) and supports the concept of water movement across the lipid matrix.

VI. CONCLUSION

Water transport across the Novikoff cell membrane exhibits the following characteristics: (1) the ratio of osmotic permeability coefficient to diffusional permeability coefficient, Pos/Pd, is close to unity, (2) the apparent E_a for osmotic water movement is significantly greater than the E_a for the self-diffusion of water, and (3) osmotic movement of water is directly related to membrane lipid fluidity (inversely related to membrane viscosity). These items are taken as evidence that, in contrast to red cell membranes, the Novikoff cell membrane does not have aqueous pores.

Table 5
RELATIONSHIP OF Lp AND THE VISCOSITY OF WATER OR MEMBRANE LIPID WITH TEMPERATURE

Temperature (°C)	Lp (× 10^{-10}) (cm dyne^{-1} sec^{-1})	η_w[a] (cP)	η_M[b] (cP)	Lp · η_w (× 10^{-10}) (cm)	Lp · η_M (× 10^{-10}) (cm)
20	7.04	1.00	4.70	7.0	33.1
27	9.04	0.85	3.65	7.7	33.0
31	11.83	0.78	2.85	9.2	33.7
40	18.69	0.65	1.80	12.1	33.6

[a] Viscosity of water, see Reference 35.
[b] Viscosity of membrane lipid, see Reference 34.

From Polefka, T. et al., *Biochim. Biophys. Acta,* 642, 79, 1981. With permission.

REFERENCES

1. **Nageli, C. and Cramer, C.,** *Pflanzenphysiologische Untersuchungen, I. Heft,* F. Schultess, Zurich, 1855.
2. **Pfeffer, W.,** *Osmotische Untersuchungen,* Engelmann, Leipzig, 1877.
3. **Overton, E.,** Ueber die allgemeinen osmotischen Eigenschaffen der Zelle, ihre vermutlichen Ursachen und ihr Bedeutung fur die, *Physiol. Vierteljahresschr. Naturforsh. Ges. Zurich,* 44, 88, 1899.
4. **Chambers, R.,** A microinjection study on the permeability of the starfish egg, *J. Gen. Physiol.,* 5, 189, 1922.
5. **Chambers, R.,** The nature of the living cell revealed by microdissection, *Harvey Lect.,* 22, 41, 1926.
6. **Sidel, V. W. and Solomon, A. K.,** Entrance of water into human red cells under an osmotic gradient, *J. Gen. Physiol.,* 41, 243, 1957.
7. **Paganelli, C. V. and Solomon, A. K.,** The rate of exchange of tritiated water across the human red cell membrane, *J. Gen. Physiol.,* 41, 259, 1957.
8. **Sha'afi, R. I., Gary-Bobo, C. M., and Solomon, A. K.,** Permeability of red cell membranes to small hydrophilic and lipophilic solutes, *J. Gen. Physiol.,* 58, 238, 1971.
9. **Naccache, P. and Sha'afi, R. I.,** Effects of PCMBS on water transport across biological membranes, *J. Cell Physiol.,* 83, 449, 1974.
10. **Garrick, R. A. and Redwood, W. R.,** Membrane permeability of isolated lung cells to nonelectrolytes, *Am. J. Physiol.,* 233, C104, 1977.
11. **Sha'afi, R. I. and Volpi, M.,** Permeability of rabbit polymorphonuclear leukocyte membrane to urea and other nonelectrolytes, *Biochim. Biophys. Acta,* 436, 242, 1976.
12. **Hempling, H. G.,** Permeability of the Ehrlich ascites cells to water, *J. Gen. Physiol.,* 44, 365, 1960.
13. **Hempling, H. G.,** Heats of activation for the exoosmotic flow of water across the membrane of leukocytes and leukemic cells, *J. Cell Physiol.,* 81, 1, 1972.
14. **Hempling, H. G., Thompson, S., and Dupre, A.,** Osmotic properties of human lymphocytes, *J. Cell Physiol.,* 93, 293, 1977.
15. **Polefka, T. G., Redwood, W. R., Garrick, R. A., and Chinard, F. P.,** Permeability of Novikoff hepatoma cells to water and monohydric alcohols, *Biochim. Biophys. Acta,* 642, 67, 1981.
16. **Polefka, T. G., Garrick, R. A., and Redwood, W. R.,** Osmotic permeability of Novikoff hepatoma cells, *Biochim. Biophys. Acta,* 642, 79, 1981.
17. **Kedem, O. and Katchalsky, A.,** A physical interpretation of the phenomenological coefficients of membrane permeability, *J. Gen. Physiol.,* 45, 143, 1961.
18. **Blok, M. C., van Deenen, L. L. M., and DeGier, J.,** Effect of the gel to liquid crystalline phase transition on the osmotic behavior of phosphatidylcholine liposomes, *Biochim. Biophys. Acta,* 433, 1, 1977.
19. **Lucké, B., Hartline, H. K., and McCutcheon, M.,** Further studies on the kinetics of osmosis in living cells, *J. Gen. Physiol.,* 14, 405, 1931.
20. **Garrick, R. A., Polefka, T. G., Cua, W. O., and Chinard, F. P.,** Water permeability of alveolar macrophages, *Am. J. Physiol.,* 251, C524, 1986.

21. **Rule, G. S., Law, P., Kruuv, J., and Lepack, J. R.,** Water permeability of mammalian cells as a function of temperature in the presence of dimethylsulfoxide: Correlation with the state of the membrane lipids, *J. Cell Physiol.*, 103, 407, 1980.
22. **Redwood, W. R., Rall, E., and Perl, W.,** Red cell membrane permeability deduced from bulk diffusion coefficients, *J. Gen. Physiol.*, 64, 706, 1974.
23. **Morariu, V. V. and Benga, Gh.,** Water diffusion through erythrocyte membranes in normal and pathological subjects: nuclear magnetic resonance investigations, in *Membrane Processes. Molecular Biology and Medical Applications*, Benga, Gh., Baum, H., and Kummerow, F. A., Eds., Springer-Verlag, Berlin, 1984, 121.
24. **Benga, Gh., Borza, V., Popescu, O., Pop, I. V., and Muresan, A.,** Water exchange through erythrocyte membranes: nuclear magnetic resonance studies on resealed ghosts compared to human erythrocytes, *J. Membr. Biol.*, 89, 127, 1986.
25. **Sha'afi, R. I. and Gary-Bobo, C. M.,** Water and nonelectrolyte permeability in mammalian red cell membranes, *Progr. Biophys. Mol. Biol.*, 26, 105, 1973.
26. **Rich, G. T., Sha'afi, R. I., Romualdex, A., and Solomon, A. K.,** Effect of osmolality on the hydraulic permeability coefficient of red cells, *J. Gen. Physiol.*, 52, 941, 1968.
27. **Andreoli, T. E. and Troutman, S. L.,** Analysis of unstirred layers in series with "tight" and "porous" lipid bilayer membranes, *J. Gen. Physiol.*, 57, 464, 1971.
28. **Garrick, R. A. and Chinard, F. P.,** Membrane permeability of isolated lung cells to nonelectrolytes at different temperatures, *Am. J. Physiol.*, 243, C285, 1982.
29. **Stein, W. D.,** *The Movement of Molecules Across Cell Membranes*, Academic Press, New York, 1967, 124.
30. **Wang, J. H.,** Self-diffusion coefficient of water, *J. Phys. Chem.*, 58, 686, 1965.
31. **Redwood, W. R. and Haydon, D. A.,** Influence of temperature and membrane composition on the water permeability of lipid bilayers, *J. Theor. Biol.*, 22, 1, 1969.
32. **Everitt, C. T., Redwood, W. R., and Haydon, D. A.,** Problem of boundary layers in the exchange diffusion of water across bimolecular lipid membranes, *J. Theor. Biol.*, 22, 20, 1969.
33. **Vieira, F. L., Sha'afi, R. I., and Solomon, A. K.,** The state of water in human and dog red cell membranes, *J. Gen. Physiol.*, 55, 451, 1970.
34. **Esko, J. D., Gilmore, R., and Glaser, M.,** Use of a fluorescence probe to determine the viscosity of LM cell membranes with altered phospholipid composition, *Biochemistry*, 16, 1881, 1977.
35. **Weast, R. C., Ed.,** *Handbook of Chemistry and Physics*, 50th ed., The Chemical Rubber Co., Cleveland, Ohio, 1970.

Chapter 7

VIRALLY INDUCED WATER AND DIVALENT CATION MOVEMENT ACROSS PLASMA MEMBRANES

Kingsley J. Micklem and Charles A. Pasternak

TABLE OF CONTENTS

I. INTRODUCTION

A. Virally Induced Alterations in Cellular Permeability

The infective cycle of viral entry, replication, and release has been extensively studied. Generally, the emphasis has rested on the molecular events through which infection by cytocidal viruses allows the viral nucleic acid to supercede the host cell genome in dictating the direction of protein synthesis.[1,2] This process usually leads to the destruction of the cell, as a consequence either of the cessation of normal cellular repair mechanisms, or of recognition and destruction by the immune system.[2] Another possible outcome, typified by infection with retroviruses, is the integration of part of the viral genome into the host genome to produce transformed cells with heritable altered behavior;[3] this usually involves loss of growth control, so that a shutdown of cellular DNA, RNA, and protein synthesis, as occurs in the first instance, is replaced by an increase in these processes.

It is self-evident that cellular lysis following cytocidal viral infection is due to a catastrophic increase in plasma membrane permeability which leads to the release of intracellular macromolecules.[4] However, permeability changes do occur which precede and are distinct from lysis.[5] Such changes are said to be demonstrable in infected cells with membranes intact by the criterion of trypan blue exclusion, early in infection.[6,7] Indeed, these changes have been postulated to be important in the initiation and maintenance of the infected state,[8,9] although not all workers agree with this interpretation.[10-15]

In addition to these permeability changes as a consequence of viral infection and replication, there are some viruses which induce permeability changes during, and as a consequence of, their entry.[16,17] These effects are restricted to certain types of paramyxoviruses and related viruses[18] which are characteristically hemolytic in vitro. Although the molecular mechanism of the increased permeability is not known, it appears to result from the fusion of the enveloped virus with the plasma membrane,[19] which causes the incorporation of the viral envelope into the membrane.[20] While it has been postulated that diffusion of viral proteins in the plane of the membrane leads to a global change in membrane permeability,[21] or that an enzymic modification occurs,[4] it is more likely that the virus envelope is already relatively leaky[22] prior to fusion, which then initiates a series of sequential permeability changes[23] that can ultimately lead to lysis.

Although these immediate permeability changes are likely to have limited physiological or pathological significance,[24] most of our knowledge regarding changes in plasma membrane function comes from the studies on these immediate changes. In particular, the induction of permeability changes by Sendai virus and their effects on the physiology of the cell[26,27] have been extensively studied.[17,28]

The work described here was originally motivated by the need to investigate the role of calcium ions in the induction and inhibition of Sendai virus-induced permeability changes. The particular interest in calcium and the work with radionuclides which preceded and validated the magnetic resonance spectroscopy is outlined in the following sections.

B. Mobilization of Calcium During Sendai Virus Treatment

Calcium ions inhibit the permeability changes caused by hemolytic paramyxoviruses.[29,30] Experiments with ^{45}Ca showed that Sendai virus also increases the uptake and release of Ca.[31] This increased "exchangeability" of calcium was particularly interesting in view of the hypothesis that calcium is involved in the mechanism of virus-induced cell fusion[32] and in membrane fusion in general.[33,34] The possibility that calcium was involved in the destabilization of the membrane prior to fusion was correlated with this "exchangeability" of calcium ions.[31] However, realization that the permeability changes included an increase in the permeability to Ca^{2+} suggested that the mobilization of calcium is a consequence and not a cause of the permeability changes.[35]

The use of radionuclide tracers gives little information concerning the localization of calcium and so the possibility of using a spectroscopic technique to localize calcium was investigated. Manganese is a potentially useful analog of calcium,[36] which because it is paramagnetic, lends itself to magnetic resonance studies. The main part of this chapter will be devoted to a discussion of the movement of Mn^{2+} into control and virus-affected cells. Two cell types have been analyzed in some detail: human erythrocytes and Lettre cells, a line of malignant cells grown as an ascitic suspension in mice. Initial experiments were performed to ensure that manganese behaved in a similar way to calcium in control and virus-affected cells, in order to validate its use as an analog.

C. Manganese as a Calcium Analog

First, manganese can replace calcium in inhibiting the permeability changes induced by the virus.[35] Second, $^{54}Mn^{2+}$ and $^{45}Ca^{2+}$ are taken up by cells, with or without virus, to the same extent and with the same kinetics and tempeature dependence; the release kinetics are also similar.[35] Third, Mn competes with ^{45}Ca for binding, indicating that Mn binds to Ca binding sites; Ca however competes only partially with ^{54}Mn, indicating that Mn may also bind to Mg binding sites.[35]

In general, it was found that manganese behaves in a similar way to calcium and one may conclude that it is a useful analog of calcium. Two magnetic resonance techniques have been used, electron paramagnetic resonance of Mn (II) and proton nuclear magnetic resonance, to investigate the localization of manganese added to cell suspensions before and after virus treatment.

II. LOCALIZATION OF CATION BINDING BY MANGANESE ELECTRON PARAMAGNETIC RESONANCE SPECTROSCOPY

A. Erythrocytes

The binding of the paramagnetic ion Mn^{2+} can be studied directly by electron paramagnetic resonance (EPR) spectroscopy. Spectral changes reflect the alteration of the environment of the Mn^{2+}. Having established that Mn^{2+} is a useful analog of Ca^{2+}, a series of experiments to characterize the normal interaction of Mn^{2+} with human erythrocytes and Lettre cells was undertaken. Erythrocytes were studied initially because they provide a simple model system and because of extensive studies on the water permeability of erythrocytes by relaxation enhancement using added Mn^{2+}.[37]

Human erythrocytes were obtained by venipuncture and rapidly diluted into Hepes-buffered saline (HBS)[19] to prevent clotting. After washing, the cells were resuspended in HBS at 10^{10} cells per mℓ and stored on ice before use. Lettre cells were prepared as described previously[38] and resuspended at 5×10^8 cells per mℓ before use. Sendai virus was prepared as before.[38] EPR spectra were recorded with a Varian E12 spectrometer at 9.2 GHz, modulation 0.4 mT. The samples were introduced into the cavity in a quartz $0.5 \times 2 \times 15$-mm flat cell and maintained at 37°C using a variable temperature accessory.

When erythrocytes are added to a solution of Mn^{2+} the six-line spectrum of aqueous Mn^{2+} (Figure 1a) is seen. Measurement of the Mn^{2+} concentration in the cell suspension and the supernatant after centrifugation by comparison of the derivative peak height with standard Mn^{2+} concentrations under the same conditions was made at several Mn^{2+} concentrations. The results in Table 1, indicate that the Mn^{2+} is excluded from the cells and that there is no loss of signal due to binding or other phenomena. This is in accord with observations which validate the technique of measuring water permeability by relaxation enhancement and require that no Mn^{2+} enters the cells.[39]

B. Lettre Cells

When such an experiment is repeated with Lettre cells, a strain of murine Ehrlich ascites

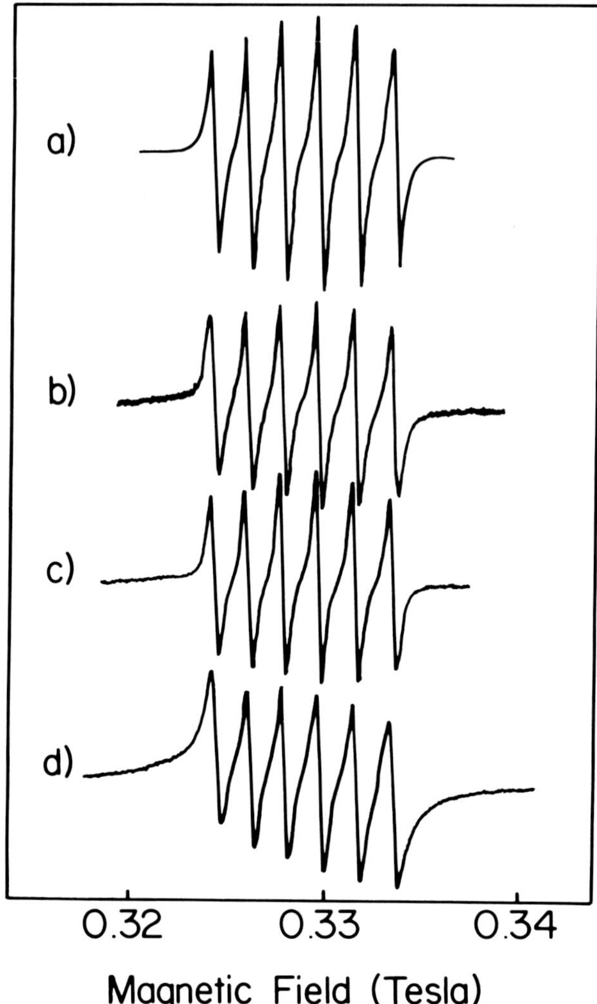

FIGURE 1. EPR spectra of cell suspensions. Spectra of 0.5 m*M* Mn^{2+} at 37°C (a) in Hepes buffer (gain 200), (b) in a suspension of 5×10^7 Lettre cells (gain 400), (c) in the supernatant of (b) (gain 400), and (d) in the pellet of (b) (gain 400). (From Getz, D., Gibson, J. F., Sheppard, R. N., Micklem, K. J., and Pasternak, C. A., *J. Membr. Biol.*, 50, 311, 1979. With permission.)

cells, the observations are quite different. The intensity of the Mn signal is reduced by an amount greater than that expected by the addition of the cells, as shown in Figure 1b and Table 1. There are two possible explanations for this behavior; broadening of part of the Mn spectrum to the point where it is undetectable and reduction or oxidation to an EPR silent state. The latter explanation was rejected for the following reasons: there is no credible redox reaction with such activity in cells, and the lost signal can be recovered by treating the cells with HCl to release bound Mn. This bound, broadened Mn is designated Mn_b.[40]

If the Lettre cell suspension is centrifuged to produce a supernatant and a pellet, measurement of the Mn^{2+} signal indicates a deviation from the expected behavior. After correction for the volume of extracellular water in the cell suspension, the expected signal intensity for the supernatant is smaller than that actually measured. The signal from the pellet is correspondingly larger than expected. This Mn^{2+} sediments with the cells and so appears

Table 1
Mn CONCENTRATIONS (m*M*) IN CELL SUSPENSIONS CALCULATED FROM THE DERIVATIVE PEAK HEIGHT REFERRED TO A CALIBRATION CURVE PREPARED WITH Mn IN HEPES BUFFER

Cells	Total Mn added	Mn conc in suspension	Mn conc in pellet	Mn conc in supernatant
Erythrocytes	2.5	2.45	0.12	3.6
	5.0	5.0	0.43	8.3
Lettre	0.5	0.36	0.21	0.43
	1.0	0.62	0.30	0.75
	2.0	1.25	0.44	1.34
Sendai virus-treated Lettre	0.5	0.17	0.28	0.24

to be cell-associated Mn^{2+} with a spectrum similar to that of aqueous Mn^{2+}. This form of cell-bound Mn^{2+} is designated Mn_b*.[40]

It is not aqueous Mn^{2+} trapped in the interstices of the cell pellet because it is present in cell pellets after centrifugation through oil, which reduces interstitial trapping. Further, the amount of trapped supernatant when estimated by measurement of the amount of trapped [^3H]insulin was typically 18%,[40] which is much lower than the proportion of the excess Mn^{2+} in the pellet (usually 50 to 60%; Table 1). Mn_b* has been identified as surface-bound Mn^{2+} by the following criteria. If Mn_b* were intracellular it would be expected to be as tightly bound as intracellular Ca^{2+}, most of which is bound to intracellular membranes.[41] Such tightly bound Mn^{2+} would exhibit a characteristic broadening of its EPR spectrum, and this is not a characteristic of Mn_b*. Also, Mn_b* is lost after only moderate dilution in Mn^{2+}-free medium, suggesting that it is only weakly bound. This release is only slightly temperature-dependent, unlike the release of $^{54}Mn^{2+}$ and $^{45}Ca^{2+}$ from intracellular pools.[35] Finally, the EPR spectrum from the Lettre cell pellet (Figure 1d), in addition to being of lower intensity than expected, has a significant component that is broadened, as indicated by the broad wings of the spectrum. A computer simulation indicated that this spectrum represents 20% Mn^{2+} with a linewidth of 1 to 1.5 mT and 80% Mn^{2+} with a linewidth of 2.4 to 4 mT, which accounts for the loss of signal in the cell pellet. This broad component is not present in the cell suspension and must arise from short-range spin-spin interactions between bound Mn^{2+} ions. The range of these effects is less than the width of the plasma membrane. It may therefore be concluded that Mn_b* is cell surface-associated Mn^{2+}. Human erythrocytes, in addition to being impermeable to Mn^{2+} as indicated by the lack of Mn_b, do not show any detectable surface-bound Mn^{2+} similar to Mn_b*, although using these techniques, such Mn^{2+} could be masked by the Mn^{2+} of the pellet interstitial medium.

Treatment of Lettre cells with 4000 hemagglutination units (HAU)/mℓ Sendai virus prior to the addition results in an increase in Mn_b with no change in Mn_b*.[40] This is quantitatively similar to the increase in $^{54}Mn^{2+}$ and $^{45}Ca^{2+}$ binding following viral treatment.[35] These results suggest that the effect of Sendai virus is to increase the permeability of the plasma membrane to divalent cations leading to an increased uptake and binding of these ions.

The principal shortcoming of EPR measurements of Mn^{2+} binding in living cells is the lack of information about the binding sites. Bound Mn^{2+}, unless in a site of high symmetry with very small zero-field splitting parameters, will give very broad, usually undetectable, signals. Attempts to produce a sharper spectrum of cell-bound Mn^{2+} by cooling to low temperatures were not successful, and in any case, disruption of the cells by freezing would make interpretation of any results obtained difficult. Weakly bound Mn^{2+}, such as Mn_b*, which is mobile over the timescale of the experiment (10^{-9} sec), can be detected but it is not possible to distinguish between a boundary layer of Mn^{2+} ions as has been described for pure phospholipid vesicles,[42] or binding to very mobile cation binding ligands. Neuraminic acid groups are a possible candidate for surface binding sites,[43] though the lack of Mn_b* in erythrocytes, and the failure of neuraminidase treatment to affect Mn_b* binding in Lettre cells,[40] argues against this.

III. LOCALIZATION OF CATION BINDING BY PROTON RELAXATION ENHANCEMENT

The water proton longitudinal or spin-lattice magnetic relaxation time (T_1) is sensitive to the magnetic environment of the proton. The presence of paramagnetic nuclei such as Mn(II) decreases the relaxation time, a process termed relaxation enhancement. The more constrained the Mn(II), for example by binding, the greater the effect.[44]

Relaxation enhancement is an established method for measuring water permeability in living cells.[37] This technique is particularly applicable to erythrocytes which do not take up the Mn^{2+} used to enhance relaxation.[39] It was intended to extend the technique to cells other than erythrocytes, and to establish whether the cation binding and localization could be determined.

A. The Origin of Multiple Proton Relaxation Rates in Cell Suspensions

The effects of exchange between different environments on the observed proton magnetic relaxation rates and population sizes have been extensively analyzed by Hazelwood.[45] If the system is made up of two compartments, such as cells (compartment A) and their suspending medium (compartment B), of fractional size P_A and P_B ($P_A + P_B = 1$), and if the exchange of water between the two compartments is rapid compared with the respective relaxation times (T_{1A}, T_{1B}), the observed relaxation time (T_1) will be the weighted average of the two relaxation times. This condition is the rapid exchange domain and the observed T_1 value is given by Equation 1.

$$\frac{1}{T_{1'}} = \frac{P_A}{T_{1A}} + \frac{P_B}{T_{1B}} \tag{1}$$

At the other extreme, in the case of negligible exchange, the observed values of T_1 and P are the same as the actual values. This is the slow exchange domain.

Between these two extremes, in the intermediate exchange domain, the observed relaxation times and apparent compartment sizes exhibit complex behavior. The intermediate exchange domain has been analyzed by Hazelwood et al.[45] for some conditions, such as very different population sizes and relaxation rates, but without considerable knowledge of the system, few parameters can be defined. It is usual to conduct experiments to measure the water permeability of erythrocytes in the slow exchange region, where residence time can be measured with some accuracy.[39]

The behavior in the intermediate exchange domain is governed by the following equations. The apparent relaxation rate of the slowly relaxing compartment $1/T_{1A}'$ and the rapidly relaxing compartment $1/T_{1B}'$ are given by

$$\frac{1}{T_{1A'}} = C_1 - C_2 \tag{2}$$

and

$$\frac{1}{T_{1B'}} = C_1 - C_2 \quad (3)$$

where

$$C_1 = \frac{1}{2}\left(\frac{1}{T_{1A}} + \frac{1}{T_{1B}} + \frac{1}{T_A} + \frac{1}{t_B}\right) \quad (4)$$

and (5)

$$C_2 = \frac{1}{2}\left[\left(\frac{1}{T_{1B}} - \frac{1}{T_{1A}} + \frac{1}{t_B} - \frac{1}{t_A}\right)^2 + \frac{4}{t_A t_B}\right]^{1/2}$$

and t_A and t_B are the residence times in compartments A and B, respectively.

The apparent sizes of the populations are given by

$$P_{B'} = \frac{1}{2} - \frac{1}{4}\left[\left(P_B - P_A\right)\left(\frac{1}{T_{1A}} - \frac{1}{T_{1B}}\right) + \frac{1}{t_A} + \frac{1}{t_B}\right]\Big/ C_2 \quad (6)$$

and

$$P_{A'} = 1 - P_{B'} \quad (7)$$

Also

$$P_{A'}/T_{1A'} + P_{B'}/T_{1B'} = P_A/\,T_{1A} + P_B/T_{1B} \quad (8)$$

and by definition,

$$P_A/t_A = P_B/t_B \quad (9)$$

With such a complex system, it is in general difficult to quantify individual parameters, even if the measured system is compatible with this model, which is not always the case (see below).

B. Generation of Two Relaxation Rates in Erythrocyte Suspensions

Proton spin-lattice relaxation times were measured at 60 MHz with a spectrometer built by R. N. Sheppard of the Chemistry Department of Imperial College, London. Measurements were made using a 90°-90° pulse sequence with 1.5-μsec pulses. The spectrometer was usually operated under the control of a computer which made replicate measurements at each pulse interval. Relaxation rates were calculated by least squares analysis of magnetization decay. Multiple rates were determined by simple subtractive deconvolution of the decay.

Human erythrocyte suspensions (10^9/mℓ, 150 μℓ) were introduced into the probe in 7-mm diameter tubes and maintained at 37°C. A series of experiments were performed in which the extracellular medium contained successively higher concentrations of Mn^{2+}. Typical results are shown in Table 2.

Table 2
MEASURED RELAXATION TIMES IN ERYTHROCYTE SUSPENSIONS AT THE INDICATED Mn CONCENTRATION

Total Mn^{2+} conc (m*M*)	T_{1A}[a]	T_{1B}[b]	P_A[c]
0	2350[d]		
0.1	800		
0.5	262		
1.25	131		
2.5	92	52	0.70
5.0	56	27	0.70

[a] The relaxation time of the cell-associated protons.
[b] The relaxation time of the medium protons.
[c] The fractional size of the cell-associated compartment.
[d] T_{1A} and T_{1B} not resolved.

As the relaxation time is reduced by the addition of Mn^{2+}, the system appears to move from the fast exchange domain into the intermediate exchange domain where two populations of protons are observed. However, closer examination of the data reveals a number of paradoxes if the system consists of freely exchanging water proton populations.

An approximate estimate of the shortest residence time in either compartment is slightly greater than the longest relaxation time at which two populations are distinguishable. In this case, it is at least 150 msec. Previous workers using T_2 measurements in the slow exchange domain have estimated the residence time of erythrocyte water protons to be between 20 msec[46,48] and 10 msec.[37,47] Such values are about an order of magnitude smaller than the residence time measured by us. Clearly, our data are inconsistent with the known water residence time.

The population with the shortest relaxation rate, in this case the extracellular medium (see below), must have a residence time considerably longer than observed relaxation time. This is because the relaxation time of this population is similar to the relaxation time of the supernatant medium. For example, at 2.5 m*M* Mn, the values are 52 msec for the medium protons in contact with cells and 51 msec when isolated as a supernatant (Table 4). Eventhough these are observed values, which within the intermediate exchange domain can vary considerably from their true values, the contribution of the population residence time must be small. A simple estimate is that the observed relaxation rate (the reciprocal of the relaxation time) is the sum of the actual relaxation rate and the exchange rate (the reciprocal of the residence time). At a conservative estimate, the exchange rate cannot be larger than 30% of the relaxation rate, giving a residence time of at least 170 msec.

The relationship between population size and residence time given in Equation 9 states that (by definition) the ratios of size to residence time in the two populations are equal. A residence time of 170 msec for the smaller population suggests an even larger residence time for the larger population.

Thus, the data indicate that the two observed proton populations are exchanging over the timescale of hundreds of milliseconds, whereas the best estimates of water exchange in this system by previous workers are an order of magnitude faster.

Even though these observations appear to be inconsistent with exchanging water protons, several other considerations indicate that water proton exchange is being observed. By

Table 3
MEASURED RELAXATION TIMES IN LETTRE CELL SUSPENSIONS AT THE INDICATED Mn CONCENTRATION

Total Mn^{2+} conc (m*M*)	T_{1A}[a]	T_{1B}[b]	P_A[c]
	3600[d]		
0.02	600	2300	0.10
0.1	315	720	0.17
0.25	170	552	0.16

[a] The relaxation time of the cell-associated protons.
[b] The relaxation time of the medium protons.
[c] The fractional size of the cell-associated compartment.
[d] T_{1A} and T_{1B} not resolved.

definition, the extracellular medium is essentially an aqueous Mn^{2+} solution and the NMR signal must come from water protons. Clearly, the cell-associated NMR signal is influenced by the extracellular water protons. As the range of direct paramagnetic relaxation enhancement is too small to influence intracellular protons from an extracellular source, then some exchange mechanism must be involved.

The cell-associated signal is a significant proportion of the total signal; typically it is slightly less than the cell volume, as a proportion of the total volume of the suspension. For example, at 2.5 m*M* Mn, the apparent cell-associated fraction is 70% in an 80% hematocrit suspension. This is much larger than would be expected, for example, for macromolecule protons. The hemoglobin concentration in erythrocytes is approximately 110 to 140 mg/mℓ which, assuming the proportion of protons in proteins is about 10%, gives an intracellular proton concentration in the 1 *M* range. By comparison, water is 110 *M* in protons.

A further consideration is that macromolecule protons have a very short spin-spin relaxation time (T_2), typically 10 to 20 μsec, which make it extremely unlikely that such protons would be detected under the conditions of measurement.

These incompatible conclusions suggest that the simple model of water proton exchange between intracellular and extracellular compartments is not applicable to this system. The only possible alternative mechanism is a process of spin exchange. Magnetic cross relaxation between solute and solvent protons, in the absence of chemical change, is known to occur.[49] It is in fact a special case in spin diffusion, similar to the transverse or spin-spin relaxation mechanism.

Such an effect has been documented in the erythrocyte system[50] and may underlie some of the early observations of "water of hydration". The slowly exchanging population in erythrocytes has been previously identified as protein,[51] although the exchange phenomenon was unexplained.

C. Generation of Two Relaxation Rates in Lettre Cell Suspensions

Suspensions (2×10^8/mℓ) of Lettre cells were used instead of erythrocytes in a series of experiments similar to those described above. The results are shown in Table 3.

Lettre cell suspensions, in a similar way to erythrocytes, exhibit two distinct proton populations when the addition of Mn enhances relaxation beyond a certain value. There is, however, a marked quantitative difference in (1) the amount of Mn needed to reveal two populations and (2) the relaxation times at this point (Table 3). For Lettre cells, the threshold is more than tenfold higher.

On the simple assumption that residence time is related to volume, i.e., assuming that erythrocytes are spherical, then Lettre cells which have a volume of about 500 μm^3 should have a residence time about five times longer than erythrocytes (volume 90 μm^3). The minimum estimate for the residence time within Lettre cells is about 700 msec, which is indeed about five times the value estimated for erythrocytes. It is, however, like the value for erythrocytes, ten times larger than expected. Although it is possible that the Lettre cell membrane may have a much lower water permeability than the erythrocyte, and that a residence time of 700 msec truly reflects the actual exchange of intracellular water with the extracellular medium, it is more likely that the cell-associated proton population is similar to that observed in erythrocytes, and that the difference between Lettre cells and erythrocytes is related to cell size.

Lettre cell suspensions differ qualitatively from erythrocyte suspensions in that the Lettre cell-associated proton signal has a shorter relaxation time than that of the extracellular medium. This is the converse of the erythrocyte suspension, for which the cell-associated signal relaxes more slowly. At first sight, the situation in Lettre cells appears paradoxical. How can Mn added to the extracellular medium enhance the relaxation of the intracellular protons to a greater extent than the extracellular water? The supposed source of the relaxation enhancement appears to be relaxing more slowly.

Lettre cells, however, differ from erythrocytes in that they bind and are permeable to Mn. This was shown by the use of ^{54}Mn[35] and Mn EPR.[40] The source of relaxation enhancement is in fact cell associated as a consequence of Mn binding to and entering the cells. Even though the amount of binding is small, the increase in relaxation enhancement due to the immobilization of the Mn is very large.[44] This result is an independent confirmation of the observation that, in contrast to Lettre cells, erythrocytes do not take up Mn.[39]

D. Measurements of Relaxation Times in Cell Suspensions, Pellets, and Supernatants

In order to assist in the analysis of the proton relaxation, measurements were made on suspensions and, after centrifugation, on the corresponding cell pellet and supernatant medium. Both erythrocytes and Lettre cells were studied at concentrations of Mn which allowed the observation of two proton populations. After a sample (150 $\mu\ell$) of a cell suspension was removed for measurement, the suspension was centrifuged for 5 min at 1000 *g* and the supernatant removed. Due to the difficulty of introducing the very viscous Lettre cell pellet into the sample tube, 80 to 90% suspensions were used. The data obtained are shown in Table 4.

1. Erythrocytes

The two experiments with erythrocytes, although suggesting an exchange rate between the cellular and extracellular compartment that is too slow to be plausibly assigned to water exchange, can however be interpreted as two exchanging compartments, with one compartment providing a source of relaxation enhancement. The following observations support this view.

In erythrocyte suspensions the shortest relaxation time (T_{1B}) originates from the extracellular medium, this signal being identified by size (observed P_B is 0.3 compared with a fractional extracellular volume of 0.25). This is consistent with aqueous Mn in the extracellular medium being the source of enhanced relaxation.

When erythrocyte suspensions are centrifuged and the pellet is examined, both relaxation times are longer, indicating that the source of relaxation enhancement, namely the supernatant medium, has been partially removed. The observed P_A does not rise to the expected 0.95, presumably because the very different relaxation rates lead to a significant reduction of the

Table 4
MEASURED RELAXATION TIMES IN ERYTHROCYTE AND LETTRE CELL SUSPENSIONS AND THE CORRESPONDING PELLETS AND SUPERNATANTS AFTER CENTRIFUGATION AT THE INDICATED Mn CONCENTRATON

Cells	Sample	Mn (m*M*)	T_{1A}[a]	T_{1B}[b]	P_A[c]
Erythrocyte	Suspension	2.5	92	52	0.70
	Pellet	2.5	180	59	0.83
	Supernatant	2.5	51[d]		
Erythrocyte	Suspension	5.0	56	27	0.70
	Pellet	5.0	148	65	0.70
	Supernatant	5.0	32		
Lettre	Suspension	0.05	335	820	0.70
	Pellet	0.05	360	205	0.68
	Supernatant	0.05	990		
Lettre	Suspension	0.5	80	250	0.12
	Pellet	0.5	110	60	0.75
	Supernatant	0.5	360		
Lettre	Suspension	2.0	37	94	0.19
	Pellet	2.0	40	17	0.82
	Supernatant		102	102	

[a] The relaxation time of the cell-associated protons.
[b] The relaxation time of the medium protons.
[c] The fractional size of the cell-associated compartment.
[d] T_{1A} and T_{1B} not resolved.

apparent size of the slowly relaxing compartment. This may result from the fact that protons leaving the larger cellular compartment return, because of the short residence time in the smaller compartment, to contribute to the signal from the larger compartment. However, the faster the relaxation rate in the smaller compartment, the less protons return unrelaxed to the larger compartment. Thus, the slowly relaxing compartment appears smaller.

For a similar reason, the apparent relaxation time in the extracellular compartment, namely the interstitial medium of the cell pellet, is slower than the corresponding suspension because fewer protons relax in the cellular compartment and are able to return to the faster relaxing compartment unrelaxed.

The supernatant medium gives a short relaxation time, indicating that the medium is the source of relaxation enhancement. This value gives a very good estimate of the actual relaxation time of the extracellular compartment in the corresponding suspensions and pellets. The unexpected observation that the actual and observed extracellular relaxation times (T_{1B}) are similar has been discussed above.

2. Lettre cells

The data for Lettre cells are markedly different. As has been mentioned above, the cell-associated proton population in Lettre cell suspensions relaxes faster, and presumably is in closer proximity to the source of relaxation enhancement, than the extracellular medium.

When Lettre cells are centrifuged to produce a more concentrated cell suspension, P_A rises as expected. However, the cell-associated protons are now relaxing more slowly than the extracellular medium. Upon centrifugation, the signal from the medium has a relaxation time four times faster than the original suspension. Clearly, unless centrifugation has caused a relocation of Mn, which is unlikely, the simple two-compartment model of the system is untenable because the source of relaxation cannot be cell associated in dilute suspensions, yet extracellular in concentrated suspensions.

3. Interpretation: a Boundary Layer?

It is known that the behavior of ^{54}Mn (and ^{45}Ca) is not markedly dependent on cell concentration, and centrifugation does not alter cation binding.[35] Although radionuclide data do not provide an explanation for the behavior of Lettre cell suspensions, the Mn EPR data do.[40]

To recapitulate, the EPR experiments indicated that in suspensions of Lettre cells, but not erythrocytes, two forms of bound Mn could be identified: Mn_b*, weakly bound in a boundary layer close to the outside of the cell surface, and Mn_b, tightly bound either in or on the surface of the cells. It is plausible to explain the behavior of the proton populations in Lettre cells by postulating that the boundary layer of Mn_b* introduces a third proton relaxation environment close to the cell surface.

Such a boundary layer would not be distinguished as a separately relaxing population from the bulk phase of the extracellular medium because of the relatively short residence time expected from such a compartment. The influence of this rapidly relaxing compartment would be relatively minor in a dilute cell suspension. The major contribution to relaxation enhancement under these conditions would be the tightly bound cell-associated Mn. However, in a more concentrated suspension, the boundary layer Mn predominates in enhancing relaxation in the extracellular medium.

There is an alternative explanation which does not invoke a boundary layer that is more conventional in its assumptions, though the unexpectedly slow exchange rate makes it less likely. It is as follows. In a simple two compartment system, altering the proportions of the compartments would change the apparent relaxation times by changing the residence time of the compartment. From Equation 9, the residence time by definition follows the population size provided no other parameters change. According to the relative sizes of the relaxation and residence times of a compartment, the observed relaxation time will change. If the residence time is much shorter than the relaxation time, the observed relaxation will follow the population size. Conversely, if the relaxation time is much shorter than the residence time, the population size will have little effect on the observed relaxation rate. Thus, reducing the proportion of extracellular medium by concentrating the cell suspension, from say 80 to 20%, would, if the residence time were shorter than the relaxation time, reduce the observed relaxation time of that compartment by about fourfold, as is observed (see Table 4).

This simple explanation predicts the expected behavior of erythrocyte suspensions and pellets also. However, in that case, centrifugation causes the apparent relaxation time of the extracellular protons to increase; clearly the effect of reducing the population size, and so the residence time, is insignificant compared with other factors influencing observed relaxation times. Indeed as the interstitial medium of the erythrocyte pellet is so small, 10% at most, then pelleting must have reduced the residence time by a factor of three without affecting the observed relaxation rate. This suggests that the residence times estimated when the behavior was discussed earlier could actually be much longer. This is even more inexplicable in terms of water proton exchange.

To clarify this, it is possible to estimate the relative magnitude of the residence time and relaxation time from the deviation of the observed relaxation time from the actual value. This can be done by using the measured value for the supernatant. As even the pellet extracellular compartment T_{1B} is similar or even larger than the supernatant T_1 (Table 4), then the residence time is clearly much longer than the relaxation time.

A similar examination of the Lettre cell data shows that the suspension extracellular T_{1B}

is somewhat shorter than the supernatant T_1. In fact, it is possible to accommodate some of the Lettre cell data in a simple model where the residence time for the extracellular medium is around 2300 msec, which is the minimum value indicated by the longest T_1 where two populations can be distinguished. Reducing the population by a factor of four gives a residence time in the pellet of around 600 msec, which clearly will influence a compartment with a T_1 of 990 msec, as is the case for 0.05 m*M* Mn (see Table 4). The rest of the data at higher Mn concentrations increasingly deviates from this model. At 2 m*M* Mn, the suspension T_{1B} of 102 msec would need to be reduced to 17 msec for a reduction in residence time of only 230 msec to 600 msec.

It must be concluded that only the model, including a boundary layer, which relaxes faster than the extracellular bulk phase, can explain why, when cells are concentrated, the extracellular protons of Lettre cell suspensions relax so much faster.

E. Localization of the Source of Relaxation Enhancement

The objective of the research described in this chapter is to ascertain whether, using magnetic resonance techniques, it is possible to localize a paramagnetic analog of calcium, and by inference calcium. The EPR results confirmed the initial binding studies with radionuclides, but were unable to localize the tightly bound Mn_b*. Intrinsically, the information on different proton compartments provided by NMR measurements is more likely to be useful for the localization of the source of relaxation enhancement and therefore of Mn. The results described for Lettre cells indicate that T_1 measurements produce data incompatible with a simple model of exchanging water protons similar to that described for T_2 with erythrocytes. Notwithstanding this, the data have provided an independent confirmation of an Mn boundary layer, as suggested by the EPR results. In this case unequivocal localization of this Mn at the cell surface has been possible. Given the working hypothesis of two apparently slowly exchanging proton compartments, namely the extracellular water and an unknown cell-associated proton population, can the cell-associated source of relaxation in Lettre cells be located? The following sections examine the data obtained so far and present further experiments designed to localize Mn in Lettre cells.

The measurements of suspensions and corresponding pellets and supernatants suggest that Mn binds to Lettre cells in such a way that the cell-associated protons are not influenced by the relaxation of the extracellular medium. Inspection of Table 4 shows that centrifugation does not significantly alter T_{1A} at the three Mn concentrations used, in spite of fourfold changes in the extracellular compartment. The small difference between the suspension T_{1B}, and the supernatant T_1, indicate that in Lettre suspensions the source of relaxation enhancement has a marked effect on the cell-associated compartment, while hardly affecting the extracellular compartment. Either the source of relaxation is within the cells, or it is at a site specifically accessible to the intracellular proton compartment. If the system consisted of two exchanging water proton populations separated by the cell plasma membrane, then such data would definitely localize the Mn within the cell membrane. However, the paradoxically slow exchange rate between extracellular water protons and the cell-associated population makes such a conclusion unsafe.

There is a possible configuration which has been proposed[40] as an alternative to Mn entry. As the system clearly exhibits long residence times, it is possible to postulate a simple model which does not involve Mn entry while remaining consistent with the experimental results. If one assumes that the cell-associated Mn is located at the cell surface it could reduce the relaxation time to such an extent that any protons entering this region would relax before leaving. In formal terms, a third proton population would exist with a relaxation time much shorter than its residence time. Although the size and geometry of such a compartment would give it a very short residence time, it is quite plausible that bound Mn could reduce the relaxation time to a smaller value. Under these conditions the two detectable compartments would be effectively isolated with respect to proton magnetization, and the rapidly relaxing compartment would act as a ''magnetization sink'' between the compartments. The observed relaxation times would then depend upon the residence times of the two compartments and the Mn concentration. Thus, the largest compartment would have the longest observed T_1. This would account for the observation that when Lettre cell suspensions are concentrated, the longest T_1 is associated with the largest compartment. Although this elegant model does explain most of the results, it does not convincingly explain why the cell proton compartment T_1 is not affected to the same extent as the extracellular compartment by a change in relative population size. Such an asymmetry in response to bound Mn suggests that Mn is at least partly intracellular.

A series of experiments conducted to establish the location of Mn in suspensions of erythrocytes and Lettre cells are described and discussed below.

F. The Use of an Impermeant Relaxation Enhancing Agent

In order to ascertain whether the entry of Mn into cells is necessary to produce the results obtained so far, an impermeant relaxing agent was used and the results compared with those obtained with Mn^{2+} alone. This is an approach similar to the use of dextran-magnetite as a relaxing agent described in another system.[52] The impermeant relaxing agent was Mn complexed to trans-1,2-diamino cyclohexane tetra acetic acid (CDTA). CDTA has a very high affinity for transition metals, the stability constant for the Mn complex being approximately 10^{17}.[25] The complex was prepared using a slight excess of CDTA and passed down an ion exchange column prior to use to eliminate any free Mn^{2+}.

The data are presented in the form of relaxation enhancement plots.[53] These are made by plotting the observed relaxation rate of a cell suspension against the relaxation rate of an aqueous solution of relaxing agent (Mn-CDTA) or $MnCl_2$ at the same concentration. Thus, behavior similar to an aqueous solution would be depicted as a straight line of unit slope.

1. Erythrocytes

Figure 2 shows the results of the data obtained with erythrocyte suspensions. The concentrations of relaxing agent and the relaxation rate are indicated. A higher concentration of Mn-CDTA than of $MnCl_2$ was required to produce the same relaxation rate. This is due to the reduced water of hydration of the chelate compared to that of the Mn^{2+} ion, thus lowering the efficiency of relaxation of the water protons.

At low concentrations of Mn, the suspension exhibits one T_1 (filled circles) quite similar to an aqueous solution (shown as diagonal line). At a critical concentration, two proton populations are distinguishable, the faster relaxing population associated with the medium (squares) and the slower relaxing population with cells (triangles). If Mn-CDTA (open symbols) is used instead of $MnCl_2$ (filled symbols), the results are essentially the same. The range of measurements to shorter T_1s was extended using Mn-CDTA. This avoids the use of excessive Mn^{2+} concentrations, which may be toxic to the cells.

It may be concluded that in erthrocyte suspensions, the data are consistent with the erythrocyte plasma membrane being as impermeable to Mn^{2+} as to Mn-CDTA.

2. Lettre Cells

A similar series of experiments was performed with Lettre cell suspensions. Figure 3 shows the results obtained using added $MnCl_2$ (filled symbols). As described before, the faster relaxing protons are cell associated (filled triangles) and the extracellular medium (filled squares) behaves similarly to an aqueous solution (indicated by the dotted line). This is comparable with the results in Table 4, showing that the medium T_1 is similar to, but slightly shorter than, the isolated supernatant. The results using Mn-CDTA however are quite different (open symbols). The protons in the medium relax rapidly (open squares) and the cell-associated protons relax slowly (open triangles). This is the reverse of the results using $MnCl_2$ and is broadly similar to the behavior of erythrocytes.

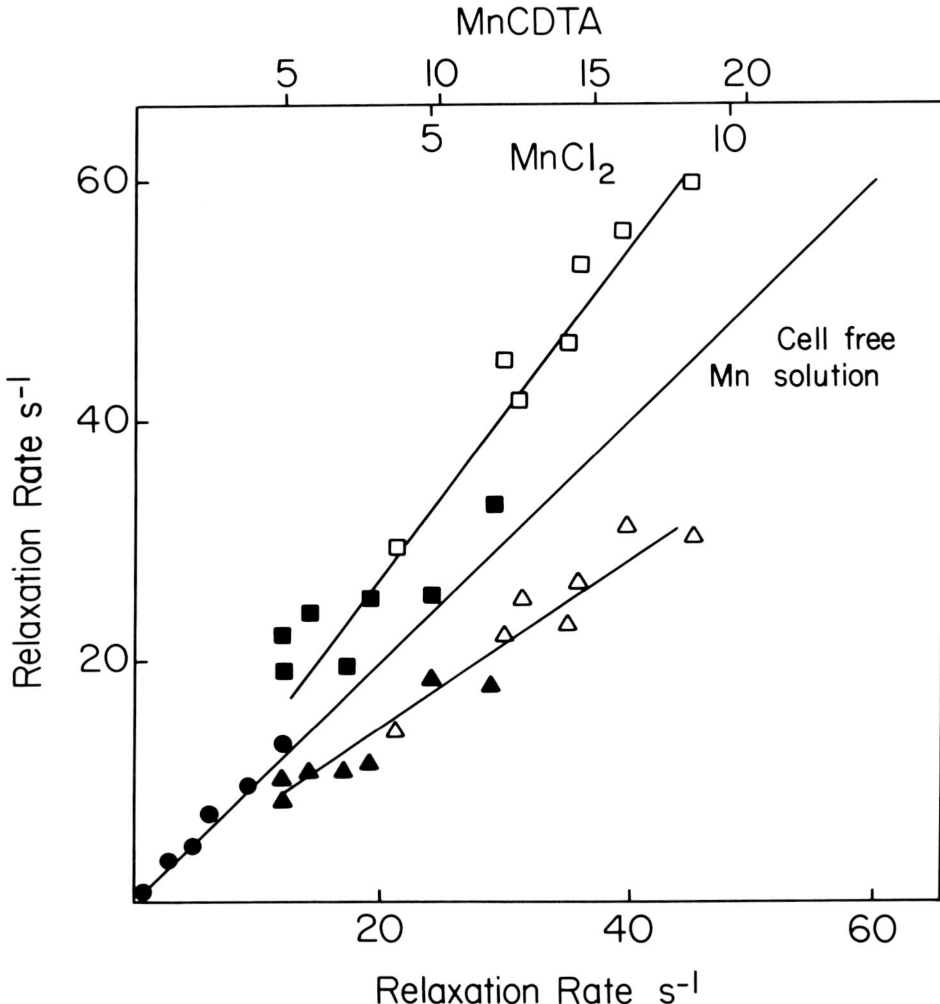

FIGURE 2. Relaxation enhancement plot of erythrocyte suspensions. Relaxation enhancement plot of relaxation rates measured in erythrocyte (abscissa) against the relaxation rate of a solution of the same Mn concentration (ordinate), in the presence of either $MnCl_2$ (closed symbols) or MnCDTA (open symbols). The equivalent Mn concentrations (in Mn) are indicated at the top. (●) Single T_1 conditions; (▲, △) cell associated T_1; (■, □) medium-associated T_1. The middle line indicates cell-free $MnCl_2$ solutions.

The Mn-CDTA complex differs from Mn^{2+} in two regards: first, it is large and impermeant; secondly, being uncharged it does not bind to Lettre cells. By eliminating the contribution of entry and binding to relaxation enhancement in Lettre cell suspensions, they behave qualitatively like erythrocytes, although they appear to exhibit slower exchange. While these results are compatible with the model for Lettre cells described above (Section II.D.3), they do not distinguish binding from entry, and so do not permit a definite localization of Mn^{2+}.

One of the original objectives of this research was to characterize the increased uptake and release of $^{45}Ca^{2+}$ during Sendai virus treatment of Lettre cells. Such alterations in Ca movement could reflect changes in membrane permeability to Ca^{2+} and/or changes in membrane bound Ca^{2+}. On the assumption that Sendai virus permeabilizes Lettre cells to Mn^{2+} and to Mn-CDTA[35] (it has been shown to do so for EDTA[68]) the experiments described below were undertaken to examine the effect of entry of a relaxation enhancing agent such as Mn-CDTA. In addition, the effect of using the divalent cation ionophore A 23187 was studied.

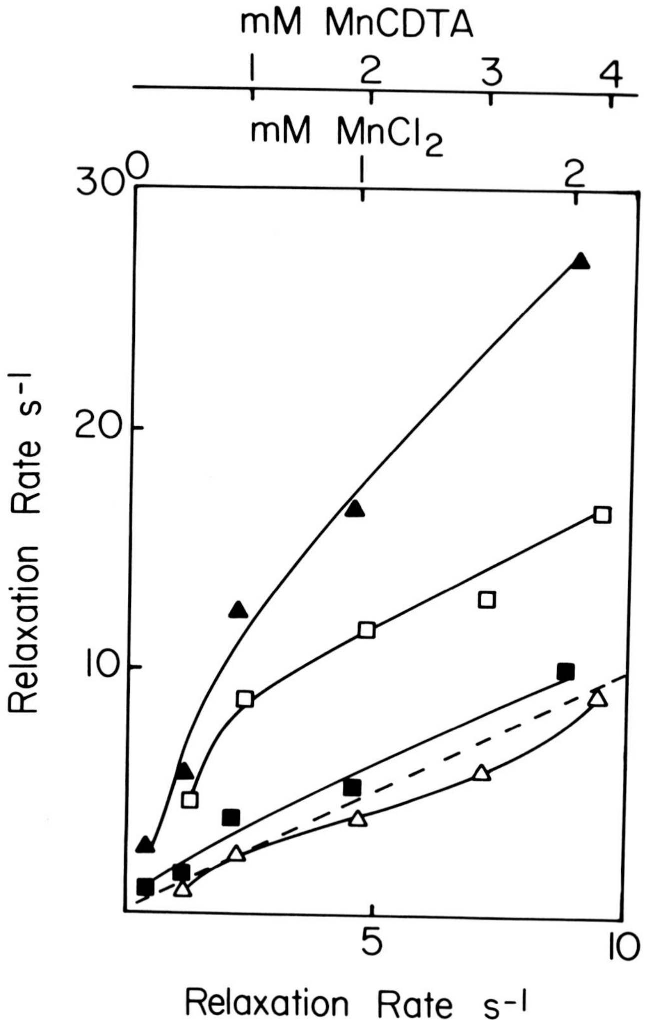

FIGURE 3. Relaxation plot of Lettre cell suspensions. Relaxation enhancement plot of relaxation rates measured in Lettre cell suspensions (abscissa) against the relaxation rate of a solution of the same Mn concentration (ordinate), in the presence of either $MnCl_2$ (closed symbols) or MnCDTA (open symbols). The equivalent Mn concentrations (in/m*M*) are indicated at the top. (▲, △) Cell-associated T_1; (■, □) medium-associated T_1; the broken line indicates cell-free $MnCl_2$ solutions.

G. Permeabilization of Lettre Cells with Sendai Virus and an Ionophore

1. Sendai Virus

Suspensions of Lettre cells (2×10^8/mℓ) containing either $MnCl_2$ or Mn-CDTA as indicated, were incubated for 10 min at 37°C with 4000 hemagglutination units (HAU)/mℓ Sendai virus. The proton relaxation times in the suspensions were measured as before. Measurements of virus suspensions with $MnCl_2$ gave relaxation times identical to the equivalent $MnCl_2$ solution, indicating that the addition of virus in itself produced no effects.

Remarkably, the effect of Sendai virus treatment prior to $MnCl_2$ addition is to cause three relaxation rates to be observed. A typical result is shown in Figure 4. Data at three Mn concentrations for suspensions and the corresponding pellets and supernatants are shown in Table 5.

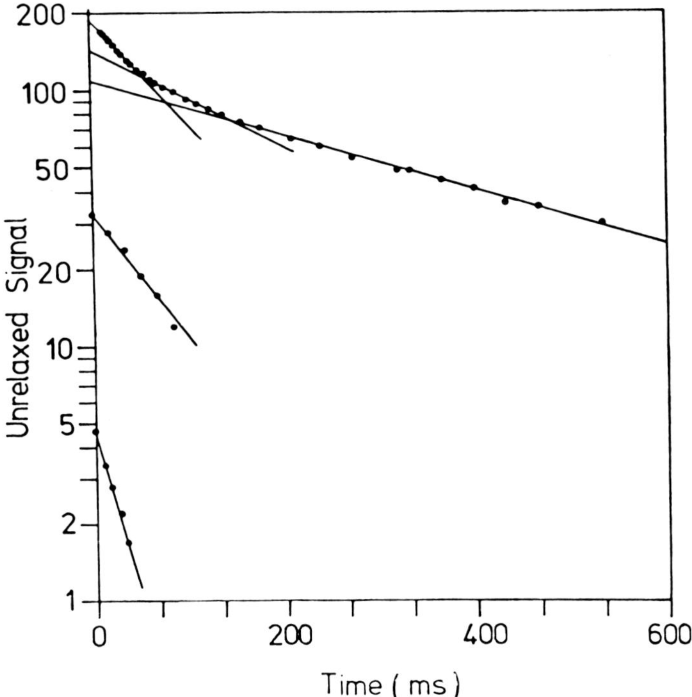

FIGURE 4. Proton relaxation in a suspension of Sendai virus-treated Lettre cells. Typical triexponential relaxation curve showing three populations of protons in a 0.1 m*M* Mn-containing suspension of Lettre cells treated with Sendai virus. The composite logarithmic plot (top) may be decomposed into three straight lines (a, b, c) by subtraction. The slopes and intercepts of these lines give the apparent relaxation times and populations. (From Getz, D., Gibson, J. F., Sheppard, R. N., Micklem, K. J., and Pasternak, C. A., *J. Membr. Biol.*, 50, 311, 1979. With permission.)

The population with the longest T_1 is presumed to be the extracellular medium and the two shorter T_1s are assumed to be cell associated. It is difficult to explain the origin of these two proton populations. As can be seen from Figure 4, they are quite distinct although the quantitation of the relaxation times and population sizes is subject to quite large errors. In particular, the sizes of the cell-associated populations vary in a nonsystematic way and probably are not significantly different.

The cell proton populations are clearly not separate water populations: first, no plausible intracellular compartments would be expected to have a residence time of at least 190 or 360 msec; second, the new population is not being resolved due to proton relaxation being enhanced from a rapid exchange kinetic domain into an intermediate domain. This second conclusion stems from comparison between the cell-associated relaxation times in untreated and virally treated Lettre cells. At 2 m*M* Mn, the cell-associated relaxation time of untreated Lettre cell suspensions is 37 msec (Table 4). Such a relaxation time is much shorter than the minimum apparent residence times of the two compartments observed in virally treated cells (190 or 369 msec). Since the enhancement of relaxation is not causing two cell-associated proton populations to be resolved, it must be due to a different localization or configuration of the Mn. A possible explanation is that a different population of Mn binding sites is occupied due to the increase in cytoplasmic Mn concentration which presumably occurs in an analogous way to Ca.[17,35] The nucleus may be the site of such bound Mn after virus treatment.

Table 5
MEASURED RELAXATION TIMES IN SENDAI VIRUS-TREATED LETTRE CELL SUSPENSIONS AND THE CORRESPONDING PELLETS AND SUPERNATANTS AFTER CENTRIFUGATION AT THE INDICATED Mn CONCENTRATION

Sample	Mn (m*M*)	$T_{1A}I$[a]	P_AI[b]	$T_{1A}II$[c]	P_AII[d]	T_{1B}[e]	P_B[f]
Suspension	0.05	360	0.12	190	0.19	710	0.69
Pellet	0.05			186[g]			
Supernatant	0.05			1655			
Suspension	0.1	130	0.21	102	0.08	422	0.71
Pellet	0.1			185			
Supernatant	0.1			1125			
Suspension	1.0	39	0.21	17	0.12	79	0.66
Pellet	1.0			12			
Supernatant	1.0			108			

[a] The longest cell-associated relaxation time.
[b] The longest cell-associated populaton.
[c] The longest cell-associated relaxation time.
[d] The longest cell-associated population.
[e] The medium proton relaxation time.
[f] The medium proton population.
[g] $T_{1A}I$, $T_{1A}II$, and T_{1B} not resolved.

The uptake of Mn by these cells is sufficient to deplete the extracellular medium, as shown by the supernatant T_1 which is much longer than the T_1 for the original Mn solution. For example, at 0.05 m*M* Mn, the supernatant, T_1, is 1655 msec and the T_1 of a 0.05-m*M* $MnCl_2$ solution is about 1000 msec. This increased uptake of Mn agrees with the results obtained with ^{54}Mn, which showed that virus treatment enhanced uptake.[35]

The cell pellet measurements show only one detectable proton population. The reason for this behavior is not clear and without knowing the origin of the multiple T_1s, it is difficult to interpret it.

A similar series of experiments was performed using the Mn-CDTA complex and the results are shown as a relaxation enhancement plot in Figure 5, together with the results using $MnCl_2$ for comparison (filled symbols). It can be seen from the scales and the position of the diagonal (dotted line) that these suspensions are showing a much greater relaxation enhancement than untreated Lettre cells. This is directly related to the binding of Mn; bound Mn is much more effective at enhancing relaxation than the aqueous ion.[44]

The results of adding Mn-CDTA to virally treated Lettre cells (open symbols) are qualitatively different and may shed some light on the origin of the multiple T_1s. At low Mn-CDTA concentrations, only two T_1s are detectable, which are not significantly different from those measured in untreated cells (compare Figure 3 with Figure 5). However, above 1 m*M* Mn-CDTA, three T_1s are detectable. The relaxation times are somewhat longer than the T_1s in the presence of $MnCl_2$, presumably because no Mn ions are binding. It can be inferred that entry of the source of relaxation enhancement is required for resolution of two cell-associated proton populations. As more Mn-CDTA than $McCl_2$ is required to resolve these T_1s, it appears that bound Mn ions are more effective than Mn-CDTA.

2. A23187

An alternative means of permeabilizing cells to Mn^{2+} is to use a cation ionophore. Lettre cells were treated with 1 μm A23187 for 5 min at 37°C in the presence of 1.5 m*M* $MnCl_2$

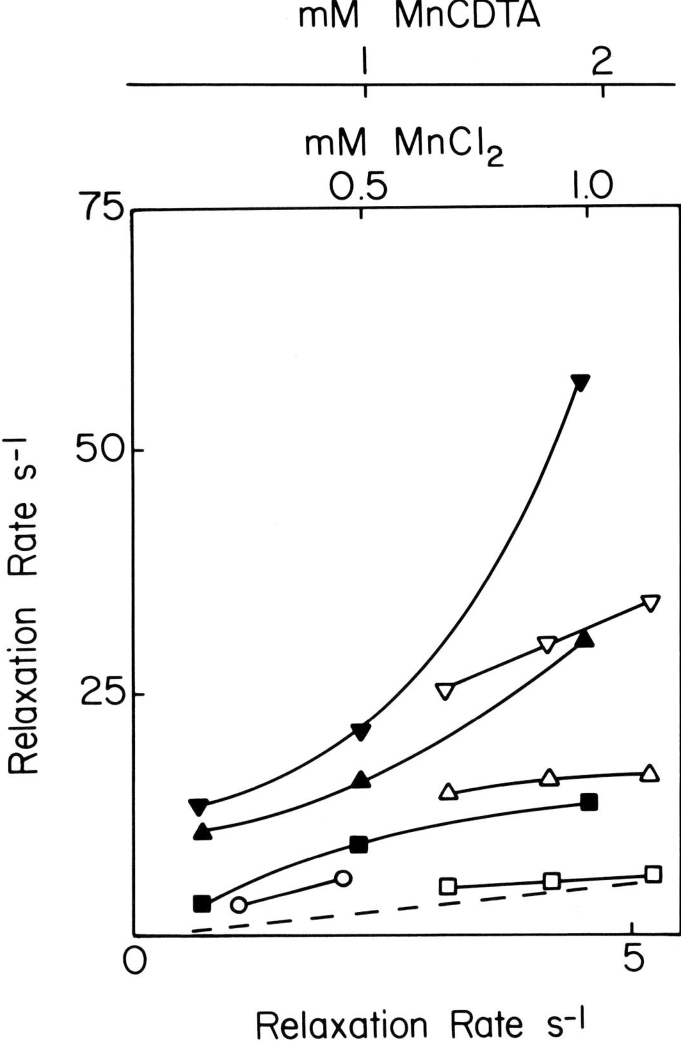

FIGURE 5. Relaxation enhancement plot of Sendai virus-treated Lettre cell suspensions. Relaxation enhancement plot of relaxation rates measured in Sendai virus-treated Lettre cell suspensions (abscissa) against the relaxation rate of a solution of the same Mn concentration (ordinate), in the presence of either $MnCl_2$ (closed symbols) or MnCDTA (open symbols). The equivalent Mn concentrations (in m*M*) are indicated at the top. (o) Single T_1 conditions; (▲, △) shortest cell-associated T_1; (▼, ▽) longest cell-associated T_1; (■, □) medium-associated T_1. The broken line indicates cell-free $MnCl_2$ solutions.

before measurement of proton relaxation times. The results shown in Table 6 are very similar to those obtained with virus treatment (Table 5, Figure 4).

3. Discussion

It must be concluded that Mn entry is required for the resolution of two cell-associated proton populations. A further conclusion can be drawn concerning the localization of Mn in untreated Lettre cells. It is clear that Mn entry due to a cytopathological change, caused either by virus or by ionophore permeabilization, results in a very characteristic separation of the intracellular protons into two measurable populations. This is not dependent on Mn^{2+}

Table 6
MEASURED RELAXATION TIMES IN AN IONOPHORE-TREATED LETTRE CELL SUSPENSION AND THE CORRESPONDING PELLET AND SUPERNATANT AFTER CENTRIFUGATION AT THE INDICATED Mn CONCENTRATION

Sample	Mn (m*M*)	$T_{1A}I$[a]	$P_{A}I$[b]	$T_{1A}II$[c]	$P_{A}II$[d]	T_{1B}[e]	P_{B}[f]
Suspension	1.5	21	0.16	44	0.27	104	0.57
Pellet	1.5			32[g]			
Supernatant	1.5			170			

[a] The longest cell-associated relaxation time.
[b] The longest cell-associated population.
[c] The longest cell-associated relaxation time.
[d] The longest cell-associated population.
[e] The medium proton relaxation time.
[f] The medium proton population.
[g] $T_{1A}I$, $T_{1A}II$, and T_{1B} not resolved.

ion binding, for it can occur when Mn-CDTA is used to enhance relaxation. Thus, if Mn entry does occur in untreated Lettre cells it results in a qualitatively different distribution to that caused by increasing the plasma membrane permeability to Mn.

This conclusion concerning the mechanism of separation of the two-intracellular compartment is different to that concerning the resolution of extracellular water protons from cell-associated protons. In this latter case, it appeared that binding of Mn^{2+} is required. It was not established whether entry of Mn, as well as surface binding, is required.

The dependence of the resolution of two cell T_1s on the qualitative localization of intracellular Mn suggests that the geometry of the relaxing agent is important. This favors a mechanism for resolving the two populations which involves the magnetic spin isolation of protons in different regions by the interposition of an environment of enhanced relaxation. This is a restatement of the hypothesis explaining separation of the extra- and intracellular compartment by a mechanism involving a boundary layer of Mn in the absence of Mn entry. While it may not be the most plausible explanation in that situation, within the cell such a mechanism may explain why just reducing the relaxation time is insufficient to resolve these populations in untreated cells.

It could be postulated that a quantitative increase in Mn ion binding to intracellular locations may have a qualitatively different effect on cellular proton behavior. This appears to be ruled out by the ability of complexed Mn to elicit this effect, albeit with reduced efficiency.

IV. CONCLUSIONS REGARDING THE BEHAVIOR OF Mn^{2+} IN CONTROL AND VIRUS-AFFECTED CELLS

Although much remains to be established concerning the localization of calcium-like cations and the behavior of the proton populations in cell suspension, several conclusions can be drawn. The combination of radionuclide tracer measurements and EPR and NMR spectroscopy have provided an analysis of a complex system.

A. Resolution of Multiple Relaxation Rates in Cell Suspensions

The very slow exchange between the proton populations detected by NMR indicates that under the conditions of measurement the proton spin-lattice relaxation times recorded are not due only to water protons. The cell-associated protons, in particular, do not appear to be intracellular water and may therefore by proteins, suggestive of Mn-protein (and hence Ca-protein) binding. Moreover, the interaction between the two main proton populations does not appear to be entirely due to water exchange across the cell plasma membrane, and may be due to a magnetic cross-relaxation phenomenon.

B. Erythrocytes

Mn(II) EPR measurements indicate that no detectable Mn binds to erythrocytes, either tightly on the plasma membrane or within the cell. Also, no loosely bound, Mn-enriched, boundary layer was detected. The proton NMR experiments indicate that there is a slowly exchanging cell-associated proton population which could be resolved by enhancing the relaxation of the extracellular water protons. The measurements on corresponding suspensions, pellets, and supernatants indicate that no Mn ions enter or bind to erythrocytes. These results are in close agreement with the data obtained from experiments using ^{54}Mn and ^{45}Ca.

C. Lettre Cells

Mn(II) EPR has identified two types of a bound Mn: a relatively immobile, tightly bound Mn, termed Mn_b, and a loosely bound, mobile boundary layer close to the external surface of the cell, termed Mn_b^*. Proton NMR experiments have demonstrated binding of Mn ions to Lettre cells and have confirmed the presence of the boundary layer. Although it is not possible to unequivocally state that Mn enters Lettre cells, this is the most probable interpretation.

D. Permeabilization of Lettre Cells

Permeabilization of cells results in the resolution of two cellular proton populations, probably by a "magnetization sink" mechanism rather than by relaxation enhancement out of a rapid exchange kinetic domain. Such resolutions may be attributed to entry of Mn as it can only be observed when entry occurs. This is in confirmation of the increased binding of radionuclides and the increase in immobilized Mn detected by Mn EPR. In addition, it unequivocally confirms that increased entry of Mn following viral treatment is the cause of the increased binding.

It is therefore clear that Mn^{2+} is a potentially useful analog with which to examine changes in intracellular Ca^{2+} by techniques that do not involve the introduction of specific Ca^{2+}-binding ligands such as the light-emitting proteins obelin and aequorin[54] or the fluorescent derivatives of EDTA.[55] Situations which lend themselves to such an investigation include hormonally, or otherwise triggered, changes in metabolism,[56] the stimulation of growth by external growth factors as well as by the activation of cellular oncogenes,[57] the contraction of all types of muscle, especially that of the heart, the alteration of direction of movement in protozoa, and many other instances in which an alteration of intracellular Ca^{2+} has been implicated or suspected to take place.

V. WATER MOVEMENT IN VIRUS-AFFECTED CELLS

The preceding sections have documented the fact that Mn^{2+}, and hence, Ca^{2+}, enters cells that have been rendered permeable by treatment with Sendai virus. Since the pore or leak created by this process is nonspecific, it is clear that other ions, such as Na^+ and K^+, will leak into and out of cells according to the gradient between cell and medium.[58] Because cells contain an excess of nondiffusible anionic proteins compared with extracellular medium, there is also a tendency for water to enter; normally this is counteracted by the operation of the Na^+ pump, which effectively maintains a lower cation concentration intracellularly compared to the external medium, as well as a negative osmotic pressure. As the capacity of the Na^+ pump to exclude Na^+ is gradually exceeded by the creation of pores in Sendai

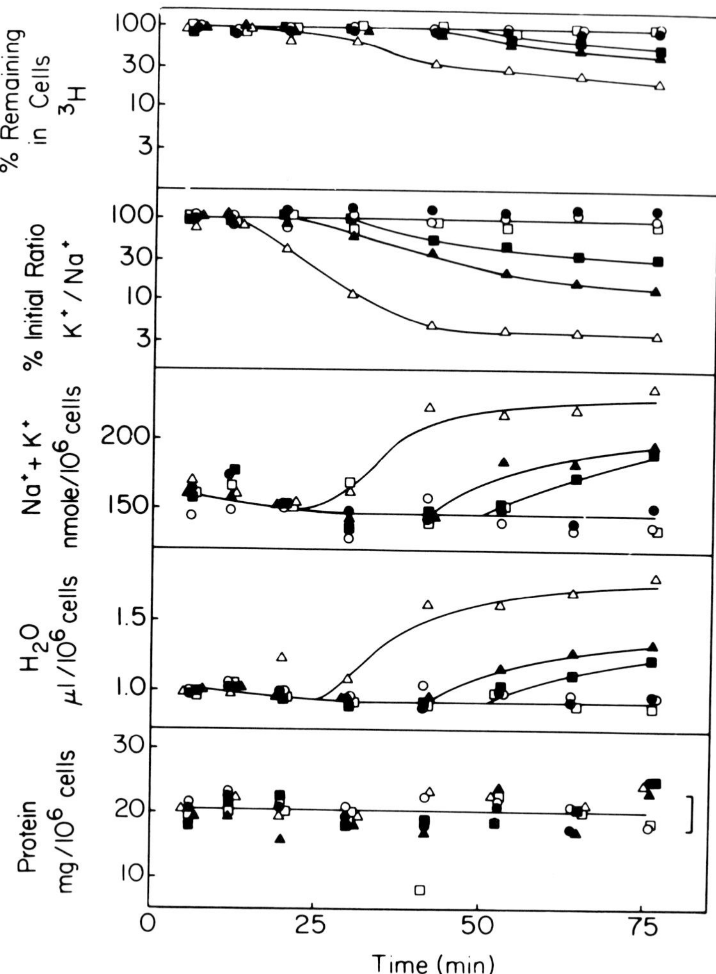

FIGURE 6. Sequential leakage of ions, water, and metabolites from virus-treated Lettre cells. Lettre cells preincubated with [^{3}H] choline were diluted into medium ($1.4 \cdot 10^7$ cells per mℓ) containing 150 m*M* NaCl, 5 m*M* KCl, 5 m*M* Hepes, and 1 m*M* $MgCl_2$ (pH 7.4), incubated at 19°C, and the content of ^{3}H, Na^+, K^+, and water in cell pellets measured by spinning samples (1 mℓ) without further dilution through oil. Sendai virus (10 HAU/mℓ), $CaCl_2$ (1 m*M*), prenylamine (1.2×10^{-5}*M*), and ouabain + furosemide (each 1 m*M*) were present as follows: prenylamine, ●-●; prenylamine + ouabain + furosemide, ○-○; virus, ■-■; virus + Ca^{2+}, □-□; virus + prenylamine , ▲-▲; virus + prenylamine + ouabain + furosemide, △-△. Note that changes in Na^+ + K^+ content (middle panel) mirror changes in water content; in cell pellets obtained as described above, a total Na^+ + K^+ content of 150 m*M* is equivalent to 1 µℓ of cell water. (From Bashford, C. L., Micklem, K. J., and Pasternak, C. A., *Biochim. Biophys. Acta,* 814, 247, 1985. With permission.)

virus-affected cells, so water will begin to enter, and cells swell. In the case of erythrocytes, it is this colloid osmotic swelling (not the creation of pores large enough to accommodate haemoglobin) that leads to hemolysis. In the case of nonerythroid cells, lysis does not necessarily ensue, as most cells have mechanisms of volume expansion,[59] as well as for recovery and repair of the membrane lesion.[60,61]

The changes in Lettre cell water content and cell volume following Sendai virus treatment illustrated in Figure 6 are difficult to correlate with the changes measured by proton NMR

(Section III.G.1). As the cell-associated proton population does not appear to be water protons, it is not likely that movement of water influences the changes observed by NMR.

VI. CONCLUSION

The experiments reviewed in this chapter document changes in divalent cations and water that take place when hemolytic viruses fuse with cells. As mentioned in Section I.A, the clinical significance of the action of such viruses on cells is likely to be limited. On the other hand, the demonstration[62-65] that similar permeability changes, including a sensitivity to external Ca^{2+}, accompanies the attack on susceptible cells of bacterial toxins such as *Staphylococcus aureus* α and δ toxins, major causative agents of bacterial infections,[66] of animal toxins such as melittin, the initiator of pain in bee stings,[67] or of the components of activated complement, makes the virally induced changes in water and divalent cations that have been described, a prototype for a much wider spectrum of interactions between cells and potentially hemolytic agents.

REFERENCES

1. **Norrby, E.,** The morphogenesis of virions, in *Textbook of Medical Virology,* Lycke, E. and Norrby, E., Eds., Butterworths, London, 1983, chap. 10.
2. **Mimms, C. A. and White, D. O.,** *Viral Pathogenesis and Immunology,* Blackwell Scientific, Oxford, 1984, 17.
3. **Weiss, R., Teich, N., Varmus, H., and Coffin, J., Eds.,** *RNA Tumor Viruses,* Cold Spring Harbor Laboratory, Cold Spring Harbor, N.Y., 1982.
4. **Kohn, A.,** Early interactions of viruses with cellular membranes, *Adv. Virus Res.,* 24, 223, 1979.
5. **Pasternak, C. A. and Micklem, K. J.,** Virally induced alterations in cellular permeability: a basis of cellular and physiological damage?, *Biosci. Rep.,* 1, 431, 1981.
6. **Carrasco, L.,** Membrane leakiness after viral infection and a new approach to the development of antiviral agents, *Nature (London),* 272, 694, 1978.
7. **Carrasco, L. and Lacal, J. C.,** Permeabilization of cells during animal virus infection, *Pharmacol. Ther.,* 23, 109, 1983.
8. **Carrasco, L. and Smith, A. E.,** Sodium ions and the shut off of host cell protein synthesis by picornavirus, *Nature (London),* 264, 807, 1976.
9. **Carrasco, L. and Smith, A. E.,** The replication of animal viruses, in: *Viral Chemotherapy: International Encyclopaedia of Pharmacology and Therapeutics,* Section 3, Vol. 2, Shugar, D. Ed., Pergamon Press, Oxford, 1984.
10. **Dawson, K. M., Stewart, A., and Stebbing, N.,** Absence of significant viral effects of β-methylene GTP on encephalomyocarditis virus infection of L cells and mice, *J. Gen. Virol.,* 45, 237, 1979.
11. **Norrie, D. H., Wolstenholme, J., Howcroft, H., and Stephen, J.,** Vaccinia virus-induced changes in $[Na^+]$ and $[K^+]$ in HeLa cells, *J. Gen. Virol.,* 62, 127, 1982.
12. **Schaeffer, A., Kuhne, J., Zibarre, R., and Koch, G.,** Poliovirus-induced alterations in HeLa cell membrane function, *J. Gen. Virol.,* 44, 444, 1982.
13. **Foster, K. A. and Micklem, K. J., Agnarsdottir, G., Lancashire, C. L., Bogomolova, N. N., Boriskin, Y. S., and Pasternak, C. A.,** Myxoviruses do not induce non-specific alterations in membrane permeability early on in infection, *Arch. Virol.,* 77, 139, 1983.
14. **Gray, M. A., Micklem, K. J., and Pasternak, C. A.,** Protein synthesis in cells infected with Semliki Forest virus is not controlled by intracellular cation changes, *Eur. J. Biochem.,* 135, 299, 1983.
15. **Gray, M. A., Austin, S. A., Clemens, M. J., Rodrigues, L., and Pasternak, C. A.,** Protein synthesis in Semliki Forest virus-infected cells is not controlled by permeability changes, *J. Gen. Virol.,* 64, 2631, 1983.
16. **Pasternak, C. A. and Micklem, K. J.,** Permeability changes during cell fusion, *J. Membr. Biol.,* 14, 293, 1973.
17. **Pasternak, C. A.,** Virally-mediated changes in cellular permeability, in *Membrane Processes: Molecular Biology and Medical Applications,* Benga, G., Baum, H., and Kummerow, R., Eds., Springer-Verlag, Berlin, 1984, chap. 8.

18. **Poste, G. and Pasternak, C. A.,** Mechanisms of virus-induced cell fusion, in *Cell Surface Reviews,* Vol. 5, Poste, G. and Nicholson, G. L., Eds., North-Holland, Amsterdam, 1978, 306.
19. **Wyke, A. M., Impraim, C. C., Knutton, S., and Pasternak, C. A.,** Components involved in virally-mediated membrane fusion and permeability changes, *Biochem. J.,* 190, 625, 1980.
20. **Apostolov, K. and Almeida, J. D.,** Interaction of Sendai virus with human erythrocytes: a morphological study of haemolysis cell fusion, *J. Gen. Virol.,* 15, 227, 1972.
21. **Okada, Y., Koseki, I., Kim, J., Maeda, Y., Hashimoto, T., Kanno, Y., and Matsui, Y.,** Modification of cell membranes with viral envelopes during the fusion of cells with HVJ, *Exp. Cell Res.,* 93, 368, 1975.
22. **Homma, M., Shimizu, K., Shimizu, Y. K., and Ishida, N.,** On the study of Sendai virus hemolysis. I. Complete Sendai virus lacking in hemolytic activity, *Virology,* 71, 41, 1976.
23. **Bashford, C. L., Micklem, K. J., and Pasternak, C. A.,** Sequential onset of permeability changes in mouse ascites cells induced by Sendai virus, *Biochim. Biophys. Acta,* 814, 247, 1985.
24. **Pasternak, C. A.,** Viruses as toxins, *Arch. Virol.,* 93, 169, 1987.
25. **Sillen, L. G. and Mantell, A. E.,** *Stability Constants and Metal Ion Complexes,* The Chemical Society, London, 1964.
26. **Forda, S. R., Gillies, G., Kelly, J. S., Micklem, K. J., and Pasternak, C. A.,** Acute membrane responses to viral action, *Neurosci. Lett.,* 29, 237, 1982.
27. **Pasternak, C. A. and Micklem, K. J.,** Altered function of excitable cells caused by paramyxoviruses, in *Viruses and Demyelinating Disease,* Mims, C. A., Cuzner, M. L., and Kelly, R. E., Eds., Academic Press, New York, 1983, 101.
28. **Pasternak, C. A.,** How viruses damage cells: alterations in plasma membrane function, *J. Biosci.,* 6, 569, 1984.
29. **Klemperer, H. G.,** Hemolysis and the release of potassium from cells by Newcastle disease virus, *Virology,* 12, 540, 1960.
30. **Pasternak, C. A. and Micklem, K. J.,** The biochemistry of virus-induced cell fusion. Changes in membrane integrity, *Biochem. J.,* 140, 405, 1974.
31. **Micklem, K. J. and Pasternak, C. A.,** Surface components involved in virally-mediated membrane changes, *Biochem. J.,* 162, 405, 1977.
32. **Poste, G. and Allison, A. C.,** Membrane fusion, *Biochim. Biophys. Acta,* 300, 421, 1973.
33. **Lucy, J. A.,** Mechanisms of chemically induced cell fusion, in *Cell Surface Reviews,* Vol. 5, Poste, G. and Nicolson, G. L., Eds., North-Holland, Amsterdam, 1978, 267.
34. **Lucy, J. A.,** Molecular aspects of membrane fusion, in *Membrane Processes: Molecular Biology and Medical Applications,* Benga, G., Baum, H., and Kummerow, F. A., Eds., Springer-Verlag, Berlin, 1984, chap. 2.
35. **Impraim, C. C., Micklem, K. J., and Pasternak, C. A.,** Calcium, cells and virus: alterations caused by paramyxoviruses, *Biochem. Pharmacol.,* 28, 1963.
36. **Bajaj, S. P., Nowak, T., and Castellino, F. J.,** Interaction of manganese with bovine prothrombin and its thrombin-mediated cleavage products, *J. Biol. Chem.,* 251, 6294, 1976.
37. **Conlon, T. and Outhred, R.,** Water diffusion permeability of erythrocytes using an NMR technique, *Biochim. Biophys. Acta,* 288, 354, 1972.
38. **Impraim, C. C., Foster, K. A., Micklem, K. J., and Pasternak, C. A.,** Nature of virally-mediated changes in membrane permeability to small molecules, *Biochem. J.,* 186, 847, 1980.
39. **Morariu, V. V. and Benga, G.,** Evaluation of a nuclear magnetic resonance technique for the study of water exchange through erythrocyte membrane in normal and pathological subjects, *Biochim. Biophys. Acta,* 469, 301, 1977.
40. **Getz, D., Gibson, J. F., Sheppard, R. N., Micklem, K. J., and Pasternak, C. A.,** Mn as a Ca probe: EPR and NMR spectroscopy of intact cells, *J. Membr. Biol.,* 50, 311, 1979.
41. **Kretsinger, R. H.,** The informational role of calcium in the cytosol, in *Advances in Cyclic Nucleotide Research,* Vol. 1, Academic Press, New York, 1979, 1.
42. **Pushkin, J. S.,** Divalent cation binding to phospholipids: an EPR study, *J. Membr. Biol.,* 35, 39, 1977.
43. **Mehrishi, J. N.,** Molecular aspects of the mammalian cell surface, *Progr. Biophys. Mol. Biol.,* 25, 1, 1972.
44. **Mildvan, A. S. and Cohn, M.,** Aspects of enzyme mechanisms studied by nuclear spin relaxation induced by paramagnetic probes, *Adv. Enzymol. Relat. Areas Mol. Biol.,* 33, 1, 1970.
45. **Hazelwood, C. F., Chang, D. C., Nicols, B. L., and Woessner, D. E.,** Nuclear magnetic resonance transverse relaxation times in skeletal muscle, *Biophys. J.,* 14, 583, 1974.
46. **Shporer, M. and Civan, M. M.,** NMR study of ^{17}O from $H_2{}^{17}O$ in human erythrocytes, *Biochim. Biophys. Acta,* 385, 81, 1975.
47. **Andrasko, J.,** Water diffusion permeability of human erythrocytes studied by a pulsed gradient NMR technique, *Biochim. Biophys. Acta,* 428, 304, 1976.

48. **Pirkle, J. L., Ashley, D. L., and Goldstein, J. H.,** Pulse nuclear magnetic resonance measurements of water exchanges across the erythrocyte membrane employing a low Mn concentration, *Biophys. J.*, 25, 389, 1980.
49. **Koenig, S. H. and Schillinger, W. E.,** Nuclear magnetic relaxation dispersion in protein solutions, *J. Biol. Chem.*, 244, 3283, 1969.
50. **Eisenstadt, M. and Fabry, M. E.,** NMR relaxation in the hemoglobin-water proton spin system in red blood cells, *J. Magn. Res.*, 17, 301, 1978.
51. **Fabry, M. E. and Eisenstadt, M.,** Water exchanges between red cells and plasma, *Biophys. J.*, 15, 1101, 1975.
52. **Ashley, D. L. and Goldstein, J. H.,** The application of Dextran-magnetite as a relaxation agent in the measurement of erythrocyte water exchange using pulsed nuclear magnetic resonance spectroscopy, *Biochem. Biophys. Res. Commun.*, 97, 114, 1980.
53. **Micklem, K. J.,** Virally-Mediated Permeability Changes, Ph.D. thesis, University of London, 1982.
54. **Campbell, A. K.,** Living light, *Trends Biochem. Sci.*, 11, 104, 1986.
55. **Tsien, R. Y.,** A non-disruptive technique for loading calcium buffers and indicators into cells, *Nature (London)*, 290, 527, 1981.
56. **Campbell, A. K.,** *Intracellular Calcium: Its Universal Role as Regulator*, John Wiley & Sons, New York, 1983.
57. **Berridge, M. J. and Irvine, R. F.,** Inositol triphosphate, a novel second messenger in cellular signal transduction, *Nature (London)*, 312, 315, 1984.
58. **Bashford, C. L., Alder, G., Micklem, K. J., and Pasternak, C. A.,** A novel method for measuring intracellular pH and potassium concentration, *Biosci. Rep.*, 1, 631, 1983.
59. **Knutton, S., Jackson, D., Graham, J. M., Micklem, K. J., and Pasternak, C. A.,** Microvilli and cell swelling, *Nature (London)*, 262, 52, 1976.
60. **Ramm, L. E., Whitlow, M. B., Koski, A. L., Shin, M. L., and Mayer, M. M.,** Elimination of complement channels from the plasma membranes of 0937, a nucleated mammalian cell line: temperature dependence of the elimination rate, *J. Immunol.*, 131, 1411, 1983.
61. **Micklem, K. J., Nyaruwe, A., and Pasternak, C. A.,** Permeability changes resulting from virus-cell fusion: temperature dependence of the contributing processes, *Mol. Cell. Biochem.*, 66, 163, 1985.
62. **Bashford, C. L., Alder, G. M., Patel, K., and Pasternak, C. A.,** Common action of certain viruses, toxins, and activated complement: pore formation and its prevention by extracellular Ca^{2+}, *Biosci. Rep.*, 4, 797, 1984.
63. **Pasternak, C. A., Bashford, C. L., and Micklem, K. J.,** Ca^{2+} and the interaction of pore-formers with membranes, *J. Biosci.*, 8, 273, 1985.
64. **Pasternak, C. A., Alder, G. M., Bashford, C. L., Buckley, C. L., Micklem, K. J., and Patel, K.,** Cell damage by viruses, toxins and complement: common features of pore-formation and its inhibition by Ca^{2+}, *Biochem. Soc. Symp.*, 50, 247, 1985.
65. **Bashford, C. L., Alder, G. M., Menestrina, G., Micklem, K. J., Murphy, J. J., and Pasternak, C. A.,** Membrane damage by haemolytic viruses, toxins, complement and other cytotoxic agents: a common mechanism blocked by divalent cations, *J. Biol. Chem.*, 261, 9300, 1986.
66. **Hedstrom, S. A. and Kronvall, D., Eds.,** Symposium on staphylococcal septicaemia and endocarditis, *Scand. J. Infect. Dis.*, Suppl. 41, 1, 1983.
67. **Prince, R. C., Gunson, D. E., and Scarpa, A.,** Stings like a bee, *Trends Biochem. Sci.*, 10, 99, 1985.
68. **Micklem, K. J. and Pasternak, C. A.,** unpublished observations.

Chapter 8

ON THE THEORY OF SOLUTE-SOLVENT COUPLING IN EPITHELIA

Jorge Fischbarg

TABLE OF CONTENTS

I. THE RULES OF THE GAME

Much as a disclaimer such as this may give rise to a feeling of "déja vu" in the readers, it must be pointed out that this article is not intended to constitute a comprehensive monograph on the question. It may instead be regarded as a historical account combined with a "topical review", all of this tinged with the particular subjective bias of the author. The choice of quotations and material to be covered is therefore a highly personal matter. Yet, in talking about and presenting the question of solute-solvent coupling to other colleagues both inside and outside the field of fluid transport, many of the points covered here were the subject of interesting discussions. It seems therefore appropriate to cover them in this chapter.

II. UNITS FOR SIMPLICITY

In what follows, frequent references will be made to osmotic permeabilities of cell membranes and tissues. However, the units in which the osmotic permeabilities are expressed by different workers are not always the same, which makes it advisable to dwell on this subject. Intuitively, the units for osmotic permeability or hydraulic permeability (or osmotic conductance or hydraulic conductance) which are easiest to grasp, are those of flow per unit generalized driving force. We will call such a permeability Lp, *a la* Kedem and Katchalsky.[1] In their nomenclature, the water (or solvent) flow, Jv, is the first derivative of a given volume per unit area, so consequently, Lp will also be expressed per unit area. Caveat researcher! For any variation on this subject, one needs to keep a careful account of the area for which Jv was originally defined. With the notions above, flow may be expressed in, say, $cm^3/cm^2/sec$, which happens to simplify to cm/sec; using consistent c.g.s. units, the driving force will be expressed in $dynes/cm^2$, i.e., a hydrostatic Δ P. For an osmotic pressure difference, the "driving force" $\Delta\pi = RT\Delta C$ has also units of force per area, and, in both cases, $Lp = Jv/\Delta P$ will have dimensions of $length^2 \times time \times mass^{-1}$. So far, so reasonably well. The problem that immediately ensues is that different authors will fancy different "driving forces". Examples: ΔC, expressed as $mOsm/cm^3$, ΔP given in mmHg, milliPascals, millibars, and so *ad nauseam.* This, to say nothing of the fact that most if not all of such units do not conform to the Système Internationale (S.I.). This explosion of creativity has usually brought either puzzled expressions or approving nods at conferences. Admittedly, some authors may occasionally prefer to camouflage their results behind intractable units. Otherwise, however, there is a convenient set of "standard" units. These are arrived at by defining a "normalized" Lp* as $Lp^* = LpRT/V_w$, where V_w is the molar volume of water. With this procedure, the units of Lp* are simply $length \times time^{-1}$; they are generally expressed in the literature as cm/sec or, if one wants to conform to the S.I., in μm/sec. The physical simplicity of this expression for a velocity is attractive; all units arising from solute and solvent masses and volumes cancel out, leaving the result to express the average velocity of the solvent flow. In addition, these units are the same as those for a permeability, which makes possible immediate numerical comparisons between osmotic and diffusional coefficients. While on this, if one is attempting to contrast osmosis vs. diffusion, the standard terminology is to label the osmotic permeability P_f (for permeability for filtration) and the diffusional one P_d. As can be gathered, Lp* is identical to P_f. Since there is no standard set of abbreviations, we will use these indistinctly.

Another useful concept that deserves presentation (if sketchy) is that of the "effective" or overall osmotic permeability P_{os}, which is simply given by $P_{os} = \sigma Lp$ (or $P_{os}^* = \sigma Lp^*$), where σ is the reflection coefficient for the solute. For further reference, the reader is advised to consult textbooks such as Davson's.[2]

III. A LITTLE HISTORY: CLAUDE BERNARD AND THE RISE, FALL, AND RISE OF LOCAL OSMOSIS

From a historical perspective, it seems useful to remember that, many years ago, the transport of water by epithelia was not connected to the notion of solute-solvent coupling. There was no apparent need for that; instead, in the early days of modern physiology, such transport was simply referred to as "osmosis". It seems instructive to reproduce one of the pertinent writings of Claude Bernard:[3]

> ...L'absorption des liquides aqueux s'exerce en général avec facilité sur toutes les surfaces où l'absorption gazeuse peut avoir lieu. La peau chez certains amimaux fail l'exception. L'absorption aqueuse y parait très-faible ou même douteuse; ce qui tient aux propriétés de l'épiderme cutané, qui se trouve imprègné de sécrétions huileuses.
>
> L'absorption des liquides, comme celle des gaz, s'accomplit en vertu d'un échange entre le liquide intra-vasculaire, ou milieu intérieur, et le milieu éxtérieur ou extra-vasculaire, de manière à constituer toujours un double courant osmotique. Les phénomènes d'endosmose ou d'osmose, découverts dans ce siècle par un physiologiste français, Dutrochet, ont une très-grande importance pour l'explication des mécanismes de l'absorption, de la sécrétion ou de l'excrétion. Toutefois, cette découverte féconde est loin d'être épuisée. Elle a été et est encore le point de départ d'une foule de travaux très-importants pour la physiologie générale, parmi lesquels il faut citer en première ligne les belles recherches de Graham sur la dialyse.
>
> La condition essentielle pour qu'il y ait diffusion ou échange entre deux milieux gazeux ou liquides, contigus ou séparés par une membrane, est qu'ils soient de constitutions différentes....

As can be seen, that ancient notion was attractively simple. An ionic concentration difference is somehow created, and water moves by osmosis according to such concentration difference. Nowadays, we would class this mechanism as a valid form of "local osmosis".

So far as one can tell, this type of explanation remained unchallenged (or, perhaps, peacefully unnoticed) until the mid-1950s. At that time, in keeping with a general upswell of interest in epithelial function, measurements of osmotic permeability of the ciliary epithelium were performed by Auricchio and Barany.[4] They noted that the value that they measured (the steady-state P_{os} for that layer) was smaller than they expected, in fact, so small that the rate of fluid transport postulated to exist across the ciliary epithelium could only be accounted for, on the basis of local osmosis, by an impossibly large ionic gradient across the tissue.

Nowadays, we would immediately suspect that unstirred layers and solute polarization (vide infra) would result in a large underestimate of the P_{os} thusly measured. In those days, however, the role of such factors was not clear, and experiments such as these cast a long shadow on local osmosis. As a result, the possible mechanism by which the transport of electrolytes (or "solute") was coupled to the transport of water (or "solvent") across epithelia began to receive greater scrutiny; this search for alternatives to local osmosis was going to be branded the study of solute-solvent coupling.

IV. PIONEERS IN MODELING. CURRAN'S DOUBLE MEMBRANE

Thus, the thought began to emerge that a more complex mechanism had to be invoked in order to explain epithelial fluid transport. The first mention of such a mechanism may have been put forward by Ballintine in the recorded discussion of a symposium on aqueous humor production.[5] He speculated that, within the complexity of an epithelium, local spaces could exist that would harbor ionic gradients much higher than could be perceived or measured from the outside of the tissue. Although he much later modestly disclaimed that comment as "armchair speculation", the idea was bound to grow. It gained fertile ground in the Biophysical Laboratory in Harvard, to which A.K. Solomon had attracted a group of gifted co-workers. Ideas involving localized gradients were to be found in two papers from that laboratory,[6,7] and another paper in Europe.[8] Eventually, these notions were put in a

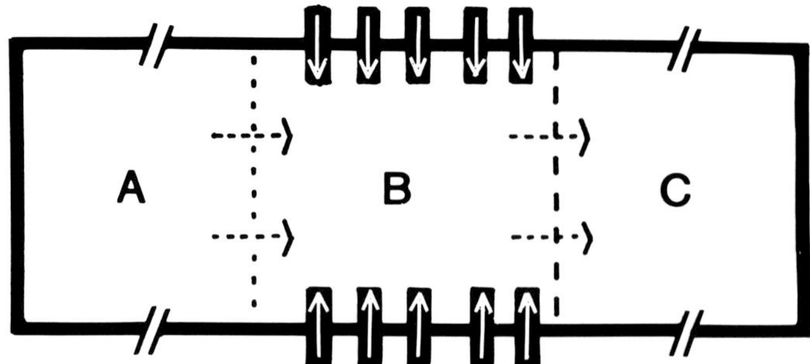

FIGURE 1. The double-membrane model. Solid arrows depict solute transport, dashed arrows signify water movement. The two membranes (A-B and B-C) have different hydraulic conductivities and reflection coefficients, and such asymmetry produces as a result the net water flow shown (from A to C).

more elaborate format by Curran and his co-workers.[9] Curran's hypothesis was to gain wide acceptance and to become famous under the name of the "double-membrane" model (Figure 1). His short description was to be enlarged by that of Patlak et al.[10] Their basic postulation was that a central compartment existed in epithelia bound by two membranes with different hydraulic permeabilities (Lp) and osmotic reflection coefficients (σ). By making the epithelium pump solute into or out of the central compartment and manipulating the values of Lp and σ, the epithelial layer could be made to display across itself any variation of coupled transports of fluid and electrolytes known to (or imagined by) man.

The next advance came when Kaye and his co-workers (followed in short order by Tormey and Diamond) proposed that the intermediate compartment of the double-membrane hypothesis was none other than the intercellular spaces.[11,12] The whole notion now fit well with established morphology, and acceptance for this scheme grew. Yet, its undisputed reign was destined to be relatively short-lived.

V. DIAMOND AND BOSSERT

Interest in this area had been awakened, and other researchers had joined the fray, notably J. Diamond. In 1964, he produced a crucial paper that put in question the double-membrane hypothesis on the basis of his results with the so-called "sweat preparation" of the gallbladder.[13] An excellent discussion of this and many other related topics can be found in the book by C.R. House.[14] Not one to leave things halfway, Diamond (in conjunction with Bossert) proceeded to develop an attractive alternative explanation for solute-solvent coupling.[15,16] Their scheme came to be known as the "standing–gradient" hypothesis.

In this work, they concentrated on the intercellular space as the primary structure responsible for fluid transport. This was in itself a significant advance over Curran's hypothesis, albeit made possible by the morphological studies cited above.[11,12] The differential equations that Diamond and Bossert advanced for this simplified system were to become standard for this formulation. They have withstood the test of time, and they reappear implicitly or explicitly even in an exceedingly detailed recent analysis such as that of Mathias.[17]

Diamond and Bossert next faced a choice of boundary conditions for their treatment. Some of the main assumptions they opted for were as follows: (1) closed junctions and (2) concentrations equal to that of the surrounding medium (C in the bulk medium or C_b in this chapter) both at the open end of the intercellular channel and inside the cells (Figure 2). However, the implementation of their analysis must have led to some interesting difficulties. Of necessity, they adopted values for the osmotic permeability of the lateral cell membrane

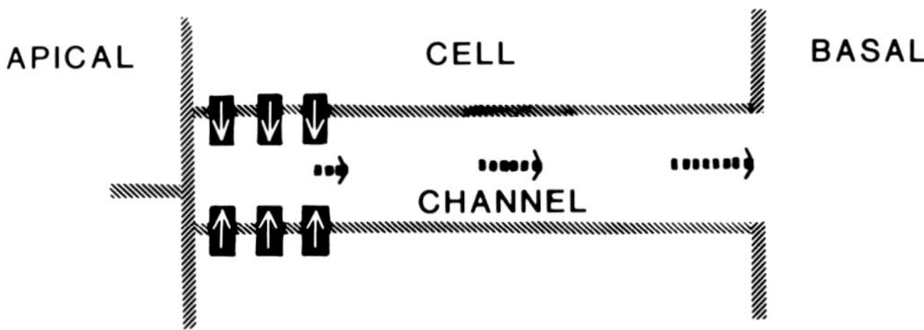

FIGURE 2. The standing-gradient model. Transport of solute into the dead end of the intercellular channel raises the local osmolarity and attracts water from the cell into the channel. A local hydrostatic pressure buildup makes the fluid flow out of the open end of the channel.

(Lp) which were in the general province of those published for permeation across whole epithelia. The values which they found most reasonable to choose were in the order of 20 to 100 μm/sec. Of course, as we now know, solving their equations with such values and the assumption above, the fluid transported turns out to be embarrassingly hypertonic. Hence, they needed to introduce an additional assumption, namely, that the rate of solute transport into the channel was restricted to the initial 1/10 bottom part of the closed dead-end channel considered. With that, things fell into place. The initial segment of the channel pumped solute, which accumulated in that neighborhood and gave a characteristic "hump" in the concentration profile curves. The concentration of solute decayed towards the open end of the channel. This characteristic concentration profile *along* the channel was what was termed a "standing-gradient" by Diamond and Bossert. The rest of the channel acted to bring osmotic equilibration in that it allowed an ample supply of solvent to flow in and dilute the high-concentration fluid initially generated at the junctional end. The accumulation of fluid in the channel increased the hydrostatic pressure in there, which propelled the fluid out through the open end. The end result: isotonic fluid transport.

VI. THE SERENE DOMINATION OF THE STANDING-GRADIENT HYPOTHESIS

This rather elegant formulation was to dominate the textbooks for years to come. It had intellectual appeal and conceptual simplicity and it explained isotonic fluid transport for a wide physiological audience. Hence, it attracted a well-earned following. As an example of this, a mathematical analysis by Segel appeared that was based on it and aimed at simplifying the rather forbidding nonlinear differential equations of Diamond and Bossert into a more accessible formulation for "everyday" easy use.[18] Such simplification has proved quite useful, judging from the number of others who have utilized it.[19-21]

VII. CLOUDS ON THE HORIZON

Yet, for all the praise and acceptance of the standing gradient model, little actual work had been done since 1968 on the substance of Diamond and Bossert's proposal. With the advent of new players, that state of affairs began to change slowly. Looking in retrospect, the first aspect that attracted the attention of skeptics was the assumption that the transport of solute was restricted to a short length of the channel. That looked peculiar; Na^+-K^+ ATPases were distributed with boring uniformity along lateral cell borders of epithelia, and the more that was learned about cell polarization in epithelia, the more precarious that assumption must have appeared. Given this type of new information being generated, the field was ripe for a critical reevaluation of these concepts in the mid-seventies.

VIII. THE STORM BREAKS OUT

At about that time, fittingly enough, three different papers appeared in which the tonicity of fluids transported by different epithelia had been calculated without assuming that transport was restricted to a small segment of the channel. These were those of Hill, Lim, and Fischbarg, and Sackin and Boulpaep.[19,20,22] All three basically concurred on the fact that the fluid transported under such circumstances was hypertonic. Sackin and Boulpaep considered a number of well-known fluid-transporting preparations in their analysis.[22] They noted several interesting aspects, i.e., that with uniform site density, the standing-gradient practically disappeared, and that there seemed to be no advantage in using the standing gradient formulation vs. the double membrane one. Hill searched the literature for suitable Lp values and geometrical parameters for a range of epithelia.[19] He concluded with a heavy indictment of the standing-gradient model, which generated hopelessly hypertonic fluid in all cases. As for us, we limited ourselves to utilize the available evidence for the corneal endothelium fluid transport system, and we verified once more that the fluid theoretically pumped with a uniform site distribution was uncomfortably hypertonic.[20]

IX. INTEREST REVIVES

All through the intervening period, the resurgence of interest in this area fueled the appearance of several other theoretical papers. King-Hele and Paulsen and Huss and Marsh, examined the implication of an open junction.[23,24] Diamond and Bossert had assumed a closed one, which they themselves qualified as an oversimplification, albeit perhaps not too serious.

The next move in this animated game came from Diamond. Characteristically, he put up a spirited defense.[25] His main theme was that all Lp values used until then had been severely underestimated, and that the geometrical parameters of epithelial layers could also have been incorrectly evaluated, given the vagaries of quantitative morphology. About the only dissonant note in that cogent writing was his curious assertion that Pedley and this author were "wrong" in our evaluation of the effects of unstirred layers on Lp measurements.[26] There was no error, of course; we merely looked at a simple ideal unstirred layer in one dimension, while Diamond was referring instead to the three-dimensional problem of solute polarization (meaning buildup or depletion) inside an intercellular convoluted epithelial channel, its neighboring cells, and adjacent layers. We would prefer not to call such a complex constellation an "unstirred layer". Practically any other name appears preferable for it, such as "intraepithelial solute polarization".

X. SO WHAT IS THE P_{os} FOR EPITHELIA?

One of the results of these intellectual encounters was a revival of interest in P_{os} measurements in epithelia. It became quite important to reevaluate those measurements and perform new ones in order to determine up to which point Hill's arguments were valid. As usual in science, there had been some straws in the wind pointing to the fact that Lp values for epithelia could be larger than previously suspected. For instance, Ullrich et al. reported in 1964 that the osmotic permeability of proximal kidney tubule was as large as 2000 μm/sec.[27] Fittingly enough, since that number was much larger than its counterparts for other epithelia, it was seemingly put aside until "the world" could understand it. Whichever the arguments for the kidney tubule, in 1972 Smulders et al. observed that the P_{os} across gallbaldder had a steep time dependence, and extrapolation to time zero of a time-variant flow curve gave a calculated P_{os} as much as ten times higher than the 50 μm/s that they reported for the steady-state.[28] Wright called attention to this observation in at least another occasion.[29]

Table 1
RECENT MEASUREMENTS OF EPITHELIAL OSMOTIC PERMEABILITIES

Membrane or tissue	P_{os} value (μm/sec)	Ref.
Necturus gallbladder, apical membrane	550	31
Necturus gallbladder, basolateral membrane	1200	31
Rabbit gallbladder, cell layer	500	36
Rabbit kidney proximal tubule, cell layer	5190—7620	35
Rabbit kidney proximal tubule, basal cell membrane	2760	37, 45
Rabbit corneal endothelium, cell	595	39
layer	711	38
Rabbit corneal endothelium, apical membrane	1420	38

Note: All P_{os} values are *uncorrected* for membrane infoldings.

Could it be that epithelial P_{os} would be large enough to retrieve the besieged standing gradient? Several laboratories set themselves to find out, having discarded (with arguable justification) Diamond's warning that the true P_{os} might be unmeasurable.[25] More on this later.

XI. THE NATIONAL INSTITUTES OF HEALTH (NIH) STEPS IN

The first ones to bring forward an answer were Spring and his co-workers. They put lavish NIH support to excellent use, and found rather large values for Necturus gallbladder P_{os} (550 μm/sec for the apical cell membrane and 1200 μm/sec for the basolateral one).[30,31] Of course, working with this tissue, they faced the problem of ''the length of the coast of England'' (or of the Malvinas, for that matter). It so happens that the length of a coastline measured with a scale depends on the size of the scale as with any fractal process. The method of Spring involved patient tracings of all outlines to measure cell perimeters and, from them, cell volume. Hence, perimeter values may change with the magnification used and, the more convoluted the cell membrane, the more the change. One alternative may be to measure cell area by capacitance, whenever feasible. Still, in spite of these difficulties, there seems to be at present no alternative but to rely on careful quantitative histology in order to normalize cell membrane P_{os} per unit membrane area.

Spring's values became an overnight success. For one thing, they confirmed Wright's and Diamond's intuitive contentions. In addition, they brought into some question the possibility that the fluid transported by leaky epithelia might traverse the intercellular junctions, which had been suggested by several laboratories.[32-34] Clearly, with such large cell membrane P_{os} values, the need to invoke a junctional route became distinctly less acute. Stimulated by Spring's findings, several laboratories (ours included) accelerated their efforts to measure cell membrane P_{os} in a variety of epithelia, and their results largely confirmed that cell membrane P_{os} were indeed larger than had been previously measured; some of these recent results are detailed in Table 1 below.

There was, however, a somewhat less desirable fallout from Spring's values. As important as it was to recognize that a route for fluid transport may go across the cell membranes, this was only one of the issues covered by the ongoing discussion about isotonic transport;

there were several other ones to consider, perhaps even more important, such as the distribution of pumping sites, cell geometry, cell tonicity, and so on. However, since large cell membrane P_{os} values were in line with one of Diamond's expectations,[25] the standing-gradient model received renewed support automatically, almost by extension. In fact, some misconceptions still had to be conquered.

Let us give an example of the difficulties that remained at this stage. As the careful readers of the paper by Sackin and Boulpaep may notice, a model with uniform site distribution yields a concentration profile nearly uniform along the length of the channel, hence, *no standing gradient* (that is, no sizable gradient *along* the channel).[22] In addition, with the "old" P_{os} values,[15] they verified that the fluid transported was, of course, hopelessly hypertonic. Rather than clearing the problem of the tonicity of the absorbate, Spring's values merely turned something very bad into something else simply bad; the fluid transported was now *only moderately hypertonic*. We will give an example, using Segel's approximation, for the standing gradient formulation, and using suitable values for *Necturus* gallbladder:

1. r = half width of intercellular spaces = 0.4 μm
2. L = length of intercellular spaces = 34 μm
3. C_b = bulk concentration = cell concentration = 210 mOsm/ℓ
4. P_{os} = lateral cell membrane osmotic permeability =
   "old" value: 50 μm/sec (or 9.1×10^{-5} cm^4/mOsm/sec)
   "new" value: 1200 μm/sec (or 217×10^{-5} cm^4/mOsm/sec).
5. D = diffusion coefficient = 1.5×10^{-5} cm^2/sec.

We will term rOs the ratio between the osmolarity of the absorbate and that of the ambient medium Cb. With Segel's formula,[18] one has "old" rOs: 5.29; "new" rOs: 1.31.

So the absorbate comes out still about 30% hypertonic, certainly too much for comfort. A similar reservation was voiced by Gonzalez et al. in 1982.[21] Redoing their Segel-like calculations for the best case for the standing gradient, we also obtain an absorbate hypertonic by 20% (they claim 30%). In conclusion, even with the advent of Spring's values, the standing-gradient model was still failing on two counts: (1) lack of a sizable standing-gradient and (2) hypertonicity of the absorbate.

XII. THE THIRD WAVE

Slowly, these problems became more apparent. A newer wave of thought appeared that eschewed the question of the standing gradient altogether and instead made practical use of the latest P_{os} in light of the notion that with such relatively high P_{os} values, a small local gradient would suffice to draw all the fluid one might desire, and hence, that an acceptable explanation for isotonic transport could be none other than *local osmosis*. At last, Claude Bernard was being partially vindicated. Yet, there was still an obstacle...

XIII. THE FALL OF THE LAST BASTION (WE HOPE): ISOTONIC CELL DETHRONED

It was not until relatively recently that a better truth began to emerge, almost by chance. The isotonic cell boundary condition of Diamond and Bossert was relaxed in several papers.[17,40-42] Why? As far as one can determine, it was done empirically. Liebovitch and Weinbaum paid close attention to the issue of what the intracellular concentration ought to be in isotonic pumping by the corneal endothelium. From simple steady-state considerations, they argued that the cell concentration had to be the average of those across the different cell membrane boundaries, weighted by the area of each boundary. They verified that the fluid transported was, alas, isotonic.[40] Almost at the same time, more NIH heavy artillery was directed at this problem, and Weinstein and Stephenson put together a comprehensive theoretical treatment of fluid pumping. The determined readers who will zero in on their Tables A1-B and A2-B will discover that, in the steady state, their cell compartments are anisotonic.[41] Similar considerations about cellular anisotonicity were made by Mathias.[17]

Curiously, although they noted it, the authors above did not overtly emphasize that they were breaking the unspoken taboo of cell isotonicity. However, this factor once again forced its way into center stage when the empirical need to deviate from cell isotonicity became quite marked during some of our recent work.[42] We studied a model for apical local osmosis in a generic epithelial cell. Somehow, the need to depart from cell isotonicity emerged here more clearly than in the precedents; one reason might have been that all our expressions were analytical, with the consequent ease for analyzing and plotting the relevant variables. We concluded that an isotonic absorbate could *never* be reached with an isotonic cell; it was simply a mathematical impossibility.[42] At this point, all previous observations seem to fall into place; for instance, we believe that the boundary condition of cell isotonicity used in the original Diamond and Bossert paper is the main problem with their treatment. The P_{os} values for the cell membranes of specialized fluid transporting epithelia would, of course, be expected to be comparatively high, but the repercussion of the P_{os} would be somewhat less drastic.

A razor's edge. Of course, Diamond and Bossert were not off by all that much. It is difficult to quarrel with an assertion that a cell is "isotonic" if its tonicity is 0.3005 rather than 0.3000 Osm/ℓ, almost a semantic difference. And yet, in this curious universe of solute-solvent coupling, a tenuous, almost imperceptible change in a boundary condition makes a very large difference in the outcome. A shift of a fraction of a mOsm/ℓ in cell concentration may spell the difference between isotonic transport or the stubbornly hypertonic flow we had referred to earlier.

XIV. RESPECT THE PIONEERS, YOUNG MEN

Presumably, that is how science moves. Some of the boundary conditions of Diamond and Bossert, quite acceptable in 1967, apparently can no longer be defended. Still, pioneers should not be criticized; it must be easier to work in this area nowadays with the benefit of *2 decades* of experimental and theoretical work by many than it may have been to decide on boundary conditions at a much earlier stage of development of this field. Most researchers in the area will acknowledge the powerful stimulation that Diamond's ideas exerted in them; this author certainly does.

XV. THE PARTY LINE AND UNDERGROUND MODELS FOR FLUID TRANSPORT

A. The Party Line

What next? Presumably, textbook writers will want to know how to account for isotonic transport (or isotonic solute-solvent coupling) in front of today's audiences. As far as this writer is concerned, a tenable model would have to include, in short:

1. A (slightly, almost imperceptibly) anisotonic cell (even if there is at present no hard evidence for it)
2. Two local osmotic gradients, one at the basolateral and another one at the apical cell membranes (there is some evidence that the intercellular spaces are hypertonic by some 3 to 5% in absorptive epithelia[43,44])
3. Reasonably high P_{os} values for the cell membranes (some recent evidence is given in Table 1 below)

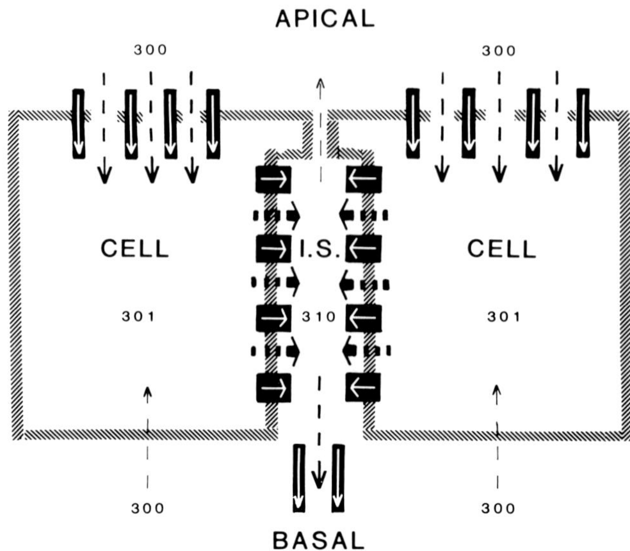

FIGURE 3. The anisotonic cell model. Osmotic flows take place across all cell membranes, driven by osmotic gradients across them. The numbers indicate the approximate milliosmolar concentrations theorized to exist in and next to an idealized epithelium. As in the standing-gradient model, there is a hydrostatic pressure buildup in the intercellular spaces; fluid is driven out predominantly through the basal end, although some loss across the apical leaky tight junction may occur. Note that fluid is "lost" back into the cell across the basal membrane; for this model to work more efficiently, the P_{os} of the basal or basolateral membrane (per unit membrane area) would have to be smaller than that of the apical membrane.

This scheme is, in a way, a compendium of several characteristics for fluid transport that have been implicitly or explicitly depicted in recent publications.[17,31,38,40-42] The similarities are not coincidental and represent a significant convergence of views on this subject.

What would one call this model? The consideration of the two gradients needed across the two regions of a cell membrane suggests a name such as "two-site transcellular local osmosis". But perhaps the most striking feature in this context is the intellectual paradox brought by the need of an anisotonic cell acting as a pivotal element to ensure isotonic flow. Hence, our particular preference goes to a name such as *"anisotonic cell model for isotonic fluid transport"*. The features in question are highlighted in Figure 3.

B. The Underground Model, Osmotic Flow Through Intercellular Junctions

This notion recurs through the literature from time to time. It would, of course, make eminent sense for leaky epithelia to use their leaky junctions to translocate fluid by osmosis through them. The problem is that solutes diffuse exceedingly fast through the leaky junctions, and their reflection coefficient for salt is hence bound to be very small, so that osmotic flows arising from a salt gradient at this level are very difficult to explain in terms of what we presently know about junctional morphology and physiology. To generate evidence to back that hypothesis has proved difficult. Yet, in retrospect, to show that the flow goes instead through cell membranes without otherwise disturbing an epithelial sheet is probably just as difficult.

A very brief review follows on evidence and opinions on the subject of osmosis through junctions. One could start arbitrarily with Whittembury's[32] calculations that lent credence to that possibility for the kidney proximal tubule. We[33] found what seemed a comparatively high P_{os} for corneal endothelium, and, given the context, suggested that it might be due to junctional osmotic flow. Hill and Hill[61,62] argued that the sucrose flux during osmosis across gall bladder was convective in nature, and that hence water flow must take place across the junctions. Diamond,[25] in reply, found objections to most prior arguments for junctional flow and gave the "party line", siding emphatically against that notion and for flow through cell membranes. Yet, unperturbed by the pendulum of debate, evidence continues to emerge (if slowly) in support of the "underground" view. Steward[63] and Whittembury et al.[64] reinforced A. E. Hill's arguments for the gall bladder. Naftalin and Tripathi[65] found large electro-osmotic coupling coefficients in intestinal epithelium and argued that such flow had to cross the junctions. Pappenheimer and his colleagues[66-68] went on to argue that paracellular transport of water and small solutes not only exists but is regulated by a dynamic process at the junctions of intestinal epithelium (as reviewed by Madara[69]). Weinstein[70] modeled the kidney proximal tubular epithelium and found that some ionic fluxes (notably Cl^-) were partly convective, and hence that osmosis through the junctions could not be excluded. In spite of this all, in a recent review (albeit written before 1988), Spring[71] evaluated the available evidence as insufficient to support flow through junctions, and favored transcellular flow.

At the time this goes to press (mid-1988), for this author the issue remains open. It will obviously turn on future experimental evidence. One alternative might be, for instance, that everyone is right and the flow traverses *both* cells and junctions. Still, it may be that transjunctional flow has been rather unduly disregarded, and that, as in some countries nowadays, the party line here too may face upheavals.

XVI. MORE ON THE OSMOTIC PERMEABILITY OF EPITHELIAL CELL MEMBRANES

One might ask what precisely are the values for the cell membrane P_{os} and the gradients one needs to account for isotonic transport. In fact, these go linked; given a certain P_{os}, a corresponding gradient becomes mandatory to account for fluid transport. However, some entertaining twists await those who travel this route.

One of the main problems is to determine the area of membrane to which the P_{os} value measured has to correspond. Typically, the value of water flows into or out of cells can be determined by microscopy, after which corrections are generally applied for the foldings and villosities of cells, which greatly increase the total area available for water flow and consequently decrease the P_{os} corrected per unit of actual cell membrane area (see Table 1). To give the gist of the problem, we will refer to Table 1, which gives a summary of some recent measurements. Among those, one may consider the P_{os} measurements of Welling et al. and of Carpi-Medina et al.[37,45] They seem a good choice, in that they both agree exceedingly well in terms of the P_{os} uncorrected for cell membrane area. Let us follow the detail of either method. From the rate of cell volume change, a certain value of P_{os} is arrived at for the basolateral membrane. Now comes the crucial step: the cell membrane area correction reduces the P_{os} value, typically by a factor of the order of 10 to 15 $\times$. However, there is a potential flaw in this type of correction. Let us picture an idealized membrane, with many foldings or a brush border (Figure 4). Let the bathing medium be made suddenly hypotonic. Water will rush into the dead-end valley spaces in-between the villi and will cross the membrane, leaving all its solute behind. In a very short time, solute will accumulate inside the valleys at a concentration much higher than that of the bathing medium, effectively nullifying the osmotic gradient just imposed in the membrane area of the valleys. In short, we will face a classical case of "solute polarization" (or solute buildup, in this case). How much membrane area has now been left for the fluid to traverse by osmosis? Lo and behold, about the "en face" or tangential area of tissue, which is what we began with before we became ambitious and applied our correction. There is, therefore, the possibility that P_{os} values *per unit of actual cell membrane area* may be larger than thought until now.

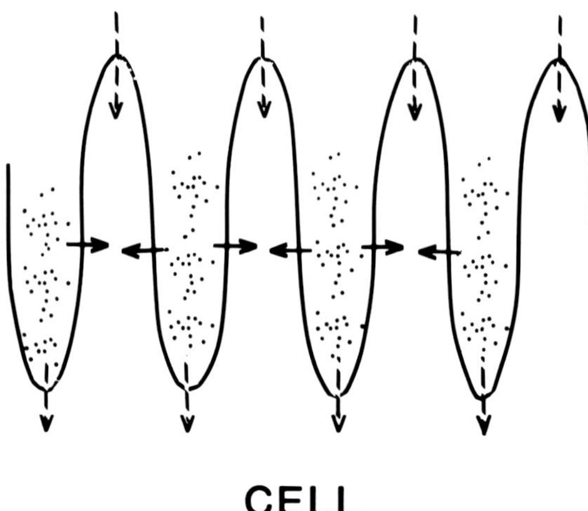

FIGURE 4. Osmotic flow into a cell across its villous membrane. The osmotic gradient is decreased by a build-up of solute concentration in the "valleys" between the villi.

XVII. TIME FOR SOME SPECULATION? THE QUESTION OF THE WATER CHANNEL

To change the subject somewhat, we turn to the route of water movement across cell membranes. For reasons such as those given above, this route emerges as crucial for the theory of water transport by cells and tissues. There is at present agreement among a majority of specialists that a water channel (or pore) exists across cell membranes, and that water largely traverses that route rather than the lipid bilayer. The origins of this story might be traced to the work of Solomon and his co-workers, who concluded that there was a pore (or channel) for water across red blood cell membranes.[49,50] Next, Macey and Farmer showed that the osmotic flow across those same cell membranes was inhibited by SH blockers.[51] Later on, this view was fueled by the spectacular discovery (by Chevalier et al.) that the antidiuretic hormone-induced increase in osmotic permeability across toad urinary bladder is accompanied by the insertion of protein aggregates into the cell membrane.[52]

The identity of this water channel is a different matter. A proposal was advanced that it might reside in the anion channel protein (band 3) of red blood cells.[53,54] However, that proposal rested on an observed inhibitory effect of 5,5′-dithiobis-(2-nitro-benzoic acid) (DTNB) on P_f,[53] and such inhibition could not be confirmed in a recent crucial study in which P_d was determined with (NMR) techniques.[55]

Another reasonable candidate for lodging the water channel is none other than the ubiquitous glucose transporter protein (band 4.5). Although not new, this view has remained stymied for years because attempts by several laboratories have failed to detect any inhibition of P_f or P_d by phloretin.[51,55,56] However, the work of Benga and his colleagues has maintained that possibility as a tenable one.[53] In a recent example, they have argued that, from the pattern of PCMBS binding to membrane protein fractions and concomitant inhibition of water diffusion in red blood cells, the site of water permeation clearly seemed to be limited to only band 3, or band 4.5, or both.[57,58] No further conclusions were drawn by them because, as they pointed out, no specific inhibitors for either anion or glucose permeation had been found that far to inhibit water transport. We have investigated this matter in the corneal endothelium and have found that several blockers of the glucose transporter also block osmotic water flow. We concluded that, in this epithelial preparation, water may traverse a glucose transporter channel. Still, the global picture is far from clear, and the identity of the water channel remains to be established.

REFERENCES

1. **Kedem, O. and Katchaksky, A.,** Thermodynamic analysis of the permeability of biological membranes to non-electrolytes, *Biochim. Biophys. Acta,* 27, 229, 1958.
2. **Davson, H.,** *A Textbook of General Physiology,* 4th ed., Williams & Wilkins, Baltimore, 1970, 548.
3. **Bernard, C.,** *Rapport sur les Progres et la Marche de la Physiologie Generale en France,* Imprimerie Imperiale, Paris, 1867.
4. **Auricchio, G. and Barany, E. H.,** On the role of osmotic water transport in the secretion of the aqueous humor, *Acta Physiol. Scand.,* 45, 190, 1959.
5. **Ballintine, E. J.,** in *Transactions of the Third Conference on Glaucoma,* Newell, F. W., Ed., Josiah Macy, Jr. Foundation, New York, 1958, 107.
6. **Durbin, R. P., Frank, H., and Solomon, A. K.,** Water flow through frog gastric mucosa, *J. Gen. Physiol.,* 39, 535, 1956.
7. **Curran, P. F. and Solomon, A. K.,** Ion and water fluxes in the ileum of rats, *J. Gen. Physiol.,* 41, 143, 1957.
8. **Heinz, V. E.,** Grandmechanismus der magensaureproduktion und deren regulation, *Klin. Physiol.,* 1, 184, 1960.
9. **Ogilvie, J. T., McIntosh, J. R., and Curran, P. F.,** Volume flow in a series-membrane system, *Biochim. Biophys. Acta,* 66, 441, 1963.
10. **Patlak, C., Goldstein, D. A., and Hoffman, J.,** The flow of solute and solvent across a two-membrane system, *J. Theor. Biol.,* 5, 426, 1963.
11. **Kaye, G. I., Wheeler, H. O., Whitlock, R. T., and Lane, N.,** Fluid transport in the rabbit gallbladder. A combined physiological and electron microscope study, *J. Cell Biol.,* 30, 237, 1966.
12. **Tormey, J. M. and Diamond, J. M.,** The ultrastructural route of fluid transport in rabbit gallbladder, *J. Gen. Physiol.,* 50, 2031, 1967.
13. **Diamond, J. M.,** The mechanism of isotonic water transport, *J. Gen.,Physiol.,* 48, 15, 1964b.
14. **House, C. R.,** *Water Transport in Cells and Tissues,* Edward Arnold Publ., London, 1974.
15. **Diamond, J. M. and Bossert, W. H.,** Standing-gradient osmotic flow. A mechanism for coupling of water and solute transport in epithelia, *J. Gen. Physiol.,* 50, 2061, 1967.
16. **Diamond, J. M. and Bossert, W. H.,** Functional consequences of ultrastructural geometry in "backwards" fluid-transporting epithelia, *J. Cell Biol.,* 37, 694, 1968.
17. **Mathias, R. T.,** Epithelial water transport in a balanced gradient system, *Biophys. J.,* 47, 823, 1985.
18. **Segel, L. A.,** Standing-gradient flows driven by active solute transport, *J. Theor. Biol.,* 29, 233, 1970.
19. **Hill, A. E.,** Solute-solvent coupling in epithelia: a critical examination of the standing-gradient osmotic flow theory, *Proc. R. Soc. London Ser. B.,* 190, 99, 1975.
20. **Lim, J. J. and Fischbarg, J.,** Standing-gradient osmotic flow: examination of its validity using an analytical method, *Biochim. Biophys. Acta,* 443, 339, 1976.
21. **Gonzalez, E., Carpi-Medina, P., and Whittembury, G.,** Cell osmotic water permeability of isolated rabbit proximal straight tubules, *Am. J. Physiol.,* 242, F321, 1982.
22. **Sackin, H. and Boulpaep, E. A.,** Model for coupling of salt and water transport. Proximal tubular reabsorption in necturus kidney, *J. Gen. Physiol.,* 66, 671, 1975.
23. **King-Hele, J. A. and Paulson, R. W.,** On the influence of a leaky tight junction on water and solute transport in epithelia, *J. Theor. Biol.,* 67, 61, 1977.
24. **Huss, R. E. and Marsh, D. J.,** A model of NaCl and water flow through paracellular pathways of renal proximal tubules, *J. Membr. Biol.,* 23, 305, 1975.
25. **Diamond, J. M.,** Osmotic water flow in leaky epithelia, *J. Membr. Biol.,* 51, 195, 1979.
26. **Pedley, T. J. and Fischbarg, J.,** The development of osmotic flow across an unstirred layer, *J. Theor. Biol.,* 70, 427, 1978.
27. **Ullrich, K. J., Ruhmrich, J., and Fuchs, G.,** Wasserpermeabilitat und transtubularer wasserfluss corticaler Nephronabschnitte bei verschiedenen diuresezustanden, *Pfluegers Arch. Gesamte Physiol. Menschen Tiere,* 280, 99, 1964.

28. **Smulders, A. P., Tormey, J. M., and Wright, E. M.,** The effect of osmotically induced water flows on the permeability and ultrastructure of the rabbit gallbladder, *J. Membr. Biol.*, 7, 164, 1972.
29. **Wright, E. M.,** in *Intestinal Ion Transport*, Robinson, J. W. L., Ed., University Park Press, Baltimore, 1976, 210.
30. **Spring, K. R., Hope, A., and Persson, B. -E.,** Quantitative light microscopic studies of epithelial fluid transport, in *Water Transport Across Epithelia*, Ussing, H. H., Bindslev, N., Lassen, N. A., and Sten-Knudsen, O., Eds., Munksgaard, Copenhagen, 1981, 190.
31. **Persson, B. -E. and Spring, K. R.,** Gallbladder epithelial cell hydraulic water permeability and volume regulation, *J. Gen. Physiol.*, 79, 481, 1982.
32. **Whittembury, G.,** Sobre los mecanismos de absorcion en el tubo proximal del rinon, *Acta Cient. Venezolana*, 3, 71, 1967.
33. **Fischbarg, J., Warshavsky, C. R., and Lim, J. J.,** Osmotic and hydraulic permeabilities across epithelial layers, *Nature (London)*, 266, 71, 1977.
34. **Hill, A. E.,** Junctional flows of salt and water: towards a unified theory of fluid transfer, in *Water Transport Across Epithelia*, Ussing, H. H., Bindslev, N., Lassen, N. A., and Sten-Knudsen, O., Eds., Munksgaard, Copenhagen, 1981, 284.
35. **Schafer, J. A., Patlak, C. S., Troutman, S. L., and Andreoli, T. E.,** Volume absorption in the pars recta. II. Hydraulic conductivity coefficient, *Am. J. Physiol.*, 234, F340, 1978.
36. **Van Os, C. H., Wiedner, G., and Wright, E. M.,** Volume flows across gallbladder epithelium induced by small hydrostatic and osmotic gradients, *J. Membr. Biol.*, 49, 1, 1979.
37. **Welling, D. J., Welling, L. W., and Ochs, T. J.,** Video measurements of basolateral membrane hydraulic conductivity in proximal tubule, *Am. J. Physiol.*, 245, F123, 1983.
38. **Fischbarg, J. and Montoreano, R.,** Osmotic permeabilities across corneal endothelium and ADH-stimulated toad urinary bladder structures, *Biochim. Biophys. Acta*, 690, 207, 1982.
39. **Klyce, S. D. and Russell, S. R.,** Numerical solution of coupled transport equations applied to corneal hydration dynamics, *J. Physiol. (London)*, 292, 107, 1979.
40. **Liebovitch, L. S. and Weinbaum, S.,** A model of epithelial water transport, *Biophys. J.*, 35, 315, 1981.
41. **Weinstein, A. M. and Stephenson, J. L.,** Models of coupled salt and water transport across leaky epithelia, *J. Membr. Biol.*, 60, 1, 1981.
42. **Fischbarg, J., Liebovitch, L. S., and Koniarek, J. P.,** A central role for cell osmolarity in isotonic fluid transport across epithelia, *Biol. Cell*, 55, 239, 1985.
43. **Simon, M. S., Curci, S., Gebler, B., and Fromter, E.,** Attempts to determine the ion concentrations in the lateral spaces between the cells of Necturus gallbladder epithelium with microelectrodes, in *Water Transport Across Epithelia. Alfred Benzon Symposium 15*, Ussing, H. H., Bindslev, N., Lassen, N. A., and Sten-Knudsen, O., Eds., Munksgaard, Copenhagen, 1981, 52.
44. **Sackin, H.,** Electrophysiology of salamander proximal tubule. II. Interspace NaCl concentrations and solute-coupled water transport, *Am. J. Physiol.*, 251, F334, 1986.
45. **Carpi-Medina, P., Lindemann, B., Gonzalez, E., and Whittembury, G.,** The continuous measurement of tubular volume changes in response to step changes in contraluminal osmolarity, *Pfluegers Arch.*, 400, 343, 1984.
46. **Kaye, G. I., Pappas, G. D., Donn, A., and Mallett, N.,** Studies on the cornea. II. The uptake and transport of colloidal particles by the living rabbit cornea in vitro, *J. Cell Biol.*, 12, 481, 1962.
47. **Parisi, M. and Bourguet, J.,** Water channels in animal cells: a widespread structure?, *Biol. Cell*, 55, 155, 1985.
48. **Macey, R. I.,** Transport of water and urea in red blood cells, *Am. J. Physiol.*, 246, C195, 1984.
49. **Goldstein, D. A. and Solomon, A. K.,** Determination of equivalent pore radius for human red cells by osmotic pressure measurement, *J. Gen. Physiol.*, 44, 1, 1960.
50. **Solomon, A. K.,** Characterization of biological membranes by equivalent pores, *J. Gen. Physiol.*, 512, 335, 1968.
51. **Macey, R. I. and Farmer, R. E. L.,** Inhibition of water and solute permeability in human red cells, *Biochim. Biophys. Acta*, 211, 104, 1970.
52. **Chevalier, J., Bourguet, J., and Hugon, J. S.,** Membrane associated particles. Distribution in frog urinary bladder epithelium at rest and after oxytocin treatment, *Cell Tissue Res.*, 152, 129, 1974.
53. **Brown, A. P., Feinstein, M. B., and Sha-afi, R. I.,** Membrane proteins related to water transport in human erythrocytes, *Nature (London)*, 254, 523, 1975.
54. **Solomon, A. K., Chasan, B., Dix, J. A., Lukacovic, M. F., Toon, M. R., and Verkman, A. S.,** The aqueous pore in the red cell membrane band 3 as a channel for anions, cations, nonelectrolytes, and water, *Ann. N.Y. Acad. Sci.*, 414, 97, 1983.
55. **Benga, G., Pop, V. I., Popescu, O., Ionescu, M., and Mihele, V.,** Water exchange through through erythrocyte membranes: nuclear magnetic resonance studies on the effects of inhibitors and of chemical modification of human membranes, *J. Membr. Biol.*, 76, 129, 1983.

56. **Owen, J. D. and Solomon, A. K.,** Control of nonelectrolyte permeability in red cells, *Biochim. Biophys. Acta,* 290, 414, 1972.
57. **Benga, G., Popescu, O., Pop, V. I., and Holmes, R. P.,** *p*-(Chloromercuri)benzenesulfonate binding by membrane proteins and the inhibition of water transport in human erythrocytes, *Biochemistry,* 25, 1535, 1986.
58. **Benga, G., Popescu, O., Borza, V., Pop, V. I., Muresan, A., Mocsy, I., Brain, A., and Wrigglesworth, J. M.,** Water permeability in human erythrocytes: identification of membrane proteins involved in water transport, *Eur. J. Cell Biol.,* 41, 252, 1986.
59. **Fischbarg, J., Liebovitch, L. S., and Koniarek, J. P.,** Evidence that water permeates cell membranes through the glucose channel, *Invest. Ophthalmol.,* Suppl. 27(Abstr.), 84, 1986.
60. **Fischbarg, J., Liebovitch, L. S., and Koniarek, J. P.,** Inhibition of transepithelial osmotic water flow by blockers of the glucose transporter, *Biochim. Biophys. Acta,* 898, 266, 1987.
61. **Hill, B. S. and Hill, A. E.,** Fluid transfer by *Necturus* gall bladder epithelium as a function of osmolarity, *Proc. R. Soc. London Ser. B,* 200, 151, 1978.
62. **Hill, A. E. and Hill, B. S.,** Sucrose fluxes and junctional water flow across *Necturus* gall bladder epithelium, *Proc. R. Soc. London Ser. B,* 200, 163, 1978.
63. **Steward, M. C.,** Paracellular non-electrolyte permeation during fluid transport across rabbit gallbladder epithelium, *J. Physiol. (London),* 322, 419, 1982.
64. **Whittembury, G., Martinez, C. V., Linares, H., and Paz-Aliaga, A.,** Solvent drag of large solutes indicates paracellular water flow in leaky epithelia, *Proc. R. Soc. London Ser. B,* 211, 63, 1980.
65. **Naftalin, R. J. and Tripathi, S.,** Passive water flows driven across the isolated rabbit ileum by osmotic, hydrostatic and electrical gradients, *J. Physiol. (London),* 360, 27, 1985.
66. **Pappenheimer, J. R. and Reiss, K. Z.,** Contribution of solvent drag through intercellular junctions to absorption of nutrients by the small intestine of the rat, *J. Membrane Biol.,* 100, 123, 1987.
67. **Pappenheimer, J. R.,** Physiological regulation of transepithelial impedance in the intestinal mucosa of rats and hamsters, *J. Membrane Biol.,* 100, 137, 1987.
68. **Madara, J. L. and Pappenheimer, J. R.,** Structural basis for physiological regulation of paracellular pathways in intestinal epithelia, *J. Membrane Biol.,* 100, 149, 1987.
69. **Madara, J. L.,** Tight junction dynamics: is paracellular transport regulated?, *Cell,* 53, 497, 1988.
70. **Weinstein, A. M.,** Modeling the proximal tubule: complications of the paracellular pathway, *Am. J. Physiol.,* 254, F297, 1988.
71. **Spring, K. R.,** Mechanism of fluid transport by epithelia, in *Handbook of Physiology—The Gastrointestinal System IV,* American Physiological Society, Washington, D. C., 1988.

Chapter 9

WATER PERMEABILITY OF AMPHIBIAN URINARY BLADDER

Jacques Bourguet, Jacques Chevalier, Mario Parisi, and Pierre Ripoche

TABLE OF CONTENTS

I. INTRODUCTION

A surprisingly sophisticated membrane traffic appears to play a pivotal role in the antidiuretic response. Observations during the last 10 years have indeed progressively led to the idea that changes in water permeability in ADH-responsive target epithelial cells do directly result from the insertion in the plasma apical membrane of new components that contain channels for water.

Although lipidic channels could also be considered, the exquisite specificity of the channels that accept water, but only water, strongly suggests that they are formed by proteins: more precisely, intrinsic proteins having access to both faces and span the whole membrane. Although these intrinsic proteins are firmly anchored to the lipidic core of the membrane and cannot be easily inactivated by removing them from the membrane and transferring them to the cytoplasm, there is actually strong experimental evidence that they do shuttle between these two locations. They are, however, not destroyed but recycled. Inactivation directly results from their removal from the apical frontier, which is the limiting barrier to water movements, and there is no need to invoke a conformational alteration of the channel. Consequently, *de novo* protein synthesis is not required to have the channels operational once again. Finally, the channels' protein never quit their liphophilic environment and are always inserted in a membrane, either in the plasma membrane or in the membrane of small cytoplasmic tubules and vesicles. They transit from one to the other place by an exo- or endocytosis-like process that protects them from the hydrophilic environment.

The present review deals with cellular mechanisms of the ADH-induced increase in water permeability, which gives rise to the so-called hydroosmotic response, in amphibian urinary bladder.

There is first a short survey of the initial steps of hormone action, starting from its binding to its receptor in the laterobasal membrane and leading to the activation of a still unidentified phosphoprotein under the influence of one of the several cAMP-dependent protein kinases that have been discovered in these cells.

We then turn to another well-known sequence, the final apical steps, and review the evidences resulting from morpho-functional studies, particularly freeze-fracture and related cytochemical approaches.

After describing the structure of the intramembrane particle aggregates involved in the hydroosmotic action, we discuss all the evidence pointing to their specificity and to their unique relationship to net water transepithelial movements in this tissue. The end of this section is devoted to the description and discussion of the hypothesis of a shuttle between cytoplasm and plasma membrane and the possibility of the reuse of the aggregates.

Although there are still important gaps in our knowledge of the events that lead from the early events in hormone action and the production of the intracellular mediator, cAMP, to the apparition of particle aggregates in the apical membrane and to the very final step of water permeability, two sections are devoted to the multiple roles of intracellular calcium and to the participation of the cytoskeleton to the response.

The last two sections review, respectively, the characteristics of the pathways for water movements and the potential biochemical correlates of some parameters whose importance has been underlined by functional and morphological studies: the potential correspondence between membrane particles, microfilaments, microtubules, granules, and other characteristic ultrastructures and cAMP-controlled phosphoproteins, the chemical nature of the coat and, most of all, of the particle aggregates.

Last but not least, attention is called to other examples of hormone-controlled transport that also involve similiar membrane traffic in their regulation: glucose transport under the control of insulin in the adipocyte[1] and proton transport in the gastrointestinal tract[2] and in renal-like epithelia.[3,4] Interestingly, they point to a more general mechanism controlling membrane transport and open the way to fruitful comparisons of these different classes of biological models.

Several other reviews have been devoted these last years to the mechanism of water permeability control of ADH. Most of them[5-7] emphasize the importance of the contribution that morphological studies have brought to the study of a strictly physiological process such as membrane transport.

II. INTRACELLULAR MEDIATION OF ADH ACTION

A. ADH Receptors and Cyclase Stimulation

The first steps of the action of vasopressin are now clearly defined. We have some confidence about the binding of the hormone to specific receptors in the basolateral membrane[8,9] and the subsequent production of an intracellular mediator.[10] Adenylate cyclase derived from epithelial cells of the urinary bladder of the toad, *Bufo marinus,* has been characterized by H.P. Bär et al.[11-13] Stoff et al.[14] have shown that the level of cAMP was increased threefold in the presence of ADH in the epithelial cells. There is also some evidence for the presence of an adenylate cyclase which does not control water flow and which, under certain circumstances, contributes to the production of cAMP of the water flow compartment.[15] Moreover, not only may there be several pools of cAMP, but the possibility that vasopressin may stimulate cAMP production in one type of cell, mitochondria-rich cells (MRC), with transfer to granular cells (GC) through intercellular channels, has been proposed.[16] This possibility is, however, still a matter of discussion. The fact that the cell swelling observed after vasopressin exposure to an hypotonic medium on the luminal side of the epithelium is confined specifically to the granular cells has suggested strongly that it is the luminal membrane of these cells which is altered by vasopressin.[17]

Methylxanthines and other compounds which inhibit cAMP phosphodiesterases and increase intracellular concentration of the intracellular mediator can also mimic the action of the hormone.[18] Interestingly, several trialkylxanthines, probably because of a better penetration into the cell, also elicit a hydroosmotic response when added to the mucosal medium.[19]

B. Post-Cyclic AMP Steps

From cyclic AMP production to the transformation of the luminal membrane, two ways were described. First, stimulation of dephosphorylation by some phosphatases.[20-23] Also, in a preparation of apical membrane of collecting ducts, Schwartz et al.[24] have described an endogenous phosphorylation of membrane proteins by a cAMP-dependent protein kinase also present in this membrane. The possibility that the dephosphorylation was posterior to the kinase stimulation has been proposed by the same authors.[25,26] There is direct evidence that a cAMP-dependent protein kinase is activated in toad or frog urinary bladder,[27,29] and it is possible that a cAMP-dependent phosphorylation of microtubule-associated protein is involved in the post-cAMP steps, like in brain.[30,31]

Some secondary information has been obtained about these first steps. The permeability change is dependent on the energy metabolism.[32,33] The hydroosmotic response to ADH can be altered by sulfhydryl-reducing agents like cystein.[34] Some other hormones and hormone-like substances also modify the response. Adrenergic agents inhibit the ADH stimulation and this inhibition can be eliminated by α-adrenergic blocking agents. Adrenal steroid hormones have a permissive effect on water permeability.[35]

An inhibitory effect of prostaglandins has been demonstrated by Orloff et al.[37,38] They show an inactivation of the cyclase but no effect on the response to cAMP.

The hypertonicity of the serosal medium can induce a hydroosmotic response comparable to the ADH one,[39,40] but the step where the hypertonicity action takes place is still unknown. It seems to be subsequent to the cAMP production.

C. Other Possible Ways

For a few years, the accent was given to possible regulation by the polyphosphoinositol cycle. In 1978, Orloff and Zusman[41] brought out direct evidence that ADH activates a phospholipase that serves to release arachidonic acid, the precursor of PGE_2. This derivative diminishes adenylate cyclase activity, the consequence of which the response of the enzyme to vasopressin is modulated. Schlondorff and Satriano,[42] using specific inhibitors and measuring the ^{32}P incorporation into phosphatidic acid and phosphatidylinositol, show that vasopressin stimulates PG synthesis in a cAMP-independent manner and also in a cAMP-dependent manner. As proposed for rat hepatocyte, a calcium-dependent, phosphoinositide-specific phosphodiesterase is not active under normal cellular conditions, but is activated upon the addition of vasopressin to the intact cell.[43] This mechanism has been also advanced in cultured mesangial cell.[44] So, these effects of vasopressin on the polyphosphoinositide breakdown are perhaps the key for a regulation system of the hydroosmotic response. This hypothesis led to some research on phorbol esters action: desensibilization of adenylate cyclase,[45] protein kinase C stimulation[46] with phosphorylation of a cytoskeleton protein[47] or of rabbit erythrocyte band 4.1. A more direct action at the apical level (exocytosis, endocytosis, and microvilli elongation) has been described by Masur.[48] A schema, proposed by Nishizuka, of a possible cascade and feedback control of cellular function could be taken into account to understand the regulation of our system.[49,50]

The last developments in the knowledge of the adenylate cyclase[51] and polyphosphoinositol systems would be a potent help for a better understanding of the mechanism underlying the hydroosmotic response to ADH.

III. MORPHOLOGICAL CORRELATES

A. Anatomy

1. At Rest

The amphibian urinary bladder is a bilobed structure that can be used as a bag[32] or mounted in chambers for water movement studies.[52,53] The tissue consists of a layer of mucosal cells supported by a basement membrane, resting on a loose layer of connective tissue with smooth muscle bundles and capillaries and covered on its contramucosal surface with a lining of serosal cells.[54,55] The thickness of the epithelial layer strongly depends on the stretching of the tissue[56] and on the existence of a hydroosmotic transepithelial flux of water[57,58] At rest, this epithelial layer is about 5 μm thick[59] while the connective tissue varies considerably throughout the bladder (30 to 100 μm), depending on the amount of blood vessels and muscle fibers.

Four morphologically distinct cell types constitute the epithelial layer of toad bladder:

GC, MRC, Goblet (or mucous) cell, and basal cell.[54] However, only the first three cell types come in contact with the lumen, the basal cells being located underneath this cell surface monolayer. Granular cells comprise 80 to 90% of the epithelial cells,[17,54,60,61] depending on the species studied (e.g., goblet cells do not exist in frog bladder) and cover most of the apical surface facing the lumen of the bladder. MRC and goblet cells have a pear shape and present only a small tip on the luminal side. As can be seen in surface views of preparations which were freeze-fractured[62,63] or stereo-scanned,[61,64-67] the epithelial cells give an appearance of a pavement structure with straight borders between the cells.

Thin-section electron microscopy shows that the mucosal membrane of the granular cells is covered with a dense cell coat of filamentous material attached perpendicular to the plasma membrane, 0.2 to 0.5 μm in length.[54,55,68] This fuzzy coat is rich in negative charges[69] and displays a high concentration of surface glycoproteins.[70-76]

Adjacent cells are linked together at their apical pole by a well-developed tight junction, the zonula occludens, about 0.4 μm in width.

On their lateral side, epithelial cells are deeply imbricated, each cell developing long infoldings into the intercellular spaces. In the absence of a net water flow, intercellular spaces are closed.[17,54,55,57,58,68,77] Desmosomes are sparsely distributed over the basolateral membranes of epithelial cells and gap junctions appeared only associated with basal cells.[78] Direct morphological evidences show that Na^+ pumps are restricted to the basolateral cell membranes of mucosal epithelial cells.[79]

Granular cells derived their name from the numerous distinctive granules which line the apical cytoplasm of the cell.[54,55,80] The major fibrillar elements of the cytoskeleton — microtubules (25 nm), actin microfilaments (5 to 7 nm), and intermediate filaments (10 nm) — have been identified as components of the cytoskeletal lattice in the apical region of these granular cells.[54,81-85]

The morphology of the epithelial cells, their organization in sheets, and their normal relationships are well preserved despite the absence of submucosa. As shown by microdissection[86] or low hydrostatic pressure and collagenase treatment,[87] the responsiveness of the isolated epithelium is unmodified and this indicates that the cellular events induced by the hormone and leading to an increase in the hydroosmotic permeability of the tissue are located in the epithelium itself.

2. ADH-Induced Modifications

Years ago, it was observed that ADH induces drastic modifications of the general aspect of amphibian urinary bladder epithelial cells. For instance, an important cell swelling was observed after vasopressin exposure with a hypotonic medium on the luminal side of the epithelium. This swelling is confined specifically to the granular cells,[17,55,58,68] but is clearly a secondary effect not essential to the action of vasopressin because exposure to ADH in the absence of an osmotic gradient results in no swelling even though there is an increase in epithelial permeability.[68,77] Concomitant with this swelling, a dilation of the intercellular spaces occurs,[55,57,58,77,88] although the mechanisms by which ADH induces such a dilation is still controversial.[89,90] Scanning electron microscopy reports[64-67,91-93] show that vasopressin action is accompanied by a dramatic change in the external surface topology of the granular cell apical membrane although there is a controversy over whether[67,92] or not[64,93] the change requires both ADH and water flow. After ADH challenge, the structure of MRC and goblet cells remains totally unchanged.[67]

Masur et al.[80] demonstrated that vasopressin induces the addition of new membrane components to the apical surface of granular epithelial cells of the toad bladder. These investigators observed that vasopressin stimulates the exocytosis of membrane-limited granules, and they suggested that the water permeability change elicited by the hormone is secondary to the release of secretory material and/or incorporation of the granule membrane into the existing apical membrane. In fact, the apical membrane area is increased (by some 40%) concomitant with the release of granules during vasopressin stimulations, at the initiation of the hydroosmotic response while endocytosis, internalization of previously surface connected membranes become significant later in the ADH response.[94-96] Pietras et al.,[97] on the other hand, reported that vasopressin provokes the release of lysosomal enzymes at the mucosal surface of the epithelium and induces a redistribution of binding sites for the lectin Concanavalin A.[73] Both processes are dependent on the cellular microtubule integrity.[73,85]

Lastly, the total amount of negatives charges of the apical surface of GC seems to be reduced in ADH-stimulated frog urinary bladder.[69]

In conclusion, it is clear that ADH induces modifications of the epithelial cells of amphibian urinary bladder, especially of the major cell type, the granular cells. However, are these changes primary events closely associated with transport features or are they secondary effects of the hormone action, not related to the mechanism of hormone action? One must ask this question since it was clearly demonstrated that ADH increases the water permeability of the apical barrier in inducing channels inside the apical membrane of granular cells. None of the preceding observations can demonstrate the presence of such pores and since these structures might lie in the plane of the membrane, it is obvious that freeze-fracture techniques can give new insights for this question.

B. Freeze-Fractured Preparations

1. Freeze-Fracture Studies of Urinary Bladder Epithelium at Rest

In contrast to most membranes previously examined using freeze fracture, the luminal membranes of the granular cells exhibit an ultrastructural characteristic which can be closely related to the low basal level of permeability of the tissue, namely a relatively low particle partition coefficient[98] observed between the P and E fracture faces. In most biological membranes, especially in nonpolarized cells, intramembrane particles (IMPs) cleave with the inner juxtaprotoplasmic leaflet (P face). In granular cells of the bladder, a comparable situation is only encountered for the basolateral membranes while, in the apical plasma membrane, most IMPs are found to cleave with the E face, i.e., on the external leaflet. Only a few particles are found on the inner leaflet so that P face appears to be particularly smooth, a feature which indicates a good continuity of the lipidic layer in the protoplasmic leaflet and, consequently, would suggest a very low permeability of that membrane. So, apical plasma membrane of granular cells is characterized, at rest, by numerous, large IMPs (about 1000 particles per μm^2, 10 to 12 nm in diameter), evenly distributed over the exoplasmic E fracture faces.[63,99,100] These large IMPs often display two to four subunits. On the complementary P face, "footprints" of E face IMP subunits can be observed codistributed with the small particles (8 nm in diameter, 600 IMPs/μm^2) that characterize P faces in those cells.[72] From time to time, a second type of granular cell was observed which lacks the large particles classically present on E face (double replicas rule out the possibility that the particles have cleaved with P face).[101]

On the other hand, mitochondria-rich cells exhibit a conventional distribution of IMPs with a higher density over the P face than the E face, both in the apical and laterobasal sides of the cell. On P face, unique rod-shaped particles are observed, to which correspond, rod-shaped depressions on E face. These IMPs are in striking contrast to the globular-shaped particles of adjacent granular cells,[62,102] and rod-shaped particles have been shown to represent a constant feature of MRC in different cell systems;[100,103,104] of dark cells in kidney-collecting tubules[105] and of chloride cells.

In the cytoplasm of granular cells, adjacent to the luminal membrane, the characteristic granules are virtually bare of particles[63,100,106] and probably have a high cholesterol content as shown by the high density of filipin-sterol complexes observed in freeze fracture.[107,108] An inverse correlation was found between the density of filipin-sterol complexes in the apical membrane and the incidence of granules in the cytoplasm. This suggests that fusion of granules with the apical plasma membrane may be responsible for the variation in the concentration of cholesterol in the apical membrane.[108]

2. Freeze-Fracture Analysis of Antidiuretic Response

No statistically significant alterations have been observed in the structure of the intercellular junctions during hormonal challenge: neither the width of the zonula occludens, the number of junctional strands, or the density of junctional meshes were significantly modified by the hormone.[109] On the contrary, the apical plasma membrane of granular cells of amphibian urinary bladder stimulated with ADH displays numerous intramembranous particle aggregates, a structure more obviously linked to water permeability variations.[99,110]

a. Fine Structure of Particle Aggregates

In glutaraldehyde-fixed preparations, ADH-induced aggregates are only observed on P face of fractured apical plasma membrane. In each aggregate, particles form alignments of different lengths which are more or less condensed in parallel arrays with a spacing of 12 nm. The precise size of the particles cannot be measured due to a shadow overlapping, but is close to 10 to 12 nm and, consequently, aggregate particles are much larger than the particles dispersed on the same membrane leaflet and characteristic of the P face of the apical membrane of granular cells.[102,111] Aggregates often appeared grouped in islets forming stellar or triangular arrangements.[101]

On the opposite complementary E face, groove arrays match the aggregates of the P face.[110,112] This groove network presents an orthogonal organization with comparable periodicity along the two axes, close to 6.5 nm.[101] In some cases, slight differences between these two complementary fracture faces are observed,[111] but some of them probably result from deformations which occur during fracture or replica formation.

b. Specificity of Aggregate Formation

The high level of organization of the aggregates suggests that these structures do not simply represent a random aggregation of intramembrane particles as has been previously observed in various membranes under a variety of circumstances.[113-116] Rather, this feature and numerous others point to the specificity of the aggregation phenomenon and to its close relationship with the ADH challenge. First, there is a complete absence of such aggregates in specimens from preparations fixed at rest, i.e., in the absence of any ADH-like agonist. This accords with the fact that the aggregate phenomenon is completely reversible: all the aggregates disappear if the agent at the origin of their formation is suppressed. The topographic repartition of the aggregates exhibits a comparable specificity. They are present only in the apical plasma membrane of granular cells, where most authors agree that the limiting barrier to transepithelial water movements is located. Second, water flow correlates with, but does not induce aggregate formation. They are observed in preparations exposed to the hormone in the absence of any gradient (same Ringer solution on both sides of the preparation) as well as in the presence of a transepithelial osmotic gradient.[102] Finally, freeze-fracture studies show that other areas of the apical plasma membrane are not modified by hydroosmotic agonists, which only trigger the apparition of the aggregates in the membrane.

c. Relation to Net Water Flow

The main problem to the physiologist is certainly to relate the particle aggregates to a specific membrane function and to determine their exact role. All the existing evidences indicate that they are linked to water permeability variations but their precise role is not yet known. Aggregates were described for the first time in association with the hydroosmotic response of frog urinary bladder to a synthetic neuropeptide, oxytocin.[99] This observation was confirmed by Kachadorian et al.[110] in toad bladder stimulated by vasopressin and a systematic study was then conducted and the following points established:

1. Aggregates are present during hydroosmotic challenge with arginine vasotocin, the antidiuretic hormone in *Rana esculenta,* as well as with cAMP, its cellular second messenger.[102,112]
2. This observation was extended to other ADH-sensitive epithelia and to other animal species. Comparable aggregates have been observed in the mammalian-collecting ducts[117] and in frog[111] and toad[118] skin. Despite some differences in the aggregate ultrastructure from one tissue to another, there is no doubt that ADH induces a specific structure into the luminal membrane of these sensitive cells.
3. Pertinent agonists and antagonists have also been employed to establish that aggregates are specifically linked to water permeability variations. It was shown, for example, that aggregates are present in the apical plasma membrane during hydroosmotic response to serosal hyperosmolarity,[102] a condition which markedly depresses trans-epithelial sodium transport. Anesthetics such as methohexital[119] were also employed to inhibit water permeability selectively without reducing the urea and sodium translocation and demonstrate that aggregates were associated with water permeability reduction.
4. Quantitative comparative studies of water permeability variation and aggregate formation were also conducted and a satisfactory correlation was observed in various conditions. First, when different agonist concentrations were employed, both the number of aggregation sites and the surface area of the aggregates were found to be directly proportional to the net water flow.[110,120] For maximal agonist concentrations, aggregates were found to cover about 1 to 2% of the total apical surface area.[101,112,119] In other experiments, aggregate appearance was monitored during the onset of the hydroosmotic response and their number was found to increase with increasing net water flow in a roughly parallel manner.[110,121-124] These studies also revealed the cellular character of the aggregation phenomenon, some cells exhibiting a high number of aggregates at a moment when other cells had none.[101]
5. Various other inhibitors of the hydroosmotic response — poisons of the cytoskeleton such as colchicine and cytochalasin B[125-127] and detergents known to extract intrinsic plasma membrane proteins[128] were also studied, both to extend this correlation and to clarify the possible roles of other cellular organites.

Although the function of the aggregates has not been demonstrated directly, the conviction has grown very strong that these structures represent the sites of apical membrane water channels. Nevertheless, under certain conditions, cellular acidification rapidly reduces the ADH-induced water flow, but the aggregate density remains unchanged.[124,129-131] This discrepancy with the preceding results seems to correspond, in fact, to a direct effect of low cellular pH causing the water channel to shift from an "open" to a "closed" state[129] (see also later in this review).

d. The "Shuttle" Hypothesis

Particle aggregates were first considered to originate from material already present in the luminal membrane by association and alignment of individual intramembrane particles. It is not difficult to imagine that an appropriate reorganization of a spectrin-like protein[132] under the influence of ADH, could lead to an organized aggregation of IMPs. However, this seems unlikely because (1) particle aggregates appear in the apical membrane very quickly, as soon as 1 min after the addition of the hormone and straight off exhibit their typical organization[101] and (2) 2.5 min after ADH stimulation, although the total number of aggregates seen per unit area of apical membrane is clearly submaximal, their size remains constant and the frequency of small aggregates is not increased as compared to what is seen at the peak of the response.[121] These observations therefore are not consistent with assembly of aggregates in the luminal plasma membrane of granular cells either from pre-existing particles or from individual particles newly inserted into the membrane.

Another alternative came when it was observed that aggregates with organization identical to that found in the luminal membrane of ADH-stimulated bladder exist at rest in cytoplasmic vesicles and tubules in granular cells.[78,133,134] These tubules, called aggrephores, have a diameter of 0.11 μm and can be up to 2 μm long,[135] vesicles probably being short segments of these tubules. Both tubules and vesicles are separate structures from the numerous cytoplasmic granules described earlier. As observed on the luminal membrane, aggregates are seen on P face, complementary groove networks on E face. The parallel linear arrays of aggregated particles appear to be disposed at an angle to the long axis of the tubule, which perhaps imparts structural stability and accounts for their uniform diameter.[135] Aggregate particles are not the subunits from which the tubule is constructed since variable smooth spaces are often seen between groupings of aggregates.

The number of aggrephores is maximal in resting cells; in toad bladder, the number has been reported to be reduced by half during ADH stimulation.[7] Such a reduction would support the hypothesis that these cytoplasmic structures move to the apical membrane in the presence of ADH and are the precursors of the apical aggregates. Wade[7] called the "Membrane Shuttle Hypothesis" the idea that the cells might vary their luminal membrane permeability by shifting aggregates back and forth between the luminal membrane and the cytoplasmic vacuoles.

Transfer of aggregates towards the apical membrane appears to involve the fusion of the carrier tubules with the plasma membrane. In a few favorable cases, tangential fractures have been obtained,[135] which exhibit tubular membranes with typical aggregates in continuity with the plasma membranes. Such pictures, however, are rare. Most frequently, fusion images can be observed in en face views of the apical membrane.[7,106,135] The images look like circular holes, the centers of which show an ice-like amorphous structure and correspond to the lumen of the tubule or vesicle fused with the membrane. In ADH-stimulated preparations, particle aggregates can frequently be seen touching the border of the fusion image. There is no doubt that such pictures indicate a site of fusion, and their frequent association with particle aggregates provides strong argument in favor of the involvement of the fusion process in the delivery of aggregates to the plasma membrane. It should be noted, however, that most of the aggregates do not directly touch the fusion images; in some cases, they are very far from them: after delivery into the apical membrane, aggregates subsequently become dispersed and behave as sites for water flow.

The incidence of fusion events is correlated with the number of aggregates that occur in the membrane after ADH stimulation.[106,135] Pretreatment with colchicine (a microtubule disrupter) has been found to decrease the number of presumptive fusion sites of aggrephores with the apical membrane[135] in addition to its inhibitory effects on the particle aggregates and water flow response.[126,127] However, colchicine has no effect when it was added 30 min after hormonal stimulation.[125,135] Thus, microtubules appear to play a role in the translocation of the aggrephores towards the apical cell surface but only in the initiation of the hormonal response prior to the incorporation of the aggregates into the plasma membrane. On the contrary, cytochalasin B (which disrupts microfilaments) has no effect on the frequency of the membrane fusion events[135] but inhibits the ADH water flow.[125-127] Therefore, once fusion has occurred, microfilaments may play a role in the movement of aggregates from the intracellular membranes to the luminal membrane.

Recent observations[106] show numerous small fusion images in control preparations unstimulated by the hormone. However, they lack emerging particle aggregates. These small fusion images may represent points of contact between the aggrephores which, in the absence of ADH, do not transfer aggregates to the luminal membrane. This possibility suggests a constant cycling of the aggrephores which are capable of making contacts with the luminal membrane but can only consolidate their contact and transfer aggregates following ADH treatment. It has been shown indeed that many tubules are positioned within 1 to 2 μm of the luminal membrane of unstimulated bladders.[137] Contact between aggrephores and membranes may therefore be a frequent occurrence.

After hormone washout, aggregates disappear from the luminal membrane. Again, one possibility considered was that the particle constituent of the aggregates is dispersed in the luminal membrane and observations of Kachadorian et al.[121] indicated that 5 min after ADH washout, the mean size of aggregates remaining in the luminal membrane was decreased significantly and the relative frequency of very small aggregates was increased while the relative frequency of large aggregates was decreased. This pattern would be consistent with dispersion of aggregates in the luminal membrane during reversal of the response. However, we never observed in our studies clouds of condensed particles which could represent such dispersion of aggregates linked to any hormone washout. This was only seen after disruption induced by a nonionic detergent.[128] Moreover, the absolute number of aggregates is reduced for all size categories after hormone washout.

An alternative process of disappearance was proposed by Wade[7] in his membrane shuttle hypothesis. The aggregates may be retrieved from the luminal membrane intact and brought back into the cytoplasm. The larger aggregates, in this hypothesis, may be retrieved faster for some reason, thus leaving behind a higher frequency of small aggregates as observed by Kachadorian et al.[121] Therefore, in such a model, aggregates are recycled so that the same patch of membrane could make repeated trips to the luminal membrane in response to subsequent exposures to hormone.

Because of the method used to study the fusion of aggrephores, all of which involve tissue fixation or fast freezing, it is difficult to distinguish between two patterns of fusion: the first is one in which a given tubule remains fused throughout the period of hormonal stimulation; the second is one in which tubules are continuously fusing and detaching from the luminal membrane during stimulation. Using colloidal gold probes or horseradish peroxidase, evidences were obtained for the cycling of aggrephores between cytoplasm and luminal membrane during the period of ADH stimulation.[95,138] Return of fused tubules to the cytoplasm appears to take place even during the initial 15-min period of stimulation when water flow and fusion events are both increasing. Apparently, some fused tubules return to their resting positions in the cytoplasm during this early phase. However, this process seems to strongly depend on the presence or absence of an osmotic gradient during ADH stimulation which therefore appears to modulate the membrane retrieval phase.[95]

IV. PATHWAYS FOR WATER MOVEMENTS

In the absence of hormonal stimulus, the mucosal or luminal border of ADH target cells can be at least 50 times less permeable to water than the serosal one. There is general agreement, based on functional and structural studies, that the final action of the hormone is to increase the water permeability of the apical membrane of granular cells (for a review, see Bourguet et al.[5]).

The water osmotic permeability of toad urinary bladders was initially measured by following the change in weight of isolated sacs[140] while unidirectional fluxes were measured employing ^{3}HHO.[52] These methods were improved by the technique introduced by Bourguet et al:[53,141] the bladder is mounted as a diaphragm between two lucite chambers and water is automatically injected to or sucked from one of those chambers to maintain a constant volume. The magnitude of this fluid movement, equivalent to the net flux, is recorded every minute. The measurement of transepithelial water permeability was further improved by an experimental approach allowing the simultaneous minute-by-minute determination of net and unidirectional movements.[142]

Two parameters are currently determined when water fluxes are measured across epithelial barriers: the diffusive permeability coefficient for water (Pd_w), calculated from ^3HHO fluxes in the absence of a water gradient and the osmotic permeability coefficient (Pf), calculated from the net water flow in the presence of an osmotic gradient. Pd_w was initially underestimated because of the presence of unstirred layers in series with the ADH sensitive membrane and relatively high values for the Pf/Pd_w ratio were initially reported. This ratio was still higher under ADH, originating the "pore enlargement hypothesis", initially proposed to explain the increase in water permeability induced by the hormone[52] (for a review, see Hebert and Andreoli[143]).

A turning point in the ideas regarding water permeability was the observation, in artificial lipid bilayers, that Pd_w is similar to Pf after correction for the unstirred layers effects. This indicates that water is moving through the lipid bilayer by a similar mechanism in the presence or in the absence of an osmotic gradient: partition and diffusion in the hydrocarbon core of the membrane.[144,145] It was also observed that it is possible to prepare pure lipid bilayers having water permeabilities similar to those observed in animal cells.[146] These results generated a new conception of ADH action: according to this hypothesis, the hormone would work, "facilitating" (through membrane fluidification?) the water movement across the lipid bilayer of the sensitive membranes. If this was true, adequate correction of unstirred layers effects would lead to a situation in which Pf will be similar to Pd_w, either before or after ADH action.[147,148]

Very quickly the "lipid fluidification hypothesis" was found unable to explain all the available experimental evidences. Any change in the lipid composition, lipid fluidity, or structure will change the membrane permeability to water and to a series of other small lipophilic or hydrophilic molecules. This is the case in artificial bilayers but not in ADH-sensitive cells[149] where the small nonspecific increase in the permeability to certain nonionic solutes cannot be compared to the increase in water permeability. Furthermore, the small increase in membrane fluidity associated with the hormonal action cannot explain the observed increase in water permeability.[150]

A. Biophysical Characterization of the ADH-Induced Water Channels

When the experimental approach developed to evaluate the effect of unstirred layers in artificial membranes was applied to the ADH-sensitive epithelia, it was observed that, in the absence of the hormone, $Pf/Pd_w = 1$.[151] After ADH action, this relation becomes significantly higher ($11 < Pf_w < 18$).[90,151-153] This qualitative change is similar to the one observed when channel-forming molecules (gramicidine, nystatin, etc.) are added to pure lipid bilayers.[154,155] In nonstimulated preparations, some water would be passing through the lipid bilayer of the apical membrane, where the hormone induces the appearance of a new water pathway,[156] almost restricted to water. Additional evidence indicating that ADH adds a new water pathway came from experiments employing the nonionic detergent NPEO[6], that differently modifies water and nonelectrolyte permeability before and after the hormonal action.[156] The ADH-associated water pathway has been described, on the basis of time-course studies, as resulting from the addition of permeability units (specific channels) that increase in number during the development of the hormonal action.[142]

ADH-induced channels seem to be specific for water (urea is excluded, as well as Na ions) and very narrow (less that 2 Å of radius)[157] (for a review, see Hays[6]). It has been suggested that single-file transport could occur across this structure and that the Pf/Pd_w ratio would indicate (as in the case of the gramicidin channel) the number of molecules that can be accommodated inside the channel.[151,153] It must be considered, however, that the existence of complex geometric arrangements in series with each aqueous channel can influence the observed results.[122,151,153]

All the experimental evidences presented in previous paragraphs support the hypothesis describing ADH action as activating a ''shuttle'' mechanism plugging and removing water channels. These channels would be present, in one of different conceivable ways, in the intramembrane particle aggregates visualized in the apical membrane using freeze-fracture studies.[99,110] When ADH-sensitive epithelial barriers are fixed with glutaraldehyde at rest or after hormonal stimulation, they remain in the low or high level of water permeability in which they were before fixation.[59,158,159] This allows the study of water permeation in a situation in which the contributions of membrane traffic processes (exoendocytosis)[95] or regulatory feedbacks[160] can be excluded. The fixative action partially reduces the osmotic and diffusive[90] increases in water permeability induced by ADH. Nevertheless, the Pf/Pd_w ratio remains similar to the one observed in nonfixed preparations.[90] This proportional reduction in Pf and Pd_w can be understood if we accept that during the fixation procedure some permeability units are lost, while the remaining ones maintain their permeability properties, the cross-linking reagent interacting with chemical groups outside the channel.

B. Water Movement Across the ADH-Sensitive Cells

It seems clearly established that water permeability is controlled at the apical membrane of the ADH target cells, probably via the insertion and removal of specific water channels. Nevertheless, it must be remarked that in the presence of an osmotic gradient, an enormous amount of water traverses the epithelial barrier. With a net flux of 3 $\mu\ell/cm^2/min$ (a frequently observed value under oxytocin) and assuming cubical cells having a side of 10 μm, the whole cell content would be removed every 20 sec. Several observations seem to indicate that this water flow is canalized from the apical regions of the target cells towards the intercellular spaces that appear swelled and distended in this situation.[77,161] The osmotic flux drastically modifies the water distribution inside the cell and it has been considered that a structural barrier, in series with the apical barrier, puts a limit to the maximal flow detectable in the presence of the hormone.[81,162,163] Going still further, it has even been proposed that this barrier could regulate water permeability in toad urinary bladder.[164] It seems evident that the presence of barriers in series with the apical membrane can affect the measured value of net flux,[122,153] as unstirred layers affect the observed unidirectional fluxes. Nevertheless, additional evidence is necessary to clarify if these observations have physiological significance.

V. CELLULAR FACTORS REGULATING THE HYDROOSMOTIC RESPONSE

A. The Role of Microtubules and Microfilaments

After the initial report showing that drugs interacting with microtubules and microfilaments inhibit the ADH and cAMP-induced osmotic water movement across the toad bladder,[81,82] it was observed that these drugs proportionally reduce water flux and the number of intramembrane particle aggregates that appear in the apical membrane.[125-127] As previously stated, these aggregates, that probably contain water channels, would be transferred from cytoplasmic vesicles to luminal membranes under hormonal stimulation.[135,165] Microtubules and microfilaments would play a crucial role in this ''membrane traffic'' process and an elaborated hypothesis (based on morphofunctional studies) has been proposed to explain their mechanism of action: microtubules would be related to the vesicles' fusion with the apical membrane while microfilaments would ''pull off'' the aggregates towards the plane of the membrane.[125,135] More direct and less speculative biochemical and structural studies have demonstrated the existence in ADH-sensitive cells of all the molecular elements (tubulin, actin) that are the supporting material of cytoskeletal functions.[54,82-84,166-168]

Colchicine, podophyllotoxin, and vinblastine inhibit the increase in the osmotic[81,83,125,169,170-174] and diffusive[174] water permeability induced by ADH. Moreover, taxol,[175] which binds specifically to cellular microtubules, as well as nocodazole or elevated hydrostatic pressure,[171,176,177] both of which are believed to induce the rapid depolymerization of microtubules, also inhibit the ADH response. On the other hand, lumicolchicine, an isomer that does not bind to tubulin, has no influence on the hydroosmotic response,[81,178] while colchicine decreases the content of assembled microtubules in the apical cytoplasm of the granular cells.[167]

Despite the previously presented evidence, the precise role that microtubules would play in the hydroosmotic response is still unknown (see Pearl and Taylor,[85] for review). Recent studies have shown that the time course of the onset of the hydroosmotic responses to ADH[127] and cAMP is slowed by colchicine, while the offset of the reaction is not affected by the alkaloid.[127,179] This observation suggests that while microtubules would play a role in aggregates' insertion, they seem not to be required for aggregates' removal from the apical membrane. Furthermore, colchicine did not affect the hydroosmotic response once fully developed.[125,127,179] This observation, indicates that microtubules are also not required for aggregates' maintenance in the apical membrane. It must be finally mentioned that disruption of as much as 75% of cytoplasmic microtubules by colchicine does not impair the hydroosmotic response to a very low dose of vasopressin, in the presence of prostaglandin synthetase inhibitors.[180]

Cytochalasin B, which blocks monomer addition to microfilaments, inhibits the hydroosmotic response to ADH and cAMP[81,82,181,182] but has no effect on sodium or urea transport. Furthermore, cytochalasin B proportionally inhibits the appearance of the aggregates in the apical membrane.[125,126] These inhibitory actions were observed when cytochalasin B was added either before or after ADH, suggesting that the maintenance, as well as the initiation of the hormonal response, appears to depend on some aspects of the microfilament action.

Nevertheless, the effects of cytochalasin seem to be less specific than those elicited by colchicine. Deep alteration of the cytoskeleton[64,82,181] is reflected in a nonspecific increase in membrane permeability to water, sodium, chloride, and urea, as well as in a reduction of membrane resistance.[84,181] These results indicate an increase in the permeability of the paracellular pathway which was confirmed with the use of extracellular markers.[127] After correcting for this effect, proportional inhibition of both diffusional and osmotic water permeability by cytochalasin B was demonstrated.[127]

The action of cytochalasin B in the presence of an ADH-induced osmotic flow results in a marked disruption of granular cell morphology.[64,125,127,162,163,178,181,182] Deep alteration of the cytoskeleton is reflected in a strong vacuolization and distortion of the epithelial structure, and it has been suggested that this may alter the path of water and contribute to the observed inhibition of the hydroosmotic response. Nevertheless, and because of its complexity, these results generated controversial interpretations;[127,163,177] see Pearl and Taylor[85] for a review.

B. The Role of Intracellular pH and Ca^{2+}

As previously stated, microtubules and microfilaments would play a crucial role in the membrane traffic process started by ADH. In view of the evidence linking Ca^{2+} with the cytoskeletal function, the possibility that cytosolic Ca^{2+} plays a role in the hydroosmotic response must be considered. Furthermore, membrane fusion events are also dependent on Ca^{2+} ion concentration. Because of this situation, several studies were devoted to localize post-cAMP steps sensitive to intracellular Ca^{2+} concentration. The obtained results must be differentiated from those describing the effects of Ca^{2+} on steps preceding the production of cAMP, mainly on the hormone-receptor interaction,[183] or in the function of the cyclase system.[12]

Most evidence on the role of intracellular Ca^{2+} is indirect, because only in recent experiments has this parameter been directly measured in isolated toad bladder cells.[185,186] Pharmacological experiments designed to manipulate intracellular Ca^{2+} concentration used Ca^{2+} ionophore A23187, X537A,[187-190] quinidine,[188,191] verapamil,[190,192,193] trifluoperazine,[193] and different Ca^{2+} blockers.[194] While most results seem to indicate that an increase in intracellular Ca^{2+} inhibits the hydroosmotic response,[183,187,188,190,191] others suggest that the response to oxytocin is a Ca^{2+}-mediated process, Ca^{2+} being critical at a step after the generation of cAMP to activate the water permeability of the apical membrane.[189] Nevertheless, non-Ca^{2+}-related effects of these drugs like prostaglandin liberation,[195,196] membrane fluidification,[195] or other side effects[192,197] can be at the origin of the reported modifications in water permeability.

A23187 has, however, a true ionophoric action in toad urinary bladder cells.[186] Moreover, the inhibitory action of the ionophore addition to the mucosal bath is post-cAMP, temperature dependent, and reminiscent of the offset of the reaction that follows the agonist withdrawl.[130] The effect can then be situated before the "temperature-dependent, rate-limiting step"[198,199] that has been related to the mechanism controlling the water channels' plug-in and removal.[124]

Between the most direct experiments relating cellular Ca^{2+} and the increase in water permeability can be mentioned those performed by Cuthbert and Wang[200] showing that ADH and cAMP modify, in opposite directions, Ca^{2+} release through the mucosal border of toad bladder epithelium. More recently, Burch and Halushka[185] have reported that ADH enhances Ca^{2+} effux in isolated toad bladder cells by releasing plasma membrane-bound Ca^{2+}. The average cytosolic-free Ca^{2+} level (as measured in isolated toad bladder cell suspensions by Quin 2 or Fura 2) does not appear to change following stimulation by vasopressin.[186] Nevertheless, the isolated state is a nonphysiological situation for epithelial cells, whose polarity is a fundamental property. Furthermore, a localized and small change in intracellular Ca^{2+}, following the hormonal stimulus, cannot be excluded.

Ausiello and Hall[201] have provided evidence that calmodulin may be involved with vasopressin stimulation of adenylate cyclase. Furthermore, trifluoperazine binds to calmodulin and also inhibits vasopressin stimulation of water flow at a post-cAMP step.[193] These and other results[190,202] indicate that calmodulin can interact with the "hormone-effect cascade" at multiple steps. It must then be concluded that we have not yet a clear picture of the role of intracellular Ca^{2+} and calmodulin in water permeability regulation.

Low serosal Na^+,[140-203] serosal acidification,[204,205] and K^+-free serosal medium[203] inhibit the hydrosmotic response to cAMP. All these effects have been interpreted as resulting in an increase in intracellular Ca^{2+} concentration. These mentioned conditions would inhibit the postulated Na^+/Ca^{2+} exchange system[187] that in turn regulates intracellular Ca^{2+} concentration.[206,207] The degree of inhibition of the response to cAMP at a given reduced Na^+ concentration was found to be a function of the serosal Ca^{2+} concentration: as serosal Ca^{2+} was lowered, the degree of inhibition was reduced.[188]

It has been previously mentioned that serosal acidification inhibits the hydroosmotic response to cAMP and its derivatives[204,205] and these effects have been attributed to a change in intracellular Ca^{2+}.[188,208] In a series of papers from this laboratory, it has been demonstrated that the inhibitory effects of medium acidification are due to a decrease in cell pH[131] and can be interpreted as related in part to an effect on adenylcyclase,[12] in part to interference with the mechanism of channel insertion and retrieval,[129] and in part to changes in the permeability state of the putative water channel that could reversibly switch from an open to a closed configuration.[129,130] The effect of medium acidification on the water channels insertion could be mediated by a change in intracellular Ca^{2+}.[130]

It has also been observed that ADH and cAMP increase intracellular pH in frog urinary bladder and that medium alkalinization can induce a reversible increase in water permeability while potentiating the response to ADH.[209,210] Nevertheless, it is unknown if the increase in cellular pH is a necessary step leading from the hormone-receptor interaction to the change in water permeability or a parallel phenomenon not directly linked to the hydroosmotic response.[211]

VI. BIOCHEMICAL CORRELATES

A. Characterization of the Apical Membrane

1. Protein Composition

Very few things are known about the composition in protein and lipid of the apical membrane of the epithelial cells of the amphibian urinary bladder. Nevertheless, some information was brought in, in different ways. By classical histochemistry[54] and cell electrophoresis,[212] it has been shown that the surfaces possess a net negative charge and that the apical surface is composed of a polysaccharide layer. The origin of the glycoproteins of the apical membrane has been described using autoradiography of the incorporation and of the migration of glucosamine, galactose, and fucose.[70]

The identification of specific proteins has been performed. Thus, the protein D,[20-22] which is dephosphorylated by ADH stimulation, has been proposed as an apical protein. Nevertheless, the exact localization of this protein is not known. By the technique of double-isotope labeling, Scott and Slatin[214] have found that ADH increases the number of "apical membrane-like proteins" available for lactoperoxidase iodination. In this case also, the apical origin of these bands is not clearly defined.

By treatment with proteases or detergents, some proteins have also been extracted from the apical surface. Ripoche et al.[215] have described the presence, in a polyacrylamide gel, of a $NPEO_6$ extract, of two bands observed especially after ADH stimulation (50 and 71 kdaltons). Bidet et al.[216] have performed the isolation of a protein of 76 kdaltons, either with proteases (endoglycosidase and endoproteinase V8) or a detergent (octylglucoside). No clear association between this band and the ADH-induced permeability modifications has been advanced.

Evidence for the presence of specific receptors has been brought by Pietras et al.,[217] who described the presence of lectin receptors, especially concanavalin A receptors, in the apical surface. The distribution of these receptors can be altered by the ADH stimulation. Brown and Orci[218] have shown the presence of coated pits in a structure comparable to the bladder epithelium, the collecting duct. A similar observation has not yet been reported in the amphibian urinary bladder.

2. Lipid Composition

A variation in the lipid composition has been suspected to explain the water permeabilization of the apical membrane. In fact, there is no precise information on the lipid composition of the apical membrane of amphibian urinary bladder because the separation of apical and basolateral membranes has not yet been achieved. The lipid composition has been determined only by Parisi et al.[219] from all cell membranes. The special role of the cholesterol has been investigated and its localization detected with filipin: it appears that the cholesterol is essentially found overlying cytoplasmic granules, which are found just beneath the luminal plasma membrane. Filipin-sterol complexes were absent from vasopressin-induced particle aggregates.[107,108]

A change in the apical membrane deformability of kidney papillary-collecting ducts has been observed by Grantham,[220] but an eventual correlation between the variation of permeability and the increased deformability was not clear. A measurement of viscosity with a fluorescent probe has also been performed by Masters et al.[150] A decrease of the microviscosity was observed but it was not possible to explain the increase of water permeability by this effect.

A small effect of vasopressin on lipophilic solutes, increasing their permeability by 30%, has been found by Pietras and Wright[174] and also by Levine and Worthington.[221] The lipid phase appears to participate in the response to vasopressin involving a change in membrane fluidity. Pietras et al.[217] show a change in the distribution of lectin bound to membrane receptors during the hormonal response.

3. Apical Surface Markers

The evolution of our knowledge of the biochemical mechanisms implicated in the ADH response was limited by the absence of a good characterization of the cell-enzyme distribution. To purify an apical membrane fraction, it is indispensable to have a good apical marker. Some information was given about a diphosphatase.[60,222,223] There is, however, no clear specificity of this enzyme for the apical membrane. Alkaline phosphatase, a classical marker enzyme of brush-border preparation, has been presented as an apical marker,[224] but this enzyme is not specifically found in plasma membranes of frog urinary bladder. It is the same thing with a neutral aminopeptidase, whose histochemical localization is very specific for the apical surface in the intestine or in the kidney of frog, but that is irregularly distributed in the frog urinary bladder.[225] Moreover, this enzyme is not found in a membrane preparation from toad urinary bladder. Some other proteins could be used as a marker, such as the protein D or the 76-kdalton protein, but the fact that their functions are unknown makes their utilization difficult.

4. Exogenous Markers

In order to get an exogenous marker, different techniques have been used with partial success. Label fracture[72,226] is a new surface-labeling technique which allows the visualization at high resolution of cytochemical or immunocytochemical markers of cell-surface sites over large cell-surface areas. Briefly, in label fracture, cell surfaces are labeled with different probes (lectins, antibodies) revealed with colloidal gold complexes, and then freeze fractured. After thawing, the specimens are *not* digested and the replicas are extensively washed with distilled water. In these preparations, the distribution of the surface label (the probe-gold complexes) is seen superimposed on the unaltered image of freeze-fractured exoplasmic fracture face (E face).

Strum and Edelman[75] and Rodriguez and Edelman[227] have performed the iodination of the apical plasma membrane with ^{125}I. The fixation of amiloride analogs has been investigated employing, for example, bromo-amiloride or phenamyl-amiloride.[228] All derivatives used did not bind with complete irreversibility.[229] ^{14}C amiloride, ^{3}H benzamyl,[230] and ^{3}HEPA[231] radioactive compounds have been prepared and used to characterize the Na^+ channel. The irreversible fixation by photolabeling has been performed.[232-234] The association of labeled molecules to the photolabeling could be a good tool to identify the apical membrane.

Some other drugs reacting with carboxyl of histidyl groups of the Na^+ channel, like E.D.Q.Q. (*N*-ethyoxycarbonyl-2-ethory-1,2-dihydroquinoline) could also be used.[235] E.D.Q.Q. can activate the carboxyl group of the Na^+ channel and induce the covalent binding of an exogenous nucleophile like glycine methyl ester which can be followed during the membrane purification. A nitroxide derivative of amiloride and a study of amiloride binding by electron paramagnetic resonance spectroscopy has been performed by Costa et al.[236] The spin-labeled amiloride seems a useful probe for the identification of the amiloride binding site, which could be used as a molecular marker for apical membranes.

The subunit alpha of the ATPase or a similar protein (cross reacting with an IgG against the subunit) has been detected in the apical surface of cultured cells. It is a possible apical marker if this protein is also present in the luminal membrane of the epithelial cells.[237]

Another type of attempt that has been applied to culture cells could be used:[238] the incorporation of vesicular stomatitis virus in the epithelial structure. This virus buds exclusively from the free (apical) surface of the cells while some other viruses acquire their envelope only from the basolateral plasma membrane. Thus, because different viruses can select specific plasma membranes, virus-infected epithelia can provide an excellent system to study plasma membrane distribution (the localization of the protein G was revealed with an IgG anti-protein G).

Another method was used by Spiegel et al.[239] working with cultured epithelial cells A6 and MDCK: apical surfaces were exposed to rhodaminyl gangliosides. Gangliosides are incorporated in the membrane and did not move to basolateral plasma membrane.

An elegant, but probably heavy, technique has been developed,[240] oriented to the identification of aggregates. The strategy was to raise antibodies against isolated cells or plasma membranes obtained from hormone-stimulated bladders. Antibodies to proteins not inserted in response to ADH were expected to bind the apical surface in all cases. Thus, the apical surface of toad urinary bladder was used as an adsorbant to extract from antisera the antibodies that bind in the absence of vasopressin. Then, the unadsorbed part of the sera was assayed to detect vasopressin-associated constituents. This technique could be used to follow the apical membrane in a procedure of isolation.

The possibility of preparing monoclonal antibodies as a probe of apical surface was improved by the work of Moberly and Fanestil[241] on amphibian kidney cells in culture. The monoclonal antibodies were produced by in vitro immunization of apical surfaces, allowing access of spleen cells only to this surface, thus limiting determinants. This immunization may provide specific monoclonal antibodies for apical membrane proteins and should yield powerful tools for characterization of epithelial cell membranes.

B. Apical Attack

1. Detergents

Attempts have been made to lightly attack the apical surface in order to destabilize the apical membrane, eventually modifying the hormonal response, or to isolate some components from this membrane to get a useful marker. First, Corriveau et al.[242] described the action of a nonionic surfactant, $NPEO_6$, on the frog urinary bladder. The authors observed a decrease of the cell coat thickness. After a 5-min incubation of the apical surface with $NPEO_6$ (0.1%), there was a drastic reduction of the water flux induced by ADH.

Some intrinsic apical proteins are removed by this detergent. This is suggested by the impossibility of separating detergent and extracted proteins without their precipitation and by the decrease in the intramembranous particle density and alteration of aggregate structure.[128] The insertion of the detergent in the membrane is not excluded.[243] The inhibition of the response to vasopressin is slowly reversible with a time course that parallels the radioactive detergent disappearance from the preparation. Even if some specific proteins have been removed from ADH-stimulated preparations,[215] further experiments will be necessary to associate these events with any one of the ADH-induced permeability modifications.

2. Bifunctional, Cleavable Agents

Some bifunctional reagents, such as dithiobispropionimidate, were used to obtain a reversible immobilization of aggregates.[244,245] Indeed, this agent offers the possibility to undergo an ulterior cleavage. It seems that this treatment is not necessary to stabilize aggregates in the membrane as observed by Hays et al.[245] in a membrane preparation.

3. Chaotropic Agents

Such agents were also used in order to remove some proteins from the apical surface (Svelto et al.[246]). No reproducible or significant difference was observed between extraction from control bladder or from vasopressin-stimulated bladder.

4. Enzymatic Attack

Glucosidase to remove the cell coat and proteases to identify some external fragment of protein was also investigated by several authors. In this way, it was difficult to identify some peptides extracted specifically when the preparation was stimulated by vasopressin.[216] Endoglycosidase H was very effective as determined by the absence of wheat germ agglutinin (WGA) binding sites in the apical surface after this action. Peptides extracted by papain, protease V8, and proteinase K have been analyzed. Some bands in SDS PAGE were particularly important; a 76-kdalton protein was clearly identified.

5. Segregation of Domains by WGA

The role of the abundant, sugar-rich fuzzy coat has not yet been clearly determined and very little is known concerning its origin, turnover, fate, and function. Using label-fracture technique, it has been shown that WGA binding sites are located on the long glycoproteins that cover the apical surface. An extremity of these glycoprotein molecules seems to be anchored in the membrane and forms the large intramembranous particles dispersed on the exoplasmic faces of freeze-fractured luminal membranes of frog bladder epithelial cells.[72] Moreover, whereas label fracture of tissue treated for 15 min with WGA and subsequently labeled with ovomucoid-gold complexes, shows uniform distribution of gold particles along the exoplasmic fracture face, prolonged exposition of urinary bladder to WGA induces drastic ultrastructural modification of the tissue and a marked reorganization of the membrane components in the exoplasmic half of the plasma membrane.[72,247] IMPs become segregated into well-defined domains, leaving intervening areas practically smooth, devoid of particles. Advantage was taken of this phenomenon to make the role of the different membrane components precise. Even after preincubation of frog or toad urinary bladder for 3 hr with WGA, ADH still induces aggregates of particles at the apical pole of granular cells. Although the intensity of the response to ADH is reduced,[248] numerous groove networks are observed in the smooth areas of E face in-between domains of condensed particles.[72] It is therefore possible to distinguish the specific labeling of these ADH-induced structures from the labeling of the cell coat itself when immunocytochemical characterization of ADH-induced aggregates is in progress.

C. Membrane Preparation

Some attempts were made at isolating membrane fractions, especially the apical membrane. First in 1966, Hays and Barland[223] described a preparation of cellular envelope. Scraped cells were homogenized in a low-ionic strength medium, then homogenate was centrifuged to separate the cell envelope from nuclear and other cellular debris. Essentially, apical sheets were observed. This technique has found an interesting development — the possibility of isolating the plasma membrane with aggregates in place, when preparations were first stimulated by cAMP or vasopressin.[245] Rodriguez and Edelman[227] have used a separation on sucrose gradient to obtain different types of membrane. A labeling of the apical proteins by ^{125}I iodination was previously performed in order to serve as a marker of the apical membrane. An endocytosis of ^{125}I and lactoperoxidase was not completely excluded in their conditions, and the formation of a network of fibrous material was constricting.

Chase and Al-Awqati[249] suggested that the fibrous material is composed of cytoskeletal proteins since, when present in high concentration, these proteins formed a gel. Two techniques were proposed to remove them: high-salt buffer suspension or keeping membrane vesicles on ice for 1 hr and low centrifugation. This preparation allowed membrane vesicles to become susceptible to being used in sodium or calcium transport measurements.[250]

Another preparation of microsomes was based on differential centrifugations, the obtained microsomal population representing all plasma membranes.[251,252] This kind of preparation was used[253,254] to study electrolyte permeability. Drastic conditions were imposed on the system in order to increase the studied signal (vesicles were suspended in a large chemical gradient) and to allow the study of a specific ion in heterogeneous populations of membrane vesicles.

Very pure apical plasma membrane vesicle preparation has not yet been routinely prepared and water permeability of heterogeneous vesicles would be not informative. A more specific preparation from toad urinary bladder is the granule purification described by Masur et al.[255] This way could be a good starting point to a better dissection of the different membrane compartments of the epithelial cell of the bladder.

D. Epithelial Cell Culture

A way for biochemical analysis could be the cell culture. Epithelial cells from the toad urinary bladder have been grown in continuous culture. Many of the cells resemble the granular cell types of the urinary bladder. Epithelia formed have transepithelial potential differences of 40 mV and resistances from 5000 to 10,000 ohms/cm^2, but vasopressin has no effect on water and Na^+ movement, while cAMP stimulates short-circuit current. No aggregates were found after cAMP and no aggrephores were described in the apical region. It is necessary to find good culture conditions to keep or to restore the characteristic of the granular cell in terms of water permeability.[256-258] Cell culture techniques have popularized the studies of renal cell function, especially with MDCK, LLC-PK1, and A6 cells.[259] They would be a powerful tool for the study of water permeability when sensibility to vasopressin would be preserved.[256,260]

ACKNOWLEDGMENTS

The authors wish to express thanks to Mrs Michèle Boileau for the preparation of the manuscript.

REFERENCES

1. **Karnieli, E., Zarnowski, J., Hissin, P. J., Simpson, I. A., Salans, L. B., and Cushman, S. W.,** Insulin-stimulated translocation of glucose transport systems in the isolated rat adipose cell, *J. Biol. Chem.*, 256, 4772, 1981.
2. **Forte, J. G., Blackj, A., Forte, T. M., Machen, T. E., and Wolosin, J. M.,** Ultrastructural changes related to functional activity in gastric oxyntic cells, *Am. J. Physiol.*, 241, G349, 1981.
3. **Gluck, S., Cannon, C., and Al-Awqati, Q.,** Exocytosis regulate urinary acidification in turtle bladder by rapid insertion of H^+ pumps into the luminal membrane, *Proc. Natl. Acad. Sci. U.S.A.*, 79, 4327, 1982.
4. **Madsen, K. M. and Tisher, C.,** Cellular response to acute respiratory acidosis in rat medullary collecting duct, *Am. J. Physiol.*, 245, F670, 1983.
5. **Bourguet, J., Chevalier, J., Parisi, M., and Gobin, R.,** Role des agrégats de particules intramembranaires induits par l'hormone antidiurétique, *J. Physiol.*, 77, 629, 1981.
6. **Hays, R. M.,** Alteration of luminal membrane structure by antidiuretic hormone, *Am. J. Physiol.*, 245, C289, 1983.
7. **Wade, J. B.,** Hormonal modulation of epithelium structure, in *Current Topics in Membranes and Transport*, Bronner, F., and Kleinzeller, A., Academic Press, New York, 1980, 123.
8. **Campbell, B. J., Woodward, G., and Borberg, V.,** Calcium-mediated interactions between the antidiuretic hormone and renal plasma membranes, *J. Biol. Chem.*, 247, 6167, 1972.
9. **Hynie, S. and Sharp, G. W. G.,** Adenyl cyclase in the toad bladder, *Biochim. Biophys. Acta*, 230, 40, 1971.
10. **Handler, J. S., Butcher, R. W., Sutherland, E. W., and Orloff, J.,** The effect of vasopressin and of theophylline on the concentration of adenosine 3′,5′-phosphate in the urinary bladder of the toad, *J. Biol. Chem.*, 240, 4524, 1965.

11. **Bar, H. P., Hechter, O., Schwartz, I. L., and Walter, R.,** Neurohypophyseal hormone-sensitive adenyl cyclase of toad urinary bladder, *Proc. Natl. Acad. Sci.*, 67, 7, 1970.
12. **Bockaert, J, Roy, C., and Jard, S.,** Oxytocin-sensitive adenyl cyclase in frog bladder epithelial cells. Role of calcium, nucleotides and other factors in hormonal stimulation, *J. Biol. Chem.*, 247, 7073, 1972.
13. **Roy, C., Bockaert, J., Rajerison, R., and Jard, S.,** Oxytocin receptor in frog bladder epithelial cells. Relationship of ^{3}H-oxytocin binding to adenylate cyclase activation, *Febs Lett.*, 30, 329, 1973.
14. **Stoff, J. S., Handler, J. S., and Orloff, J.,** The effect of aldosterone on the accumulation of adenosine 3′,5′-cyclic monophosphate in toad bladder epithelial cells in response to vasopressin and theophylline, *Proc. Natl. Acad. Sci. U.S.A.*, 69, 805, 1972.
15. **Flores, J., Witkum, P. A., Beckmab, and Sharp, G. W. G.,** Stimulation of osmotic water flow in toad bladder by prostaglandin E1. Evidence for different compartments of cyclic AMP, *J. Clin. Invest.*, 56, 256, 1955.
16. **Goodman, D. B. P., Bloom, F. E., Battenberg, E. R., Rasmussen, H., and Davis, W. L.,** Immunofluorescent localization of cyclic AMP in toad urinary bladder: possible intercellular transfer, *Science*, 188, 1023, 1975.
17. **Dibona, D. R., Civan, M. R., and Leaf, A.,** The cellular specificity of the effect of vasopressin on toad urinary bladder, *J. Membr. Biol.*, 1, 79, 1969.
18. **Handler, J. S. and Orloff, J.,** The mechanism of action of antidiuretic hormone, in *Handbook of Physiology, Renal Physiology,* American Physiological Society, Washington, D.C., 1973.
19. **Favard, P.and Bourguet, J.,** Effects of pentoxyphylline and various methylxanthine derivatives on transepithelial water permeability of frog urinary bladder, *Scand. J. Lab. Invest.*, 41(Suppl. 156), 305, 1981.
20. **De Lorenzo, R. J., Walton, K. G., Curran, P. F., and Greengard, P.,** Regulation of phosphorylation of a specific protein toad bladder membrane by antidiuretic hormone and cyclic AMP and its possible relationship to membrane permeability changes, *Proc. Natl. Acad. Sci. U.S.A.*, 70, 880, 1973.
21. **De Lorenzo, R. J. and Greengard, J. P.,** Activation by adenosine 3′,5′-monophosphate of a membrane-bound phosphoprotein phosphatase from toad bladder, *Proc. Natl. Acad. Sci. U.S.A.*, 70, 1831, 1973.
22. **Ferguson, D. R. and Twite, B. R.,** Effects of vasopressin on toad bladder membrane proteins: relationship to transport of sodium and water, *J. Endocrinol.*, 61, 501, 1974.
23. **Walton, K. G., De Lorenzo, R. J., Curran, P. F., and Greengard, P.,** Regulation of protein phosphorylation and sodium transport in toad bladder, *J. Gen. Physiol.*, 65, 153, 1975.
24. **Schwartz, I. L., Schlatz, L. J., Kinne-Saffran, E., and Kinne, E.,** Target cell polarity and membrane phosphorylation in relation to the mechanism of action of antidiuretic hormone, *Proc. Natl. Sci. U.S.A.*, 71, 2595, 1974.
25. **Schwartz, I. L., Huang, C. J., Reisman, L., Scalettar, E., Wyssbrod, H., Cort, J. H., Roth, L. B., Li, H. C., and Ripoche, P. A.,** Cyclic and mediated dephosphorylation in renal collecting duct epithelial cell plasma membrane: a substrate level phenomenon, *Inserm*, 85, 71, 1979.
26. **Jurgensen, S. R., Chock, P. B., Taalor, S., Vandenheede, J. R., and Merlevede, W.,** Inhibition of the Mg^{2+}-ATP-dependent phosphoprotein phosphatase by the regulatory subunit of cAMP-dependent protein kinase, *Proc. Natl. Acad. Sci. U.S.A.*, 82, 7565, 1985.
27. **Jard, S. and Bastide, F.,** A cyclic AMP-dependent protein-kinase from frog bladder epithelial cells, *Biochim. Biophys. Res. Commun.*, 39, 559, 1970.
28. **Kirchberger, M. A., Schwartz, I. L., and Walter, R.,** Cyclic 3′,5′-AMP-dependent protein kinase activity in toad bladder epithelium, *Proc. Soc. Exp. Biol. Med.*, 140, 657, 1972.
29. **Schlondorff, D. and Franki, N.,** Effect of vasopressin on cyclic AMP-dependent protein kinase in toad urinary bladder, *Biochim. Biophys. Acta,* 628, 1, 1980.
30. **Vallee, R. B., Dibartolomeis, M. J., and Theurkauf, W. E.,** A protein kinase bound to the projection portion of MAP2 (microtubule-associated protein 2), *J. Cell Biol.*, 90, 568, 1981.
31. **Sloboda, R. D., Rudolph, S. A., Rosenbaum, J. L., and Greengard, P.,** Cyclic AMP-dependent endogenous phosphorylation of a microtubule associated protein, *Proc. Natl. Acad. Sci. U.S.A.*, 72, 177, 1975.
32. **Bentley, P. J.,** The effects of neurohypophyseal extracts on water transfer across the water of isolated urinary bladder of the toad *Bufo marinus, J. Endocrinol.*, 17, 201, 1958.
33. **Handler, J. S., Petersen, M., and Orloff, J.,** Effect of metabolic inhibitors on the response of the toad bladder to vasopressin, *Am. J. Physiol.*, 211, 1175, 1966.
34. **Handler, J. S. and Orloff, J.,** Cysteine effect on toad bladder response to vasopressin, cyclic AMP, and theophylline, *Am. J. Physiol.*, 206, 505, 1964.
35. **Handler, J. S., Preston, A. S., and Orloff, J.,** Effect of adrenal steroid hormones on the response of the toad's urinary bladder to vasopressin, *J. Clin. Invest.*, 48, 823, 1969.
36. **Handler, J. S., Preston, A. S., and Orloff, J.,** The effect of aldosterone on glycolysis in the urinary bladder of the toad, *J. Biol. Chem.*, 244, 3194, 1969.

37. **Orloff, J., Handler, J. S., and Bergstrom, S.,** Effect of prostaglandin (PGE1) on the permeability response of toad bladder to vasopressin, theophylline and adenosine 3′,5′-monophosphate, *Nature (London)*, 205, 397, 1965.
38. **Grantham, J. J. and Orloff, J.,** Effect of prostaglandin E1 on the permeability response of the isolated collecting tubule to vasopressin, adenosine 3′,5′ monophosphate and theophylline, *J. Clin. Invest.*, 47, 1154, 1968.
39. **Bentley, P. J.,** Physiological properties of the isolated frog bladder in hyperosmotic solutions, *J. Comp. Biochem. Physiol.*, 12, 233, 1964.
40. **Ripoche, P., Bourguet, J., and Parisi, P.,** The effect of hypertonic media on water permeability of frog urinary bladder, *J. Gen. Physiol.*, 61, 110, 1973.
41. **Orloff, J. and Zusman, R.,** Role of prostaglandin E (PGE) in the modulation of the action of vasopressin on water flow in the urinary bladder of the toad and mammalian kidney, *J. Membr. Biol.*, 297, 304, 1978.
42. **Schlondorff, D. and Satriano, J. A.,** Interactions of vasopressin, cAMP, and prostaglandins in toad urinary bladder, *Am. J. Physiol.*, 248, F454, 1985.
43. **Seyfred, M. A. and Wells, W. W.,** Subcellular site and mechanism of vasopressin-stimulated hydrolysis of phosphoinositides in rat hepatocytes, *J. Biol. Chem.*, 259, 7666, 1984.
44. **Troyer, D. A., Kreisberg, J. I., Schwertz, D. W., and Venkatachalam, M. A.,** Effects of vasopressin on phosphoinositides and prostaglandin production in cultured mesangial cells, *Am. J. Physiol.*, 249, F139, 1985.
45. **Kelleher, D. J., Pessin, J. E., Ruoho, A. E., and Johnson, G. L.,** Phorbol ester induces desensitization of adenylate cyclase and phosphorylation of the beta-adrenergic receptor in turkey erythrocytes, *Proc. Natl. Acad. Sci. U.S.A.*, 81, 4316, 1984.
46. **Yorio, T., Quist, E., and Masaracchia, R. A.,** Calcium/phospholipid-dependent protein kinase and its relationship to antidiuretic hormone in toad urinary bladder epithelium, *Biochem. Biophys. Res. Commun.*, 133, 717, 1985.
47. **Ling, E. and Sapirstein, V.,** Phorbol ester stimulates the phosphorylation of rabbit erythrocyte band, *Biochem. Biophys. Res. Commun.*, 120, 291, 1984.
48. **Masur, S. K., Sapirstein, V., and Rivero, D.,** Phorbol myristate acetate induces endocytosis as well as exocytosis and hydroosmosis in toad urinary bladder, *Biochim. Biophys. Acta*, 821, 286, 1985.
49. **Nishizuka, Y.,** The role of protein kinase C in cell surface signal transduction and tumour promotion, *Nature (London)*, 308, 693, 1984.
50. **Nishizuka, Y.,** Turnover of inositol phospholipids and signal transduction, *Science*, 225, 1365, 1984.
51. **Rodbell, M.,** Programmable messengers: a new theory of hormone action, *Trends in Biochemical Sciences*, 10, 461, 1985.
52. **Hays, R. M. and Leaf, A.,** Studies on the movement of water through the isolated toad bladder and its modification by vasopressin, *J. Gen. Physiol.*, 45, 905, 1962.
53. **Bourguet, J. and Jard, S.,** Un dispositif automatique de mesure et d'enregistrement du flux net d'eau à travers la peau et la vessie des amphibiens, *Biochim. Biophys. Acta*, 88, 442, 1964.
54. **Choi, J. K.,** The fine structure of the urinary bladder of toad, *Bufo marinus*, *J. Cell Biol.*, 16, 53, 1963.
55. **Peachey, L. J. and Rasmussen, H.,** Structure of the toad's urinary bladder in relation to its physiology, *J. Biophys. Biochem. Cytol.*, 10, 529, 1961.
56. **Gfeller, E. and Walser, M.,** Stretch induced changes in geometry and ultrastructure of transporting surfaces of toad bladder, *J. Membr. Biol.*, 4, 16, 1971.
57. **Pak Poy, R. F. K. and Bentley, P. J.,** Fine structure of the epithelial cells of the toad urinary bladder, *Exp. Cell Res.*, 20, 235, 1960.
58. **Carasso, N., Favard, P., and Valérien, J.,** Variations des ultrastructures dans les celluls epitheliales de la vessie de crapaud, après stimulation par l'hormone neurohypophysaire, *J. Microsc.*, 1, 143, 1962.
59. **Jard, S., Bourguet, J., Carasso, N., and Favard, P.,** Actions de divers fixateurs sur la permeabilite et l'ultrastructure de la vessie de grenouille, *J. Microsc.*, 5, 31, 1966.
60. **Keller, A. R.,** A histochemical study of the toad urinary bladder, *Anat. Rec.*, 147, 367, 1963.
61. **Danon, D., Strum, J. M., and Edelman, I. S.,** The membrane surfaces of the toad bladder: scanning and transmission electron microscopy, *J. Membr. Biol.*, 16, 279, 1974.
62. **Wade, J. B.,** Membrane structural specialization of the toad urinary bladder revealed by the freeze-fracture technique. II. The mitochondria rich cell, *J. Membr. Biol.*, 29, 111, 1976.
63. **Wade, J. B., Discala, J. A., and Karnovsky, M. J.,** Membrane structural specialization of the toad urinary bladder revealed by freeze fracture technique. I. The granular cell, *J. Membr. Biol.*, 22, 385, 1975.
64. **Davis, W. L., Goodman, D. B. P., Martin, J. H., Matthews, J. L., and Rasmussen, H.,** Vasopressin induced changes in the toad urinary bladder epithelial surface, *J. Cell Biol.*, 61, 544, 1974.
65. **Ferguson, D. R. and Heap, P. F.,** The morphology of the toad urinary bladder: a stereoscopic and transmission electron microscopical study, *Z. Zellforsch.*, 109, 297, 1970.
66. **Mia, A. J., Tarapoom, N., Carnes, J., and Yorio, T.,** Alteration in surface substructure of frog urinary bladder by calcium ionophore, verapamil and antidiuretic hormone, *Tissue Cell*, 15, 737, 1983.

67. **Spinelli, F., Grosso, A., and De Sousa, R. C.,** The hydrosmotic effect of vasopressin: a scanning electron microscope study, *J. Membr. Biol.*, 23, 139, 1975.
68. **Jard, S., Bourguet, J., Favard, P., and Carasso, N.,** The role of intercellular channels in the transepithelial transfer of water and sodium in the frog urinary bladder, *J. Membr. Biol.*, 4, 124, 1971.
69. **Pisam, M., Ripoche, P., and Rambourg, A.,** Etude ultrastructurale des charges négatives de la surface apicals de l'epithelium vesical de la grenouille: leurs modifications sous l'action de l'ocytocine, *C.R. Acad. Sci. Paris,* 271, 105, 1970.
70. **Pisam, M. and Ripoche, P.,** Renouvellement des glycoprotéines de surface: étude radioautographique de l'épithélium de la vessie de grenouille, *J. Microsc.*, 17, 261, 1973.
71. **Pisam, M. and Ripoche, P.,** Redistribution of surface macromolecules in dissociated epithelial cells, *J. Cell Biol.*, 71, 907, 1976.
72. **Chevalier, J., Pinto Da Silva, P., Ripoche, P., Gobin, R., Wang, X. Y., Grossetete, J., and Bourguet, J.,** Structural and cytochemical differentiation of membrane elements of the apical membrane of amphibian urinary bladder epithelial cells, *Biol. Cell,* 55, 181, 1985.
73. **Pietras, R. J.,** Vasopressin-induced redistribution of binding sites for concanavalin A at the surface of epithelial cells from urinary bladder, *Nature (London),* 264, 774, 1976.
74. **Rodriguez, H. J. and Edelman, I. S.,** Isolation of radio-iodinated apical and basal-lateral plasma membranes of toad bladder epithelium, *J. Membr. Biol.*, 45, 215, 1979.
75. **Strum, J. M. and Edelman, I. S.,** Iodination (125I) of the apical plasma membrane of toad bladder epithelium: electron-microscopic autoradiography and physiological effects, *J. Membr. Biol.*, 14, 17, 1973.
76. **Fujimoto, T. and Ogawa, K.,** Cell membrane polarity in dissociated frog urinary bladder epithelial cells, *J. Histochem. Cytochem.*, 31, 131, 1983.
77. **Carasso, N., Favard, P., Bourguet, J., and Jard, S.,** Role du flux net d'eau dans les modifications ultrastructurales de la vessie de grenouille stimulée par l'ocytocine, *J. Microsc.*, 5, 519, 1966.
78. **Wade, J. B.,** Membrane structural specialization of the toad urinary bladder revealed by the freeze-fracture technique. III. Location, structure and vasopressin dependence of intramembranous particle arrays, *J. Membr. Biol.*, Special issue, 281, 1978.
79. **Mills, J. W. and Ernst, S. A.,** Localization of sodium pump sites in frog urinary bladder, *Biochim. Biophys. Acta,* 375, 268, 1975.
80. **Masur, S. K., Holtzman, E., and Walter, R.,** Hormone-stimulated exocytosis in toad urinary bladder. Some possible implications for turnover of surface membranes, *J. Cell Biol.*, 52, 211, 1972.
81. **Taylor, A., Mamelak, M., Reaven, E., and Maffly, R.,** Vasopressin: possible role of microtubules and microfilaments in its action, *Science,* 181, 347, 1973.
82. **Carasso, N., Favard, P., and Bourguet, J.,** Action de la cytochalasine B sur la reponse hydrosmotique et l'ultrastructure de la vessie urinaire de la grenouille, *J. Microsc. Paris,* 18, 383, 1973.
83. **Kraehenbuhl, J. P., Pfeiffer, J., Rossier, M., and Rossier, B. C.,** Microfilament rich cells in toad bladder epithelium, *J. Membr. Biol.*, 48, 167, 1979.
84. **Pearl, M. L. and Taylor, A.,** Actin filaments and vasopressin-stimulated water flow in toad urinary bladder, *Am. J. Physiol.*, 245, C28, 1983.
85. **Pearl, M. and Taylor, A.,** Role of the cytoskeleton in the control of transcellular water flow by vasopressin in amphibian urinary bladder, *Biol. Cell,* 55, 163, 1985.
86. **Parisi, M., Ripoche, P., Bourguet, J., Carasso, N., and Favard, P.,** The isolated epithelium of frog urinary bladder. Ultrastructural modifications under the action of oxytocin, theophylline and cyclic AMP, *J. Microsc.*, 8, 1031, 1969.
87. **Bourguet, J., Lemonnier, R., Carasso, N., and Favard, P.,** Ultrastructure et perméabilité à l'eau de l'épithélium isolé de la vessie de grenouille, *J. Microsc. Biol. Cell,* 23, 139, 1975.
88. **Grantham, J. J., Cuppage, F. E., and Fanestil, D.,** Direct observation of toad bladder response to vasopressin, *J. Cell Biol.*, 48, 695, 1971.
89. **Dibona, D. R. and Civan, M. M.,** Pathways for movement of ions and water across toad urinary bladder. I. Anatomic site of transepithelial shunt pathways, *J. Membr. Biol.*, 12, 101, 1973.
90. **Parisi, M., Merot, J., and Bourguet, J.,** Glutaraldehyde fixation preserves the permeability properties of the ADH-induced water channels, *J. Membr. Biol.*, 86, 239, 1985.
91. **Lefurgey, A. and Tisher, C. C.,** Time course of vasopressin induced formation of microvilli in granular cells of toad urinary bladder, *J. Membr. Biol.*, 61, 13, 1981.
92. **Mills, J. W. and Malick, L. E.,** Mucosal surface morphology of the toad urinary bladder. Scanning electron microscope study of the natriferic and hydro-osmotic response to vasopressin, *J. Cell Biol.*, 77, 598, 1978.
93. **Dratwa, M., Lefurgey, A., and Tisher, C. C.,** Effect of vasopressin and serosal hypertonicity on toad urinary bladder, *Kidney Int.*, 16, 695, 1979.
94. **Gronowicz, G., Masur, S. K., and Holtzman, E.,** Quantitative analysis of exocytosis and endocytosis in the hydroosmotic response of toad bladder, *J. Membr. Biol.*, 52, 221, 1980.

95. **Masur, S. K., Cooper, S., and Rubin, M. S.,** Effect of an osmotic gradient on antidiuretic hormone-induced endocytosis and hydroosmosis in the toad urinary bladder, *Am. J. Physiol.*, 247, F370, 1984.
96. **Masur, S. K., Holtzman, E., Schwartz, I. L., and Walter, R.,** Correlation between pinocytosis and hydroosmosis induced by neurohypophyseal hormones and mediated by adenosine 3′,5′-cyclic monophosphate, *J. Cell Biol.*, 49, 582, 1971.
97. **Pietras, R. J., Seeler, B. J., and Szego, C. M.,** Influence of antidiuretic hormone on release of lysosomal hydrolase at mucosal surface of eptihelial cells from urinary bladder, *Nature (London)*, 257, 493, 1975.
98. **Satir, P. and Satir, B.,** Design and function of site-specific particle arrays in the cell membrane, in *Control of Cell Purification*, Cold Spring Harbor Symp., Cold Spring Harbor, New York, 1974, 233.
99. **Chevalier, J., Bourguet, J., and Hugon, J. S.,** Membrane associated particles: distribution in frog urinary bladder epithelium at rest and after oxytocin treatment, *Cell Tissue Res.*, 152, 129, 1974.
100. **Orci, L., Humbert, F., Ammerdt, M., Grosso, A., De Sousa, R. C., and Perreler, A.,** Patterns of membrane organization in toad bladder epithelium: a freeze fracture study, *Experientia*, 31, 1335, 1975.
101. **Chevalier, J., Parisi, M., and Bourguet, J.,** Particle aggregates during antidiuretic action. Some comments on their formation, *Biol. Cell*, 35, 207, 1979.
102. **Bourguet, J., Chevalier, J., and Hugon, J. S.,** Alterations in membrane associated particle distribution during antidiuretic challenge in frog urinary bladder epithelium, *Biophys. J.*, 16, 627, 1976.
103. **Brown, D., Ilic, V., and Orci, L.,** Rod shaped particles in the plasma membrane of the mitochondria-rich cell of amphibian epidermis, *Anat. Rec.*, 192, 269, 1978.
104. **Brown, D. and Montesano, R.,** Membrane specialization in the rat epididymis. 2. Rod-shaped intramembrane particles in the apical (mitochondria-rich) cell, *J. Cell Sci.*, 45, 187, 1980.
105. **Humbert, F., Pricam, C., Perrelet, A., and Orci, L.,** Specific plasma membrane differentiations in the cells of the kidney collecting tubules, *J. Ultrastruct. Res.*, 52, 13, 1975.
106. **Hays, R. M., Chevalier, J., Gobin, R., and Bourguet, J.,** Fusion images and intramembrane particle aggregates during the action of antidiuretic hormone: a rapid-freeze study, *Cell Tissue Res.*, 240, 433, 1985.
107. **Orci, L., Montesano, R., and Brown, D.,** Heterogeneity of toad bladder granular cell luminal membranes. Distribution of filipin-sterol complexes in freeze-fracture, *Biochim. Biophys. Acta*, 601, 443, 1980.
108. **Stetson, D. L. and Wade, J. B.,** Ultrastructural characterization of cholesterol distribution in toad bladder using filipin, *J. Membr. Biol.*, 74, 131, 1983.
109. **Chevalier, J. and Bourguet, J.,** Jonctions serrées et antidiurèse: étude par cryofracture, in *Epithelial Transport in the Lower Vertebrates*, Lahlou, B., Ed., Cambridge University Press, London, 1980, 91.
110. **Kachadorian, W. A., Wade, J. B., and Di Scala, V. A.,** Vasopressin induced structural change in toad bladder luminal membrane, *Science*, 190, 67, 1975.
111. **Chevalier, J., Adragna, N., Bourguet, J., and Gobin, R.,** Fine structure of intramembranous particle aggregates in ADH-treated frog urinary bladder and skin: influence of glutaraldehyde and N-ethyl maleimide, *Cell Tissue Res.*, 218, 595, 1981.
112. **Wade, J. B., Kachadorian, W. A., and Discala, V. A.,** Freeze-fracture electron microscopy: relationship of membrane structural features to transport physiology, *Am. J. Physiol.*, 232, F77, 1977.
113. **Mc Intyre, J. A., Karnovsky, M. J., and Gilula, N. B.,** Intramembranous particle aggregation in lymphoid cells, *Nature (London) New Biol.*, 245, 147, 1973.
114. **Mc Intyre, J. A., Gilula, N. B., and Karnovsky, M. D.,** Cryoprotectant induced redistribution of intramembranous particle aggregation in lymphoid cells, *J. Cell Biol.*, 60, 192, 1974.
115. **Pinto Da Silva, P.,** Translational mobility of the membrane intercalated particles of human erythrocyte ghosts. pH dependent, reversible aggregation, *J. Cell Biol.*, 53, 777, 1972.
116. **Speth, V. and Wunderlich, F.,** Membranes of tetrahymena. II. Direct visualization of reversible transitions in biomembrane structure induced by temperature, *Biochim. Biophys. Acta*, 291, 621, 1973.
117. **Harmanci, M. C., Kachadorian, W. A., Valtin, H., and Discala, V. A.,** Antidiuretic hormone-induced intramembranous alterations in mammalian collecting ducts, *Am. J. Physiol.*, 235, F440, 1978.
118. **Brown, D., Grosso, A., and De Sousa, R. C.,** Isoproterenol-induced intramembrane particle aggregation and water flux in toad epidermis, *Biochim. Biophys. Acta*, 596, 158, 1980.
119. **Kachadorian, W. A., Levine, S. D., Wade, J. B., Discala, V. A., and Hays, R. M.,** Relationship of aggregated intramembranous particles to water permeability in vasopressin treated toad urinary bladder, *J. Clin. Invest.*, 59, 576, 1977.
120. **Harmanci, M. C., Stern, P., Kachadorian, W. A., Valtin, H., and Discala, V. A.,** Vasopressin and collecting duct intramembranous particle clusters: a dose-response relationship, *Am. J. Physiol.*, 239, F560, 1980.
121. **Kachadorian, W. A., Casey, C., and Discala, V. A.,** Time course of ADH-induced intramembranous particle aggregation in toad urinary bladder, *Am. J. Physiol.*, 234, F461, 1978.
122. **Parisi, M., Merot, J., Ripoche, P., Chevalier, J., and Bourguet, J.,** Biophysical characterization of the ADH-induced water channel, in *Water and Ions in Biological Systems*, Vasiliescu, V., Pullman, A., and Packer, L., Eds., Plenum Press, New York, 1986.

123. **Bourguet, J., Chevalier, J., and Parisi, M.,** On the role of intramembranous particle aggregates in the hydroosmotic action of antidiuretic hormone, in *Water Transport Across Epithelia,* Ussing, H. H., Bindslev, N., Lassen, N. A., and Sten-Knudsen, O., Eds., Alfred Benzon Symp. 15, Munksgaard, Copenhagen, 1981, 404.
124. **Chevalier, J., Parisi, M., and Bourguet, J.,** The rate-limiting step in hydrosmotic response of frog urinary bladder, a freeze-fracture study at different temperatures and medium pH, *Cell Tissue Res.,* 228, 345, 1983.
125. **Kachadorian, W. A., Ellis, S. J., and Muller, J.,** Possible role for microtubules and microfilaments in ADH action on toad urinary bladder, *Am. J. Physiol.,* 236, F14, 1979.
126. **Chevalier, J., Bourguet, J., and Hugon, J. S.,** Actions combinées de la colchicine et de la cytochalasine B sur la perméabilité à l'eau et la distribution des particules intramembranaires de la vessie de grenouille, *Proc. Int. Union Physiol. Sci.,* 13, 135, 1977.
127. **Parisi, M., Pisam, M., Merot, J., Chevalier, J., and Bourguet, J.,** The role of microtubules and microfilaments in the hydrosmotic response to antidiuretic hormone, *Biochim. Biophys. Acta,* 817, 333, 1985.
128. **Chevalier, J., Bourguet, J., and Parisi, M.,** New evidence on the role of intramembranous particle aggregates as the ADH induced water pathways: the effect of a low HLB surfactant, cemulsol 6 NP-E06, in *Hormonal Control of Epithelial Transport,* Inserm Symp. Ser., Paris, 85, 147, No. 129, 1979.
129. **Parisi, M. and Bourguet, J.,** Effects of cellular acidification on the ADH-induced intramembranous particle aggregates, *Am. J. Physiol.,* 246, C157, 1984.
130. **Parisi, M. and Bourguet, J.,** Different steps in the regulation by intracellular (Ca^{++}) and pH of the ADH induced hydroosmotic response (résumé), *Biophys. J.,* 41(2), 161a, 1983.
131. **Parisi, M., Montoreano, R., Chevalier, J., and Bourguet, J.,** Cellular pH and water permeability control in frog urinary bladder: a possible action on the water pathway, *Biochim. Biophys. Acta,* 648, 267, 1981.
132. **Elgsaeter, A., Shotton, D. M., and Branton, D.,** Intramembrane particle aggregation in erythrocyte ghosts. II. The influence of spectrin aggregation, *Biochim. Biophys. Acta,* 426, 101, 1972.
133. **Chevalier, J., Parisi, M., and Bourguet, J.,** Particle aggregates during antidiuretic action. Some comments on their formation, *Biol. Cell,* 35, 207, 1979.
134. **Humbert, F., Montesano, R., Grosso, A., De Sousa, R. C., and Orci, L.,** Particle aggregates in plasma and intracellular membranes of toad bladder granular cells, *Experientia,* 33, 1364, 1977.
135. **Muller, J., Kachadorian, W. A., and Discala, V. A.,** Evidence that ADH-stimulated intramembrane particle aggregates are transferred from cytoplasmic to luminal membranes in toad bladder epithelial cells, *J. Cell Biol.,* 85, 83, 1980.
136. **Sasaki, J., Tilles, S., Condeelis, J., Carboni, J., Meiteles, L., Franki, N., Bolon, R., Robertson, C., and Hays, R. M.,** Electron-microscopic study of the apical region of the toad bladder epithelial cell, *Am. J. Physiol.,* 247, C268, 1984.
137. **Beauwens, R., Rentmeesters, M., Bergmann, P., and Dratwa, M.,** Inhibition of vasopressin-induced water flow across toad bladder wall by cationized albumin, *Arch. Int. Physiol. Biochim.,* 90, 41, 1982.
138. **Ding, G., Franki, N., and Hays, R. M.,** Evidence for cycling of aggregate containing tubules in toad urinary bladder, *Biol. Cell,* 55, 213, 1985.
139. **Beauwens, R., Te Kronnie, G., Snauwaert, J., and In'T Veld, P. A.,** Polycations reduce vasopressin-induced water flow by endocytic removal of water channels, *Am. J. Physiol.,* 250, C729, 1986.
140. **Bentley, P. J.,** The effects of ionic changes on water transfer across the isolated urinary bladder of the toad, *Bufo marinus, J. Endocrinol.,* 18, 327, 1959.
141. **Fischbarg, J., Lim, J. J., and Bourguet, J.,** Adenosine stimulation of fluid transport across rabbit corneal endothelium, *J. Membr. Biol.,* 35, 95, 1977.
142. **Parisi, M., Bourguet, J., Ripoche, P., and Chevalier, J.,** Simultaneous minute by minute determination of unidirectional and net water fluxes in frog urinary bladder, *Biochim. Biophys. Acta,* 556, 509, 1979.
143. **Hebert, S. C. and Andreoli, T. E.,** Water permeability of biological membranes. Lessons from antidiuretic hormone-responsive epithelia, *Biochim. Biophys. Acta,* 650, 267, 1982.
144. **Hanai, T. and Haydon, D. A.,** The permeability to water of bimolecular lipid membranes, *J. Theor. Biol.,* 11, 270, 1966.
145. **Cass, A. and Finkelstein, A.,** Water permeability of thin lipid membranes, *J. Gen. Physiol.,* 50, 1765, 1967.
146. **Finkelstein, A.,** Water and non-electrolyte permeability of lipid bilayer membranes, *J. Gen. Physiol.,* 68, 127, 1976.
147. **Hays, R. M. and Franki, N.,** The role of water diffusion in the action of vasopressin, *J. Membr. Biol.,* 2, 263, 1970.
148. **Parisi, M. and Piccinni, Z.,** The penetration of water into the epithelium of toad urinary bladder and its modification by oxytocin, *J. Membr. Biol.,* 12, 227, 1973.
149. **Finkelstein, A.,** Nature of the permeability increase induced by the antidiuretic hormone (ADH) in toad urinary bladder and related tissues, *J. Gen. Physiol.,* 68, 137, 1976.

150. **Masters, B. R., Yguerabide, J., and Fanestil, D. D.,** Microviscosity of mucosal cellular membranes in toad urinary bladder: relation to antidiuretic hormone action on water permeability, *J. Membr. Biol.,* 40, 179, 1978.
151. **Parisi, M. and Bourguet, J.,** The single file hypothesis and the water channels induced by antidiuretic hormone, *J. Membr. Biol.,* 71, 189, 1983.
152. **Levine, S. D., Jacoby, M., and Finkelstein, A.,** The water permeability of toad urinary bladder. I. Permeability of barriers in series with the luminal membranes, *J. Gen. Physiol.,* 83, 529, 1984.
153. **Levine, S. D., Jacoby, M., and Finkelstein, A.,** The water permeability of toad urinary bladder. II. The value of Pf/Pd (w) for the ADH-induced water permeation pathway, *J. Gen. Physiol.,* 83, 543, 1984.
154. **Holtz, R. and Finkelstein, A.,** The water and non-electrolyte permeability induced in the thin lipid membranes by the polyene antibiotics nystatin and amphotericin B, *J. Gen. Physiol.,* 56, 125, 1970.
155. **Rosenberg, P. A. and Finkelstein, A.,** Water permeability of gramicidin A-treated lipid bilayer membranes, *J. Gen. Physiol.,* 72, 341, 1978.
156. **Parisi, M., Ripoche, P., Chevalier, J., and Bourguet, J.,** A low HLB surfactant (NP-E06) differently modifies water, sodium, urea and nicotinamide permeation in frog urinary bladder, in *Hormonal Control of Epithelial Transport,* Inserm Symp. Series, Paris, 85, 289, 1979.
157. **Hays, R. M., Carvounis, C. P., Franki, N., and Levine, S. D.,** Water permeation in epithelial tissues: current concepts, *Inserm,* 85, 281, 1979.
158. **Eggena, P.,** Glutaraldehyde-fixation method for determining the permeability to water of the toad urinary bladder, *Endocrinology,* 91, 240, 1972.
159. **Eggena, P.,** Effect of glutaraldehyde on hydroosmotic response of toad bladder to vasopressin, *Am. J. Physiol.,* 244, C37, 1983.
160. **Parisi, M., Ripoche, P., Prevost, G., and Bourguet, J.,** Regulation by ADH and cellular osmolarity of water permeability in frog urinary bladder: a time course study, *Proc. N.Y. Acad. Sci.,* 372, 144, 1981.
161. **Dibona, D. R.,** Cytoplasmic involvement in ADH-mediated osmosis across toad urinary bladder, *Am. J. Physiol.,* 245, C297, 1983.
162. **Grosso, A., Spinelli, F., and De Sousa, R. C.,** Cytochalasin B and water transport. A scanning electron microscope study of the toad urinary bladder, *Cell Tissue Res.,* 188, 375, 1978.
163. **Hardy, M. A. and Dibona, D. R.,** Microfilaments and the hydroosmotic action of vasopressin in toad urinary bladder, *Am. J. Physiol.,* 243, C200, 1982.
164. **Kachadorian, W. A.,** Regulation of ADH-stimulated water flow at a post-luminal barrier in toad bladder, *Biol. Cell,* 55, 225, 1985.
165. **Wade, J. B., Stetson, D. L., and Lewis, J. A.,** ADH action: evidence for a membrane shuttle mechanism, *Ann. N.Y. Acad. Sci.,* 372, 106, 1981.
166. **Wilson, L. and Taylor, A.,** Evidence for involvement of microtubules in the action of vasopressin in toad urinary bladder. II. Colchicine binding properties of toad bladder epithelial cell tubulin, *J. Membr. Biol.,* 40, 237, 1978.
167. **Reaven, E., Maffly, R., and Taylor, A.,** Evidence for involvement of microtubules in the action of vasopressin in toad urinary bladder. III. Morphological studies on the content and distribution of microtubules in bladder epithelial cells, *J. Membr. Biol.,* 40, 251, 1978.
168. **Ausiello, D. A., Corwin, H. L., and Hartwig, J. H.,** Identification of actin-binding protein and villin in toad bladder epithelia, *Am. J. Physiol.,* 246, F101, 1984.
169. **Yuasa, S., Urakabe, S., Kimura, G., Shirai, D., Takamitsu, Y., Orita, Y., and Abe, H.,** Effect of colchicine on the osmotic water flow across the toad urinary bladder, *Biochim. Biophys. Acta,* 413, 277, 1975.
170. **Hardy, M. A., Jr., Montoreano, R., and Parisi, M.,** Colchicine dissociates the toad *(Bufo arenarum)* urinary bladder responses to antidiuretic hormone and to serosal hypertonicity, *Experientia,* 31, 803, 1975.
171. **Brady, R. J., Parsons, R. H., and Coluccio, L. M.,** Nocodazole inhibition of the vasopressin-induced water permeability increase in toad urinary bladder, *Biochim. Biophys. Acta,* 646, 399, 1981.
172. **Burch, R. M. and Halushka, P. V.,** Inhibition of prostaglandin synthesis antagonizes the colchicine induced reduction of vasopressin-stimulated water flow in the toad urinary bladder, *Mol. Pharmacol.,* 21, 142, 1982.
173. **Palmer, L. and Lorenzen, M.,** Antidiuretic hormone-dependent membrane capacitance and water permeability in the toad urinary bladder, *Am. J. Physiol.,* 244, F195, 1983.
174. **Pietras, R. J. and Wright, E. M.,** Non-electrolyte probes of membrane structure in ADH-treated toad urinary bladder, *Nature (London),* 247, 222, 1974.
175. **Leeds, V. and Taylor, A.,** Is microtubule treadmilling involved in the water permeability response to vasopressin in toad urinary bladder?, *J. Physiol. (London),* 332, 114, 1982.
176. **Hays, R. M., Franki, N., and Ross, L. S.,** Effect of metabolic inhibitors on vasopressin-stimulated transport systems in the toad bladder, *J. Supramol. Struct.,* 10, 175, 1979.
177. **Coluccio, L. M., Brady, R. J., and Parsons, R. H.,** Pressure effects on the ADH-induced initiation of water flow in toad bladder, *Am. J. Physiol.,* 244, F547, 1983.

178. **Taylor, A.,** Role of microtubules and microfilaments in the action of vasopressin, in *Disturbances in Body Fluid Osmolality,* Andreoli, T. E., Grantham, J. J., and Rector, F. C., Jr., Eds., American Physiology Society, Bethesda, Md., 1977, 97.
179. **Valenti, G., Chevalier, J., Hugon, J. S., and Bourguet, J.,** Membrane structure and water permeability in renal like epithelia, in *Endocrine Regulation in Salt and Water Transport,* Kirsch and Lahlou, Eds., Vol 1, 8th ESCP Conf., S. Karger, Basel, 1987, 67.
180. **Baron, D. A., Burch, R. M., Halushka, P. V., and Spicer, S. S.,** Blockade of colchicine-induced inhibition of vasopressin-stimulated osmotic water flow: failure to influence microtubule formation, *Am. J. Physiol.,* 249, F464, 1985.
181. **Davis, W. L., Goodman, D. B. P., Jones, R. G., and Rasmussen, H.,** The effects of cytochalasin B on the surface morphology of the toad urinary bladder epithelium: a scanning electron microscope study, *Tissue Cell,* 10, 451, 1978.
182. **De Sousa, R. C., Grosso, A., and Rufener, C.,** Blockade of the hydrosmotic effect of vasopressin by cytochalasin B, *Experientia,* 30, 175, 1974.
183. **Petersen, M. J. and Edelman, I. S.,** Calcium inhibition of the action of vasopressin on the urinary bladder of the toad, *J. Clin. Invest.,* 43, 583, 1964.
184. **Bockaert, J., Roy, C., and Jard, S.,** Oxytocin-sensitive adenyl cyclase in frog urinary bladder epithelial cells. Role of calcium, nucleotides and other factors in hormonal stimulation, *J. Biol. Chem.,* 247, 7073, 1972.
185. **Burch, R. M. and Halushka, P. V.,** ADH or theophylline-induced changes in intracellular free and membrane-bound calcium, *Am. J. Physiol.,* 247, F939, 1984.
186. **Taylor, A., Pearl, M., and Crutch, B.,** The role of cytosolic free calcium and its measurement in vasopressin-sensitive epithelium, *J. Mol. Physiol.,* 8, 43, 1985.
187. **Taylor, A. and Windhager, E. E.,** Possible role of cytosolic Ca^{++} and Na^{+}-Ca^{++} exchange in regulation of transepithelial Na^{+} transport, *Am. J. Physiol.,* 236, F507, 1979.
188. **Taylor, A., Eich, E., Pearl, M., and Brem, A.,** Role of cytosolic Ca^{++} and Na-Ca exchange in the action of vasopressin, *Inserm Symp. Ser.,* 85, 167, 1979.
189. **Hardy, M. A.,** Intracellular calcium as a modulator of transepithelial permeability to water in frog urinary bladder, *J. Cell Biol.,* 76, 787, 1978.
190. **Svelto, M. and Casavola, V.,** Evidence for the role of calcium in the hydroosmotic response to antidiuretic hormone in frog skin, *Pfluegers Arch.,* 402, 166, 1984.
191. **Arruda, J. A. L. and Sabatini, S.,** Effect of quinidine on Na^{+}, H^{+} and water transport by the turtle and toad bladders, *J. Membr. Biol.,* 55, 141, 1980.
192. **Humes, D. and Weinberg, T. M.,** Ionophore A 23187 induced reduction in toad urinary bladders epithelial cells oxidative phosphorylation and viability, *Pfluegers Arch.,* 388, 217, 1980.
193. **Humes, H. D., Simmons, C. F., and Brenner, B. M.,** Effects of verapamil on the hydroosmotic response to antidiuretic hormone in toad urinary bladder, *Am. J. Physiol.,* 239, F250, 1980.
194. **Levine, S. D., Levin, D. N., and Schlondorff, D.,** Calcium flow-independent actions of calcium channel blockers in toad urinary bladder, *Am. J. Physiol.,* 244, C243, 1982.
195. **Forrest, J. P., Schneider, C. J., and Goodman, D. B. P.,** Role of prostaglandin E2 in mediating the effects of pH on the hydroosmotic response to vasopressin in toad urinary bladder, *J. Clin. Invest.,* 69, 499, 1982.
196. **Erlij, D., Gerstein, L., Sherba, G., and Schoen, H. F.,** Role of prostaglandin release in the response of tight epithelia to Ca^{2+} ionophores, *Am. J. Physiol.,* 250, C629, 1986.
197. **Epstein, P. M., Fiss, K., Hachiso, R., and Andrenyak, D. M.,** Interaction of calcium antagonist with cyclic AMP phosphodiesterases and calmodulin, *Biochem. Biophys. Res. Commun.,* 105, 1142, 1982.
198. **Bourguet, J.,** Influence de la température sur la cinétique de l'augmentation de la perméabilité à l'eau de la vessie de grenouille sous l'action de l'ocytocine, *J. Physiol. (Paris),* 58, 476, 1966.
199. **Bourguet, J.,** Cinétique de la perméabilisation de la vessie de grenouille par l'ocytocine. Role du 3'-5'-adenosine monophosphate cyclique, *Biochim. Biophys. Acta,* 150, 104, 1968.
200. **Cuthbert, A. W. and Wong, P. Y. D.,** Calcium release in relation to permeability changes in toad bladder epithelium following antidiuretic hormone, *J. Physiol. (London),* 241, 407, 1974.
201. **Ausiello, D. A. and Hall, D.,** Regulation of vasopressin-sensitive adenylate cyclase by calmodulin, *J. Biol. Chem.,* 256, 9796, 1981.
202. **Grosso, A., Cox, J. A., Malnde, A., and De Sousa, R. C.,** Evidence for a role of calmodulin in the hydroosmotic action of vasopressin in toad bladder, *J. Physiol. (Paris),* 78, 270, 1982.
203. **Davis, W. L., Goodman, D. B. P., and Rasmussen, H.,** Ionic and metabolic requirements for the hydroosmotic reponse to antidiuretic hormone in toad urinary bladder, *J. Membr. Biol.,* 41, 225, 1978.
204. **Gulyassy, P. F. and Edelman, I. S.,** Hydrogen-ion dependence of antidiuretic action of vasopressin, oxytocin and deaminooxytocin, *Biochim. Biophys. Acta,* 102, 185, 1965.
205. **Gulyassy, P. F. and Edelman, I. S.,** Effect of pH and theophylline on uptake, elution and antidiuretic action of cyclic AMP, *Am. J. Physiol.,* 212, 740, 1967.

206. **Frindt, G., Windhager, E., and Taylor, A.,** Hydroosmotic response of collecting tubules to ADH or cAMP at reduced peritubular sodium, *Am. J. Physiol.*, 243, F503, 1979.
207. **Lee, C. D., Taylor, A., and Windhager, E. E.,** Cytosolic calcium ion activity in epithelial cells of Necturus kidney, *Nature (London)*, 287, 859, 1980.
208. **Hardy, M. A.,** Ionic requirements of the transepithelial water flow induced by hypertonicity, *Inserm*, 85, 311, 1979.
209. **Parisi, M., Wietzerbin, J., and Bourguet, J.,** Intracellular pH, transepithelial pH gradients and the ADH-induced water channels, *Am. J. Physiol.*, 244, F712, 1983.
210. **Parisi, M., Chevalier, J., and Bourguet, J.,** Influence of mucosal and serosal pH on antidiuretic action in frog urinary bladder, *Am. J. Physiol.*, 237, F483, 1979.
211. **Parisi, M. and Wietzerbin, J.,** Cellular pH and the ADH-induced hydrosmotic response in different ADH target epithelia, *Pfluegers Arch.*, 402, 211, 1984.
212. **Lipman, K. M., Dodelson, R., and Hays, R. M.,** The surface charge of isolated toad bladder epithelial cells, *J. Gen. Physiol.*, 49, 501, 1966.
213. **Liu, A. Y. C. and Greengard, P.,** Aldosterone-induced increase in protein phosphatase activity of toad bladder, *Proc. Natl. Acad. Sci. U.S.A.*, 71, 3869, 1974.
214. **Scott, W. N. and Slatin, S. L.,** Alterations in lactoperoxidase catalyzed radio-iodination of membrane proteins associated with vasopressin-induced changes in tissue permeability to water, *Biochem. Biophys. Res. Commun.*, 91, 1038, 1979.
215. **Ripoche, P., Parisi, M., Chevalier, J., Rivas, E., and Bourguet, J.,** Detergent extraction of membrane proteins related to the action of antidiuretic hormone, *Biochim. Biophys. Acta*, 693, 497, 1982.
216. **Bidet, S., Berthonaud, V., Gobin, R., Chevalier, J., Bourguet, J., and Ripoche, P.,** Apical material extracted from amphibian urinary bladder by enzymes and detergent treatment, *Biol. Cell*, 55, 191, 1985.
217. **Pietras, R. J., Naujokaitis, P. J., and Szego, C. M.,** Surface modifications evoked by antidiuretic hormone in isolated epithelial cells: evidence from lectin probes, *J. Supramol. Struct.*, 3, 391, 1975.
218. **Brown, D. and Orci, L.,** Vasopressin stimulates formation of coated pits in rat kidney collecting ducts, *Nature (London)*, 302, 253, 1983.
219. **Parisi, M., Gauna, A., and Rivas, E.,** Water permeability and lipid composition of toad urinary bladder: the influence of temperature, *J. Membr. Biol.*, 26, 335, 1976.
220. **Grantham, J. J.,** Vasopressin: effect on deformability of urinary surface of collecting duct cells, *Science*, 168, 1093, 1970.
221. **Levine, S. D. and Worthington, R. E.,** Amide transport channels across toad urinary bladder, *J. Membr. Biol.*, 26, 91, 1976.
221. **Novikoff, A. B. and Heus, M.,** A microsomal nucleoside diphosphatase, *J. Biol. Chem.*, 238, 710, 1963.
223. **Hays, R. M. and Barland, P.,** The isolation of the membrane of the toad urinary bladder epithelial cell, *J. Cell Biol.*, 31, 209, 1966.
224. **Rubin, M. S., Gronowicz, G., Copper, S., and Masur, S. K.,** Comparison of isolated dense subapical granules (DSAG) and surface membranes (SM) of toad urinary bladder epithelial cells, *J. Biol. Chem.*, 95, 263a, 1982.
225. **Caroff, A., Berthonaud, V., and Ripoche, P.,** Neutral aminopeptidase in amphibian kidney, small intestine and frog urinary bladder, *Fed. Proc. Fed. Am. Soc. Exp. Biol.*, 44, 1900, 1985.
226. **Pinto Da Silva, P. and Kan, F. W. K.,** Label fracture. A method for high resolution labeling of cell surfaces, *J. Cell Biol.*, 99, 1156, 1984.
227. **Rodriguez, H. J. and Edelman, J. J.,** Radio-iodination of plasma membranes of toad bladder epithelium, *J. Membr. Biol.*, 45, 185, 1979.
228. **Garvin, J. L., Simon, S. A., Cragoe, R. J., Jr., and Mandel, L. J.,** Phenamil: an irreversible inhibitor of sodium channels in the toad urinary bladder, *J. Membr. Biol.*, 87, 45, 1985.
229. **Barbry, P., Frelin, C., Vignet, P., Cragoe, E. J., Jr., and Lazdunski, M.,** ^{3}H-phenamil, a radiolabelled diuretic for the analysis of the amiloride-sensitive Na^+ channels in kidney membranes, *Biochem. Biophys. Res. Commun.*, 135, 25, 1986.
230. **Aceves, J. and Cuthbert, A. W.,** Uptake of ^{3}H benzamil at different sodium concentrations. Inferences regarding the regulation of sodium permeability, *J. Physiol. (London)*, 295, 491, 1979.
231. **Vignet, P., Jean, T., Barbry, P., Frelin, C., Fine, L. G., and Lazdunski, M.,** ^{3}H-ethylpropylamiloride, a ligand to analyze the properties of the Na^+/H^+ exchange system in the membranes of normal and hypertrophied kidneys, *J. Biol. Chem.*, 260, 14120, 1985.
232. **Cobb, M. H. and Scott, W. N.,** Irreversible inhibition of sodium transport by the toad urinary bladder following photolysis of amiloride analogs, *Experientia*, 37, 68, 1981.
233. **Cuthbert, A. W., Fanestil, D. D., Herrera, F. C., and Pryn, S. J.,** Irreversible inhibition of epithelial sodium channels by ultraviolet irradiation, *Br. J. Pharmacol.*, 77, 431, 1982.
234. **Benos, D. J. and Mandel, L. J.,** Irreversible inhibition of sodium entry sites in frog skin by a photosensitive amiloride analog, *Science*, 199, 1205, 1978.

235. **Park, C. S. and Fanestil, D. D.,** Functional groups of the Na^+ channel: role of carboxyl and histidyl groups, *Am. J. Physiol.*, 245, F716, 1983.
236. **Costa, C. J., Kirschner, L. B., and Cragoe, E. J., Jr.,** Identification of apical membranes from tight epithelia using spin-labeled amiloride and electron paramagnetic resonance spectroscopy, *J. Membr. Biol.*, 82, 49, 1984.
237. **Rossier, B. C.,** Biosynthesis of Na^+, K^+-Atpase in amphibian epithelial cells, *Curr. Top. Membr. Transp.*, 20, 125, 1984.
238. **Rodriguez-Boulan, E. and Sabatini, D. D.,** Asymmetric budding of viruses in epithelial monolayers. Model system for study of epithelial polarity, *Proc. Natl. Acad. Sci. U.S.A.*, 75, 5071, 1978.
239. **Spiegel, S., Blumenthal, R., Fishman, P. H., and Handler, J. S.,** Gangliosides do not move from apical to basolateral plasma membrane in cultured epithelial cells, *Biochim. Biophys. Acta*, 821, 310, 1985.
240. **Wade, J. B., Guckian, V., Koeppen, I.,** Development of antibodies to apical membrane constituents associated with the action of vasopressin, *Curr. Top. Membr. Trans.*, 20, 217, 1984.
241. **Moberly, J. B. and Fanestil, D. D.,** Monoclonal antibodies produced by in vitro immunization as probes of apical surface determinants of amphibian kidney cells (A6), *Biol. Cell*, 55, 38a, 1985.
242. **Corriveau, M., Bourguet, J., and Hugon, J. S.,** Morphological modifications of the frog bladder induced by non ionic detergents (*Rana pipiens* L.), *Virchows Arch. B*, 14, 69, 1973.
243. **Ripoche, P., Parisi, M., Bourguet, J., and Schwartz, I. L.,** Effects of a very hydrophobic non-ionic surfactant (NP-E06) on the apical border of frog urinary bladder, *Inserm*, 85, 85, 1979.
244. **Rapoport, J., Kachadorian, W. A., Muller, W. A., Franki, N., and Hays, R. M.,** Stabilization of vasopressin-induced membrane events by bifunctional imidoesters, *J. Cell Biol.*, 89, 261, 1981.
245. **Hays, R. M., Bourguet, J., Satir, B. H., Franki, N., and Rapoport, J.,** Retention of antidiuretic hormone-induced particle aggregates by luminal membranes separated from toad bladder epithelial cells, *J. Cell Biol.*, 92, 237, 1982.
246. **Svelto, M., Tauc, M., Hays, R. M., and Bourguet, J.,** Attack by urea of the apical surface of amphibian urinary bladder. Influence on antidiuretic response, *Biophys. J.*, 37, 161a, 1982.
247. **Tauc, M., Bourguet, J., Favard, N., and Favard, P.,** Marked ultrastructural alterations of the apical membrane accompany the inhibition of hydroosmotic response by wheat germ agglutinin, *J. Cell Biol.*, 99, 284a, 1984.
248. **Bourguet, J., Gobin, R., Favard, N., and Favard, P.,** Inhibition of antidiuretic action by wheat germ agglutinin. A time course study, *Biol. Cell*, 55, 33a, 1985.
249. **Chase, H. S. and Al-Awqati, Q.,** Calcium reduces the sodium permeability of luminal membrane vesicles from toad bladder. Studies using a fast-reaction apparatus, *J. Gen. Physiol.*, 81, 643, 1983.
250. **Chase, H. S. and Al-Awqati, Q.,** Regulation of the sodium permeability of the luminal border by intracellular sodium and calcium, *J. Gen. Physiol.*, 77, 693, 1981.
251. **Labelle, E. and Valentine, M. E.,** Inhibition by amiloride of $^{22}Na^+$ transport into toad bladder microsomes, *Biochim. Biophys. Acta*, 601, 195, 1980.
252. **Labelle, E. F. and Eaton, D. C.,** Sulfhydryl reagents affect Na^+ uptake into toad bladder membrane vesicles, *J. Membr. Biol.*, 71, 39, 1983.
253. **Garty, H. and Edelman, I. S.,** Analysis of hormonal regulation of sodium transport in toad bladder, *J. Gen. Physiol.*, 81, 785, 1983.
254. **Garty, H., Rudy, B., and Karlish, S. J. D.,** A simple and sensitive procedure for measuring isotope fluxes through ion-specific channels in heterogenous populations of membrane vesicles, *J. Biol. Chem.*, 258, 13094, 1963.
255. **Masur, S. K., Cooper, S., Massardo, G., Gronowicz, G., and Rubin, M. S.,** Isolation and characterization of granules of the toad bladder, *J. Membr. Biol.*, 89, 39, 1986.
256. **Johnson, J. P., Steele, R. E., Perkins, F. M., Wade, J. B., Preston, A. S., Green, S. W., and Handler, J. S.,** Epithelial organization and hormone sensitivity of toad urinary bladder cells in culture, *Am. J. Physiol.*, 241, F129, 1981.
257. **Handler, J. S.,** Toad urinary bladder epithelial cells in culture. Maintenance of epithelial structure, sodium transport and response to hormones, *Proc. Natl. Acad. Sci. U.S.A.*, 76, 4151, 1979.
258. **Handler, J. S., Preston, A. S., and Steele, R. E.,** Factors affecting the differentiation of epithelial transport and responsiveness to hormones, *Fed Proc. Fed. Am. Soc. Exp. Biol.*, 43, 2221, 1984.
259. **Handler, J. S., Perkins, F. M., and Johnson, J. P.,** Studies of renal cell function using cell culture techniques, *Am. J. Physiol.*, 238, F1, 1980.
260. **Perkins, F. M. and Handler, J. S.,** Transport properties of toad kidney epithelia in culture, *Am. J. Physiol.*, 241, C154, 1981.

Chapter 10

WATER MOVEMENTS ACROSS CORNEAL LIMITING LAYERS

Jorge Fischbarg

TABLE OF CONTENTS

I. SOME ANATOMY AND PHYSIOLOGY OF THE CORNEA

Much of what will follow applies largely to rabbit corneas. It is the corneal preparation most exhaustively characterized; in more recent years, evidence has also been gathered from higher mammals, but, for the mechanisms we will refer to, the picture in higher mammals is the same as in rabbits, so it seems sensible to concentrate on the largest and most consistent body of evidence available from rabbits.

The cornea of the eye is a complex tissue, with morphological and physiological characteristics in consonance with the functions it has to perform for the organism (Figure 1). On one hand, being the outermost frontal defense of the fragile ocular globe, it has evolved to display a large degree of mechanical resilience. On the other hand, as if this requirement would not be taxing enough, it is in the optical path and must therefore remain transparent.

This remarkable layer has evolved to display all those required attributes. Mechanical strength is achieved thanks to sheer thickness (ca. 0.5 mm) and to the peculiar juxtaposition of collagen lamellae in its stroma held in place by a ground substance largely made of mucopolysaccharides. At the same time, against one's intuitive expectations, this arrangement is transparent; the details of why this is so have been reviewed in several excellent textbooks, such as the chapters by D. M. Maurice in *The Eye*, or the pertinent chapter in the book by Hogan et al.[1,2,3] One consequence of this arrangement is that the polysaccharides of the stromal ground substance happen to need to be maintained in an unstable state of permanent relative dehydration, otherwise they would imbibe water and the stroma would swell and become opaque. That explains why the cornea has two limiting epithelial layers, and why the transfers of water across each one of them are of central importance for the visual function.

The limiting layers in question (Figure 1) are (1) the anterior epithelium, a 50-μm thick, three- to five-layered stratified tissue known as the ''corneal epithelium'' (or epithelium in what follows) and (2) the posterior epithelium, a 4- to 7-μm thick monocellular layer of flat cells known as the ''corneal endothelium'' (or endothelium in what follows). This last is somewhat of a misnomer; this layer is a typical transporting epithelium, and whatever morphological similarities it has with vascular endothelia (single layer of flat cells, a leaky paracellular pathway) are somewhat incidental. Still, its name as ''endothelium'' reigns supreme in the literature, and it seems needlessly confusing to attempt to change it.

While on the subject of corneal cells, a few more words seem in order. Although sometimes nearly forgotten, the stroma contains a good number of specialized fibroblasts (corneal fibrocytes), albeit suitably dispersed. In terms of sheer mass of cells, they add up to a number comparable to that of the epithelial cells. The endothelial cells do not even come into mention in this regard; they represent less than one tenth of the epithelial mass. However, they are metabolically very active, and their nutritional requirements are the highest per unit mass in the cornea.

To continue with the peculiarities of the cornea, it is one of the exceedingly few completely avascular organs. The evolutionary reasons are obvious: vascularized corneas are as opaque as comparable pieces of connective tissue. However, this dictates one more evolutionary imposition: for all practical purposes, all nutrients for the corneal cells must come from sources other than local blood supply. Not surprisingly, nutrients reach corneal cells mostly through the paracellular pathway across the *endothelial* layer.

As for the paracellular pathway across the *epithelium*, exactly the opposite applies. As it forms the outer barrier of the eye, it avoids permeation of bacteria and extraneous solutes across it. Sure enough, the epithelium is a tight layer, with tight junctions in a zonular apparatus at its most superficial (distal) layer in contact with the tear film.

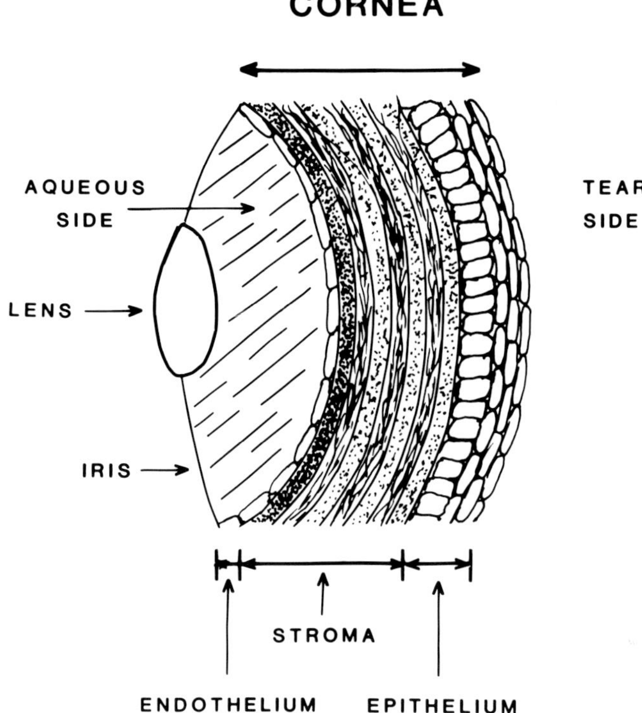

FIGURE 1. Schematic diagram of the cornea showing the layers that compose it. The scale is arbitrary; actual thicknesses (for the rabbit) are epithelium, ~50 μm; stroma, ~375 μm; endothelium, ~5 μm. For simplicity, the basement membranes for the epithelium (Bowman's membrane) and for the endothelium (Descemet's membrane) are not shown.

II. WATER MOVEMENTS AND LIMITING EPITHELIAL LAYERS

With these requirements listed, we can now begin to analyze the water relations of the limiting layers. On one hand, access of water and small nutrients is needed; on the other hand, at the same time, the stroma is to remain partially dehydrated. These two requirements are, in principle, mutually contradictory. Still, evolution has provided mechanisms to comply with them, namely, active fluid transport mechanisms in both layers which continuously pump water out of the stroma and the "leaky" paracellular pathway across the corneal endothelium to let the water and nutrients in.

In spite of its much smaller mass (or perhaps thanks to it?), the endothelium carries the brunt of the fluid transporting activity. Estimates vary, but apparently the epithelium would not pump more than 30% of what the endothelium transports. We will therefore start with the last one.

A. The Endothelium

The corneal endothelium is known to play a key role in maintaining corneal transparency. The connection between endothelial function and corneal hydration and transparency is very well established and has been dealt with before at some length, as in the early papers of Mishima and his co-workers, of Maurice, and the reviews by D. M. Maurice, by Riley, by Waring et al., and by Fischbarg et al.[1,2,4-11] Hence, these issues will be mentioned here only peripherally. Briefly stated, the endothelium is engaged in continuous pumping from stroma to aqueous of fluid which leaks into the stroma by several routes, predominantly across the endothelium itself. As in other epithelia, the transport of fluid appears to be secondary to the transport of electrolytes. As it is also characteristic of other epithelia engaged in fluid transport, the endothelium is leaky and allows relatively large molecules (up to some 45 Å in diameter, cf. Kaye et al.)[12] to pass across it, along with water and electrolytes. Other fluid transporting epithelia, such as the kidney proximal tubule or the gallbladder, are also leaky, although not quite to that degree. Of course, parenthetically, one important exception to that quasi-rule may be the ciliary epithelium, which, according to recent evidence, has been termed an electrically tight epithelium.[13]

B. Rate of Endothelial Pump and Leak

Active transport of water across the endothelium was suspected and suggested, notably by Mishima and his co-workers, since the late 1960s.[4] This research then moved ahead in earnest after a paper by Maurice appeared in 1972 with a convincing demonstration of net movement of fluid across the isolated endothelium from stroma to aqueous against a hydrostatic pressure gradient.[7] The magnitude of this fluid transport is still being revised; in the report cited, it was given as 3 to 12 in range, and 6 $\mu\ell/hr/cm^2$ in the average.[7] Later on, it was reported to be about 4.5 $\mu\ell/hr/cm^2$ when the preparation was bathed with a solution containing 5 m*M* adenosine, and 3.7 $\mu\ell/hr/cm^2$ when adenosine was absent.[14] More recently, we have increased our figure to 10 $\mu\ell/hr/cm^2$ in the absence of adenosine but with 6.9 m*M* glucose present.[15] As for its ambient requirements, it has been reported that the fluid transport depends on the presence of Na^+, K^+, and HCO_3^- in the solution, that it does not depend on $[Cl^-]$, and that it is inhibited by the presence of ouabain (10^{-5}) in solution.[16-18] For reference, the rate of fluid movement has been measured by at least two different methods: (1) microscopy, be that specular microscopy or pachometry, in which a microscope with epi-illumination is utilized to follow the change in thickness of the stroma under conditions such that water can flow only across the endothelium[6,16] and (2) flow across a de-epithelialized corneal preparation mounted in modified Ussing chambers.[7,14]

Aside from this metabolically dependent fluid transport, independently of it, a passive leak of fluid in the opposite direction (from aqueous to stroma) across the endothelium has been recognized since the 1960s.[1,5] It is driven by a hydrostatic pressure difference existing between the aqueous chamber and the stromal interstice. This happens because, as mentioned above, the stromal mucopolysaccharides are relatively dehydrated in the physiological condition, and their tendency to imbibe water can be measured as a negative hydrostatic pressure (''imbibition pressure'').[19] From all we know, this ''leak'' flow traverses the endothelial leaky tight junctions. One demonstration of this consists of treating the endothelial layer with Ca-free solutions, which disrupts the junctions and greatly increases this passive flow.[12]

To be noted, in a healthy cornea in vivo, the total thickness remains constant. Under this condition, no net water movement takes place across the endothelium; there is a dynamic balance in which the imbibition pressure (negative pressure) of the stroma creates a leak of magnitude equal to that of the rate of fluid transport by the endothelium. This implies that circulation of water may take place at the cellular scale; whatever fluid is pumped by the cells from the intercellular spaces into the aqueous may leak back driven across the junctions by the existing hydrostatic pressure difference (stromal imbibition pressure).

III. THE EPITHELIUM AND STROMA

The epithelium plays somewhat different roles in helping corneal deturgescence. First and foremost, it constitutes a sizable barrier for water flows induced by a hydrostatic pressure difference (or hydraulic flows, in short). This becomes important under conditions in which the intraocular pressure increases, such as in glaucoma; in fact, damage to the epithelium can occur then, which is not surprising in view of the above. The epithelium also plays a minor role in pumping some water out of the stroma. How much it does so under physiological conditions remains a question. The epithelial cells can pump electrolytes and fluid in either direction across it, much like some of the ileal epithelium. While transport in the undisturbed layer may proceed from the tears into the stroma, the functional apparition of a chloride permeation pathway (chloride channels) in the apical (tear-facing) membrane under stimulation by catecholamines or cAMP results in a reversal such that fluid and electrolytes are then pumped from stroma to tears.[20-24] A similar reversal appears to be the basis for the bidirectional mode of fluid transfer across small intestine (i.e., normal absorption vs. diarrhea). In fact, the role of Cl^- channels in such reversal, which was established for the corneal epithelium, turned out to be quite similar when studied later in the intestine. Having said all this, however, it should be noted that the amount of fluid pumped by the epithelium out of the stroma, even under maximum stimulation of the reversal, is relatively small. How small? In the mid-1970s calculations put it at some 5% of the amount pumped by the endothelium.[22,24,25] Therefore, it is possible that the functional role of the epithelium for fluid transport may be that of a small correction for regulation, or a reserve capacity available in emergencies.

IV. OSMOTIC PERMEABILITY OF THE LIMITING LAYERS

The osmotic permeabilities (P_{os}) of the corneal epithelium and endothelium are important for the understanding of the basic mechanisms of water permeation across them. As an example, in several models for transendothelial and transcorneal water transport,[26-28] values for these osmotic permeabilities are utilized. We will therefore present some recent measurements, together with an evaluation of them.

What is the endothelial osmotic permeability? We have been examining that central question since 1973. At first, we employed specular microscopy, and essentially redid the experiments of Mishima and Hedbys, in which the endothelial thickness was determined manually by an operator.[29] Those data were given in Figure 10 of Fischbarg and Figure 4 of Fischbarg et al.[10,30] The slope of the osmotic flow vs. osmotic gradient curve represents the osmotic permeability of the endothelium as a whole. Yet, that slope was not constant; at the lowest gradients (10 mOsm/ℓ), the slope was very steep and extrapolation to the origin yields a P_{os} of some 600 μm/sec. True, for the other osmotic gradients utilized, the slope is much less steep, and a line drawn across most points yields instead a P_{os} of some 200 μm/sec. However, since extrapolation has its dangers, in those papers we emphasized the lower value because it seemed the most consistent one. Incidentally, that value came close to that reported by Mishima and Hedbys. Still, the question of why the slope was not constant constituted a puzzle, and the matter remained under exploration. When an automatic technique for fluid-flow measurements introduced by Bourguet and Jard in 1964 was improved so as to detect volume changes of 1 to 3 nℓ, we focused again on that question.[14,31] With this technique, the steady-state flows induced by a sucrose gradient were reported in Figure 3 of Fischbarg et al.[30] Once more, the slope was not linear. Its steepest region yielded a P_{os} of 150 to200 μm/sec, consistent with the low value discussed above, but the nonlinearity puzzle persisted. The next development came from exploratory experiments performed in early 1977.[56] When the osmotic gradient was set up using molecules larger than sucrose such as polyethylene glycol, the flows obtained were much larger and implied a P_{os} of some 400 to 600 μm/sec.

This issue was hence reinvestigated. In experiments in which we utilized dextran of 40,000 mol wt, the high transendothelial P_{os} reappeared; with glucose absent from the balanced salts (BS) medium, it was as high as 451 ± 84 μm/sec.[32] By then, Klyce and Russell had reported a transendothelial P_{os} of 595 μg/sec.[28] They had performed osmotic changes at the epithelial side of corneas and, based on their theoretical model, had computed the transendothelial P_{os}

which best fit their results. Although such calculation depended on the correctness of the model employed, it all seemed promising in that their model explained satisfactorily a number of other corneal hydration dynamic characteristics.

By early 1980, the results suggested that, as argued in general for P_{os} measurements by Diamond,[33] some of the P_{os} values that others and ourselves had reported might have been underestimated because of insufficient time resolution in the techniques and because of solute polarization inside, between, and around cellular structures. Theoretical work in this laboratory had shown that simple unstirred layers could not account for P_{os} errors of more than two- or threefold;[34,35] on the other hand, epithelial structures could in principle exhibit varying apparent P_{os} due to a number of complex morphological changes combined with concentration shifts in confined spaces. The question of the transendothelial P_{os} still required an unequivocal answer. And, of course, the separate P_{os} of the apical and basolateral endothelial membranes was not known.

In order to confront such problems, we developed an optical method.[36] It was based on a modification of an earlier technique in which specular microscopy had been successfully applied to the study of corneal thickness by Klyce and Maurice and by Klyce and Russell.[37,38] With our modifications, the method could be utilized for two different purposes: (1) the measurement of global transepithelial P_{os}, i.e., the P_{os} of the entire layer and (2) the measurement of the P_{os} of the apical cell membranes of an epithelium.

In the first variant, the position of a tissue border can be followed continuously for a short time after an osmotic challenge, which allows an excellent graph of tissue thickness vs. time; the initial slope of this graph constitutes the P_{os} one is after. With this method, the P_{os} measured for ADH-stimulated, glutaraldehyde-fixed frog urinary bladder was of the same order that had been measured by other procedures by other laboratories. After this validation, we proceeded to determine the P_{os} of both the epithelial and endothelial layers. As it turned out, the endothelial P_{os} that we measured (711 ± 34 μm/sec, n = 7) is quite similar to the values given by Klyce and Russell.[28] They reported two series of six experiments from which their theoretical model allowed a computation of the transendothelial P_{os} of (in median and range) 740 (600 to 770) and 450 (340 to 740) μg/sec. Our 1982 P_{os} value is somewhat higher than the average in their report, but agrees with the value computed from their first series.

For comparison, the P_{os} across the corneal epithelium was determined with the same procedure above[36] and the value found was 137 ± 30 μg/sec (r = 5; errors are SEM). There was a historical precedent for such measurements in the value of 88 ± 30 μg/sec reported by Mishima and Hedbys.[29] Again, as one might have predicted, the new value was somewhat larger.

A preliminary assessment of these results told us that both layers had relatively high P_{os}. Beyond this, however, one has to ask what is the P_{os} of *each of the cell membranes* that the water must traverse (rather than that of the total layer). As a guide, an approximate calculation for such cell membrane P_{os} can be made as follows: if one assumes that, ideally, the water flow traverses two cell membranes in series (apical and basolateral), and if one further simplifies the problem by assuming that the areas and the P_{os} of both cell membranes are comparable, it follows that, to a first approximation, *each* of those cell membranes ought to have a P_{os} of twice the value measured for the *total* tissue layer. That entails values of 1400 μm/sec for the endothelial cell membranes and 250 μg/sec for the epithelial cell membranes. Comparisons with other epithelial cells are difficult because of their convoluted shape; however, red blood cells, which are easier to deal with because of their known area, display P_{os} values of the order of 200 to 400 μm/sec. One might say then, in principle, that the P_{os} of the cell membranes of the corneal epithelium would be in the range for "normal" cells, while the cell membrane P_{os} for the endothelial cells would display a P_{os} several times larger than that.

We investigated this point by utilizing the second variant of our procedure. In this case, scans are made of the endothelial cell thickness as fast as the motorized microscope stage can travel; data are, of course, being fed to a computer. We could determine one cell thickness point every 0.33 sec. The transition that the osmotic challenge introduces in the cell thickness is very rapid, taking only some 2 to 3 sec. The apparent apical cell membrane osmotic permeability (P_{os}) was determined graphically by eye from the initial slope of such curves. Its value was 1420 ± 160 μm/sec (n = 5).[36]

On one hand, this value gives one pause, because it is the highest ever reported for a cell membrane. On the other hand, there is no reason why all cells should have the same P_{os}, and tissues with a higher degree of specialization could exhibit these very large values. In addition, of course, this value is curiously close to what one would expect from the total transendothelial layer P_{os} (711 μm/sec) and the simple-minded calculation given above.

How valid is this very large figure? Could we be underestimating or overestimating P_{os}? Arguments can be given for both. Probably the most serious technical pitfall one faces is the possibility that the lateral cell membrane may be "recruited" for osmotic flow during the transient change in light intensity which would lead to an overestimate of P_{os}. With this scenario, assuming a change in the apical bathing solution from isotonic to hypotonic, at the same time that the water rushes into the cell through the apical membrane, salt could diffuse out of the intercellular spaces across the open junctions, leaving the spaces hypotonic and giving as a result an additional osmotic flow from spaces to cell (cf. Boulpaep's comment in the discussion of the Macy Symposium).[38] From our calculations, this effect is minor; diffusion through the junction is slower than the water onrush into the cell. However, a more conclusive answer will require further experimental work. In addition, flow from the intercellular spaces while the cell is swelling leads to quick space collapse and cessation of this secondary flow.

V. THE ROUTE OF WATER TRANSPORT

There is good evidence that the hydraulic pathway can be identified with the paracellular channels, that is, intercellular spaces and junctions in series. To begin with, the value of the hydraulic conductance Lp is just what one calculates for such a system assuming Poiseuille-like laminar flow of water between parallel plates, as recognized more than a decade ago.[10] In addition, when the junctional complexes are disrupted in Ca-free solutions (with some EDTA added), the hydraulic conductance increases drastically, as one would expect.[12]

The higher P_{os} values detected lately require some refinements to the conclusions derived from our earlier results. *Circa* 1977, the highest P_{os} found for the endothelium was in the order of 200 μm/sec, which was not too far from the hydraulic Lp for that layer (~120 μm/sec).[30] This prompted us to speculate on whether these two pathways might not be the same.[30] At this point, the P_{os} value for the layer has been found to be larger, of the order of 600 to 700 μm/sec.[28,36] That may apparently put in question our conclusion[30] about the two pathways (osmotic and hydraulic) being the same. However, if one re-examines Figure 2 of our 1977 paper,[30] the water flow can be seen to change very rapidly at lower values of applied hydrostatic pressure difference ΔP. For that ΔP range, in that Figure the hydraulic Lp is therefore very large, namely, 1.8×10^{-10} cm^3 sec^{-1} $dyne^{-1}$, or 2511 μm/sec. As can be seen, this value is of the order of the P_{os} that we attributed to the apical cell membrane. In a word, the possibility remains open that the high hydraulic conductance pathway (the junctions) and the high osmotic permeability pathway may be one and the same. All we can tell at present is that the layer is very permeable to water, whether the driving force is osmotic or hydrostatic pressure. Whether osmosis traverses the cells, the junctions, or a combination of both remains to be established.

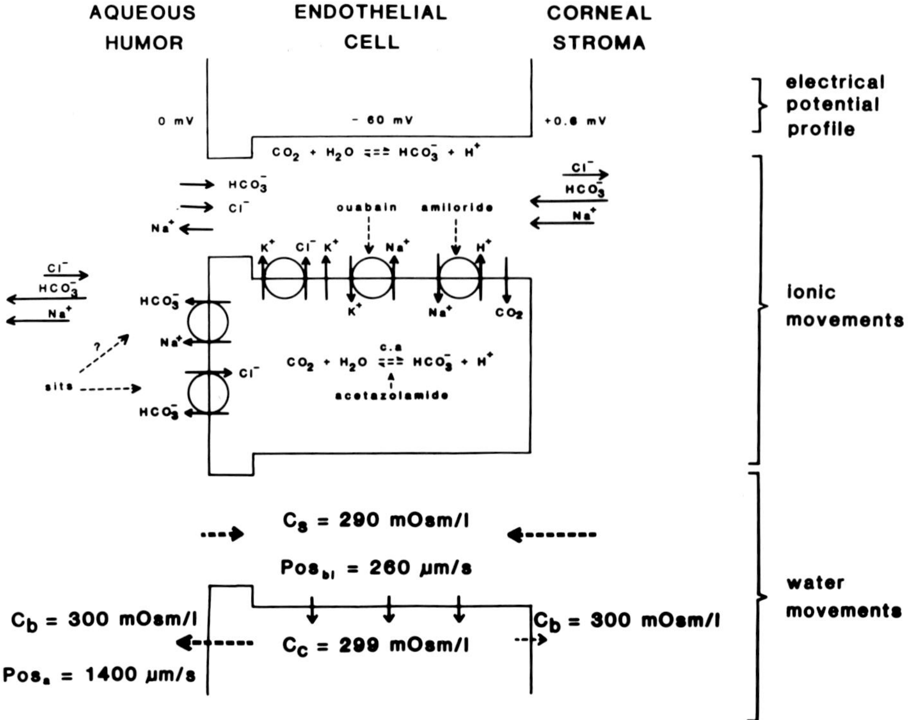

FIGURE 2. Fluid and electrolyte transport mechanisms postulated to exist across the apical and basolateral membranes of the corneal endothelium. Solid lines represent either active transport or coupled movements of electrolytes. Dashed lines represent water movements. P_{os}: osmotic permeability; c: cell; s: intercellular spaces; a: apical; bl: basolateral.

VI. ENDOTHELIAL FLUID TRANSPORT

The results reviewed so far (plus those to be gathered in the future) would be expected to fit into a global picture that would ultimately describe in detail how the endothelium transports fluid and electrolytes across it. A model presenting a partial scheme for the electrolyte movements was advanced by us in 1974;[18] subsequently, more recent updates have appeared.[11,43] In addition, except for minor modifications and extensions, many of the elements of such models have been adopted by another laboratory (Jentsch et al.).[44] As for the mechanism of fluid transport, one can single out several papers that have dealt with that subject for the endothelium.[26,27,45] From those, we will adhere most closely to the scheme of Liebovitch and Weinbaum,[27] albeit with the benefit of more recent P_{os} values.[36] We will reproduce here the basic elements of that model, as described in our 1985 review[11] plus some additions. To be sure there is an important caveat to be made. This model is *merely one of the possibilities* still open nowadays, and it illustrates *only* how transcellular osmosis may proceed. There are apparently no comparable models for the possibility of transjunctional osmotic flow in this preparation.

We will refer for this purpose to the scheme in Figure 2. The polarity of the transendothelial potential difference that others and ourselves reported years ago (aqueous negative)[10,17,18,46,47] cannot correspond to an electrogenic transport of Na^+ in the same direction as that of the fluid transport. The polarity observed could instead be explained by an electrogenic transport of HCO_3^- across this preparation, as suggested originally by Barfort and Maurice.[47] However, the route and mechanism for such transport remain unclear. One possibility is that there could be an apical carrier (cotransporter) driven by the HCO_3^- electrochemical gradient that would ferry both HCO_3^- and Na^+ in a ratio favoring the anion.[44,48] For correct stoichiometry, however, some additional ionic movements would also have to be postulated. One suitable scheme would include an HCO_3^- for Cl^- exchanger at the apical membrane, and passive Cl^- and K^+ diffusion out of the cell across the lateral membrane. Another possibility is that HCO_3^- would leave the cell by facilitated diffusion across the apical membrane, and the electric field across the junctions would then result in a current carried mostly by Na^+.[11] In both schemes, a net transfer of $NaHCO_3$ from stroma to aqueous would be achieved.

For our main subject in this review, water transport, the elements that matter in the scheme of Figure 2 are two: (1) the hypotonicity that the transport mechanism across the lateral cell membranes would create in the intercellular spaces and (2) the fact that the cell concentration would be slightly hypotonic to comply with steady-state constraints on water and electrolyte fluxes (Liebovitch and Weinbaum, Fischbarg et al.).[27,48] The scheme for fluid transport in Figure 2 is basically the same of that of Liebovitch and Weinbaum;[27] the main difference lies in the P_{os} and osmolarity values we utilize. Actually, however, a model much like that had been advanced by Shapiro and Candia in 1973.[45] In their pioneering scheme, they envisioned most water movements in the same directions as Figure 2 has them, and by the same routes. The only difference is that they attributed water exit from the cell across the apical membrane to a hydrostatic pressure gradient across it; nowadays, that step is generally thought to be due to osmotic forces. Thus, in other papers, Lim and Fischbarg, Mayes and Hodson and Mayes have described water transport solely in terms of local osmotic gradients in and around the endothelium.[26,49,50]

A. The Choice of Osmolarities

To return to the model for endothelial fluid transport, one key element to consider has to do with the osmolarity values to be attributed to the different compartments and layers. We have performed some simple calculations that yielded the estimates shown in Figure 2. For these calculations, we start from the following values:

1. Ambient concentration Cb = 300 mOsm/ℓ (of particles).
2. Apical membrane osmotic permeability P_{osa} = 1400 μm/sec = 2.52×10^{-3} cm^4/mOsm sec.[36]
3. Total trans-tissue osmotic permeability P_{ost} = 711 μm/sec = 1.28×10^{-3} cm^4/mOsm sec.[28,36]
4. Total rate of fluid transport ≃ rate of flow across the apical membrane; Jv_a: 9 μℓ/hr cm^2 = 2.5×10^{-6} 1 cm/sec (neglecting back-leak across the leaky intercellular junctions).[7,15]
5. Apical reference "en face" area A_a = 1 cm^2.
6. Basal refetence "en face" area A_b = 1 cm^2.
7. Lateral area = 4.5 cm^2 (per cm^2 of "en face" area; as calculated by us from the electronmicrographs given in Hirsch et al.[51]

We assume that the basolateral membrane is homogeneous, and therefore, that its P_{os} is uniform. We calculate this basolateral membrane osmotic permeability $P_{os_{bl}}$ by reasoning that the total osmotic permeability of the tissue results from the series juxtaposition of the apical and basolateral membranes, with their permeabilities weighted by their areas, e.g.,

$$P_{os_t} x A_a = (P_{os_a} x A_a \times P_{os_{bl}} x A_{bl})/(P_{os_a} \times A_a + P_{os_{bl}} \times A_{bl}) \quad (1)$$

or

$$r = A_{bl}/A_a \tag{2}$$

$$P_{os_t} = (P_{os_a} \times r \times P_{os_{bl}})/(P_{os_a} + r \times P_{os_{bl}} \tag{3}$$

$$P_{os_{bl}} = P_{os_t} \times P_{os_a}/r \times (P_{os_a} - P_{os_t}) = 263\ \mu m/sec$$

$$= 4.7 \times 10^{-4}\ cm^4/mOsm/sec \tag{4}$$

The cell concentration is determined by the fact that at the apical membrane we know the P_{os_a} and the total translayer rate of fluid transport, from which we argue that the apical concentration gradient ΔC must be at least the value given by:

$$\Delta C_a = Jv_a/P_{os_a} = 1\ mOsm/\ell \tag{5}$$

That is, the cell osmolarity is 299 mOsm/ℓ.

Next comes the osmolarity of the intercellular spaces (C_s). For this, we will assume that, as noted by Liebovitch and Weinbaum,[27] the integral of the water flow over the lateral area is larger than the similar integral over the apical area, the difference being given by water flow in the ''wrong'' direction across the basal membrane, driven by the gradient at that location. Let us assume that such flow is simply

$$Jv_b = \Delta C_b \times Lp_{bl} = 0.47 \times 10^{-6}\ cm/sec \tag{6}$$

In the steady state, the flow across the lateral membrane is the sum of the flows through the apical plus basal membranes, or

$$Jv_l = Jv_a + Jv_b = 3.0 \times 10^{-6}\ cm/sec \tag{7}$$

(referred to the ''en face'' or tangential area of the tissue). The gradient across the lateral membrane is, then

$$C_l = Jv_l/P_{os_{bl}} = 6.3\ mOsm/\ell \tag{8}$$

Adding the apical plus lateral gradients, we conclude that the intercellular spaces are hypotonic to the medium, in average, by

$$\Delta C_s = \Delta C_a + \Delta C_l = 7.3\ mOsm/\ell \tag{9}$$

In actual terms, one would expect the hypotonicity of the lateral spaces to be maximal at about half-way along them, and to decrease towards the open apical and basal ends of the spaces. We performed simulations[57] utilizing the Diamond-Bossert differential equations for advection-diffusion along the intercellular spaces,[52] and basically confirmed the calculation above; typically, the highest hypotonicity we generated was some 10 mOsm/ℓ at about two fifths of the channel, going from apical to basal.

Calculations similar to the ones above had been performed previously by Persson and Spring,[53] for the Necturus gallbladder fluid transport system, and the order of magnitude of their results was the same as given here. In fact, there is some recent evidence from two laboratories working in similar tissues that the intercellular spaces differ in tonicity from the ambient medium by some 3 to 5%.[54,55] The tonicity of the cell is a different matter; a difference of a fraction of a milliosmol, such as postulated, may well be impossible to detect for the time being. This may not be a totally satisfying state of affairs, but after all, some of these questions, including that of a complete experimental demonstration of these mechanisms, may have to wait for an answer.

VII. THE GLOBAL CORNEAL MODEL

In closing, it is worth emphasizing that many of our remarks above on transendothelial and transepithelial water transfers apply to particular conditions in which these layers are isolated and operating under in vitro conditions. The situation in vivo is more complex. Thus, in vivo and in the steady state, as mentioned above, there is net transtissue water transfer across the endothelium because of the presence of a partially dehydrated stroma which aspirates back as much fluid as the endothelium is pumping. A model for water relations in the cornea has been put together by Klyce and Russell, which incorporates the interactions of all corneal layers in an ensemble.[28] It is quite successful in predicting the time course of variations in the corneal thickness that ensues when the steady state is affected, such as after inhibition of the ionic transport mechanisms by exposure to cold temperatures. It also shows that the stroma can lodge unstirred layers which may result in anti-intuitive results during induced pertubations of the steady state.

REFERENCES

1. **Maurice, D. M.,** The cornea and sclera, in *The Eye,* Vol. 1, Davson, H., Ed., Academic Press, New York, 1969, 489.
2. **Maurice, D. M.,** The cornea and sclera, in *The Eye,* Vol. 1B, Davson, H., Ed., Academic Press, New York, 1983, 1.
3. **Hogan, M. J., Alvarado, J. A., and Weddell, J. E.,** *Histology of the Human Eye,* W. B. Saunders, Philadelphia, 1971.
4. **Mishima, S. and Kudo, T.,** In vitro incubation of rabbit cornea, *Invest. Ophthalmol.,* 6, 329, 1967.
5. **Trenberth, S. M. and Mishima, S.,** The effect of ouabain on the rabbit corneal endothelium, *Invest. Ophthalmol.,* 7, 44, 1968.
6. **Mishima, S.,** Corneal thickness, *Surv. Ophthalmol.,* 13, 57, 1968.
7. **Maurice, D. M.,** The location of the fluid pump in the cornea, *J. Physiol.,* 221, 43, 1972.
8. **Riley, M. V.,** Transport of ions and metabolites across the corneal endothelium, in *Cell Biology of the Eye,* McDevitt, D. S., Ed., Academic Press, New York, 1985, 32.
9. **Waring, G. O., III, Bourne, W. M., Edelhauser, H. F., and Kenyon, K.,** The corneal endothelium. Normal and pathologic structure and function, *Ophthalmology,* 89, 531, 1982.
10. **Fischbarg, J.,** Active and passive properties of rabbit corneal endothelium, in the Proc. of the Int. Symp. on Transport and the Eye, *Exp. Eye Res.,* 15, 615, 1973.
11. **Fischbarg, J., Hernandez, J., Liebovitch, L. S., and Koniarek, J. P.,** The mechanism of fluid and electrolyte transport across corneal endothelium: critical revision and uptake of a model, *Curr. Eye Res.,* 4, 351, 1985.
12. **Kaye, G. I., Sibley, R. C., and Hoefle, F. B.,** Recent studies on the nature and function of the corneal endothelial barrier, *Exp. Eye Res.,* 15, 585, 1973.
13. **Burstein, N. L. and Fischbarg, J.,** Electrical potential, resistance, and fluid secretion across isolated ciliary body, *Exp. Eye Res.,* 39, 771, 1984.
14. **Fischbarg, J., Lim, J. J., and Bourguet, J.,** Adenosine stimulation of fluid transport across rabbit corneal endothelium, *J. Membr. Biol.,* 35, 95, 1977.
15. **Fischbarg, J. and Masters, B. R.,** unpublished results.

16. **Dikstein, S. and Maurice, D. M.,** The metabolic basis to the fluid pump in the cornea, *J. Physiol. (London),* 221, 29, 1972.
17. **Hodson, S.,** The regulation of corneal hydration by a salt pump requiring the presence of sodium and bicarbonate ions, *J. Physiol. (London),* 236, 271, 1974.
18. **Fischbarg, J. and Lim, J. J.,** Role of cations, anions and carbonic anhydrase in fluid transport across rabbit corneal endothelium, *J. Physiol. (London),* 241, 647, 1974.
19. **Hedbys, B. O., Mishima, S., and Maurice, D. M.,** The imbibition pressure of the corneal stroma, *Exp. Eye Res.,* 2, 99, 1963.
20. **Zadunaisky, J. A., Lande, M. A., Chalfie, M., and Neufeld, A. H.,** Ion pumps in the cornea and their stimulation by epinephrine and cyclic-AMP, *Exp. Eye Res.,* 15, 577, 1973.
21. **Candia, O. A.,** The active translocation of Cl and Na by the frog corneal epithelium, in *Chloride Transport in Biological Membranes,* Zadunaisky, J. A., Ed., Academic Press, New York, 1982, 232.
22. **Klyce, S. D.,** Cl transport in rabbit cornea, in *Chloride Transport in Biological Membranes,* Zadunaisky, J. A., Ed., Academic Press, New York, 1982, 199.
23. **Klyce, S. D. and Wong, R. K. S.,** Site and mode of adrenaline action on chloride transport across the rabbit corneal epithelium, *J. Physiol. (London),* 266, 777, 1977.
24. **Klyce, S. D. and Crosson, C. E.,** Transport processes across the rabbit corneal epithelium: a review, *Curr. Eye Res.,* 4, 323, 1985.
25. **Huff, J.,** personal communication.
26. **Lim, J. J. and Fischbarg, J.,** Standing-gradient osmotic flow: examination of its validity using an analytical method, *Biochim. Biophys. Acta,* 443, 339, 1976.
27. **Liebovitch, L. S. and Weinbaum, S.,** A model of epithelial water transport: the corneal endothelium, *Biophys. J.,* 35, 315, 1981.
28. **Klyce, S. D. and Russell, S. R.,** Numerical solution of coupled transport equations applied to corneal hydration dynamics, *J. Physiol. (London),* 292, 107, 1979.
29. **Mishima, S. and Hedbys, B. O.,** The permeability of the corneal epithelium and endothelium to water, *Exp. Eye Res.,* 6, 10, 1967.
30. **Fischbarg, J., Warshavsky, C. R., and Lim, J. J.,** Pathways for hydraulically and osmotically-induced water flows across epithelia, *Nature (London),* 266, 71, 1977.
31. **Bourguet, J. and Jard, S.,** Un dispositif automatique de mesure et d'enregistrement du flux net d'eau a travers la peau et la vessie des amphibiens, *Biochim. Biophys. Acta,* 88, 442, 1964.
32. **Fischbarg, J., Hofer, G., and Koatz, R. A.,** Priming of the fluid pump by osmotic gradients across rabbit corneal endothelium, *Biochim. Biophys. Acta,* 603, 198, 1980.
33. **Diamond, J. M.,** The epithelial junction: bridge, gate and fence, *Physiologist,* 20, 10, 1977.
34. **Pedley, T. J. and Fischbarg, J.,** The development of osmotic flow through an unstirred layer, *J. Theor. Biol.,* 70, 427, 1978.
35. **Pedley, T. J. and Fischbarg, J.,** Unstirred layer effects in osmotic water flow across gallbladder epithelium, *J. Membr. Biol.,* 54, 89, 1980.
36. **Fischbarg, J. and Montoreano, R.,** Osmotic permeabilities across corneal endothelium and ADH-stimulated toad urinary bladder structures, *Biochim. Biophys. Acta,* 690, 207, 1982.
37. **Klyce, S. D. and Maurice, D. M.,** Automatic recording of corneal thickness in vitro, *Invest. Ophthalmol.,* 15, 550, 1976.
38. **Fischbarg, J.,** The paracellular pathway in the corneal endothelium, in *The Paracellular Pathway,* Bradley S. E. and Purcell, E. F., Eds., Josiah Macy, Jr. Foundation, New York, 1982, 307.
39. **Fischbarg, J., Liebovitch, L. S., and Koniarek, J. P.,** Evidence that water permeates cell membranes through the glucose channel, *Invest. Ophthalmol. Suppl.,* 27 (Abstr.), 84, 1986.
40. **Fischbarg, J., Liebovitch, L. S., and Koniarek, J. P.,** Inhibition of transepithelial osmotic water flow by blockers of the glucose transporter, to be published.
41. **Fischbarg, J.,** On the theory of solute-solvent coupling in epithelia, this volume, chap. 8.
42. **Macey, R. I. and Farmer, R. E. L.,** Transport of water and urea in red blood cells, *Am. J. Physiol.,* 246, C195, 1984.
43. **Fischbarg, J. and Lim, J. J.,** Fluid and electrolyte transports across corneal endothelium, in *Current Topics in Eye Research,* Vol. 4, Zadunaisky, J. A., Ed., Academic Press, New York, 1984, 201.
44. **Jentsch, T. J., Keller, S., Koch, M., and Wiederholt, M.,** Evidence for coupled transport of bicarbonate and sodium in cultured bovine corneal endothelial cells, *J. Membr. Biol.,* 81, 189, 1984.
45. **Shapiro, M. P. and Candia, O. A.,** Corneal hydration and metabolically dependent transcellular passive transfer of water, *Exp. Eye Res.,* 15, 659, 1973.
46. **Fischbarg, J.,** Potential difference and fluid transport across rabbit corneal endothelium, *Biochim. Biophys. Acta,* 228, 1972b.
47. **Barfort, P. and Maurice, D.,** Electrical potential and fluid transport across the corneal endothelium, *Exp. Eye Res.,* 419, 11, 1974.

48. **Fischbarg, J., Liebovitch, L. S., and Koniarek, J. P.,** A central role for cell osmolarity in isotonic fluid transport across epithelia, *Biol. Cell,* 55, 239, 1985.
49. **Mayes, K. R. and Hodson, S.,** Local osmotic coupling to the active trans-endothelial bicarbonate flux in the rabbit cornea, *Biochim. Biophys. Acta,* 1514, 286, 1978.
50. **Mayes, K. R.,** Further evidence for local osmotic coupling in the rabbit cornea, *Biochim. Biophys. Acta,* 600, 831, 1980.
51. **Hirsch, M., Renard, G., Faure, J.-P., and Pouliquen, Y.,** Study of the ultrastructure of the rabbit corneal endothelium by the freeze-fracture technique: apical and lateral junctions, *Exp. Eye Res.,* 25, 277, 1977.
52. **Diamond, J. M. and Bossert, W. H.,** Standing-gradient osmotic flow. A mechanism for coupling of water and solute transport in epithelia, *J. Gen. Physiol.,* 450, 2061, 1967.
53. **Persson, B. E. and Spring, K. R.,** Gallbladder epithelial cell hydraulic water permeability and volume regulation, *J. Gen. Physiol.,* 79, 481, 1982.
54. **Simon, M., Curci, S., Gebler, B., and Fromter, E.,** Attempts to determine the ion concentrations in the lateral spaces between the cells of Necturus gallbladder epithelium with microelectrodes, in *Water Transport Across Epithelia,* XV Benzon Symposium, Ussing, H. H., Bindslev, N., Lassen, N. A., and Sten-Knudsen, O., Eds., Munksgaard Publ., Copenhagen, 1981, 52.
55. **Sackin, H.,** Electrophysiology of salamander proximal tubule II. Interspace NaCl concentrations and solute-coupled water transport, *Am. J. Physiol.,* 251, F334, 1986.
56. **Warshavsky, C. R. and Fischbarg, J.,** unpublished results.
57. **Hernandez, J. and Fischbarg, J.,** unpublished results.

Chapter 11

WATER PERMEABILITY OF THE GILL EPITHELIUM: SALINITY AND TEMPERATURE RELATIONS

Peter M. Taylor

TABLE OF CONTENTS

I. INTRODUCTION

A. The Structure and General Functions of Gills

Gills (alternatively branchiae) may be defined broadly as outpushings of the body surface of an aquatic animal which are specialized primarily to maximize the respiratory exchange of dissolved gases (oxygen and carbon dioxide).[1] Specializations of branchial structure serve to incorporate a large exchange surface area and a short transintegumental diffusion distance.[2] Gills are characteristic of fishes, larval (and certain adult) amphibians, and many coelomate invertebrates. Although customarily considered in terms of a respiratory function, gills in general are adaptable, multifunctional structures which may play important roles in osmoregulation, locomotion, and feeding. Indeed, in certain invertebrates, structures classed as gills serve for dissolved-gas exchange no more than the remainder of the body surface, while in others, localized areas of the body wall may become specialized as respiratory plaques without elaboration into recognizable gills. Gills and lungs may be found together in animals specialized for a bimodal existence and in larvae undergoing emergent metamorphosis.

There is a tremendous interspecific range of complexity in branchial structure, although it is possible in the great majority of cases to identify a highly perfused lamella as the common functional unit (Figures 1 and 2). The basic lamella consists of two permeable sheets of thinned integument (2 to 10 μm in thickness), separated by regularly spaced trabecular elements which form either a network of capillaries or a series of hemocelic lacunae. In addition, there is generally a well-defined marginal canal affording separation of afferent (deoxygenated) and efferent (oxygenated) circulatory fluid. In certain decapod crustaceans trabeculae are absent and the lamellar circulation is channeled by a perforate median septum.[3] It is common for lamellae to be arranged compactly on supporting filaments which may form an arborescent gill or be borne on a branchial arch. In fishes, filaments are generally termed primary lamellae and the respiratory lamellae themselves are termed secondary lamellae. The branchial designs serve to maximize the lamellar area available for gas exchange and produce, in consequence, a major site for diffusional water movements between animal and medium. It has been demonstrated for fishes and crustaceans that the gills are the principal site of water exchange, representing over 90% of the exchange surface area.[4-7]

B. Gills and Osmoregulation: The Physiological Importance of Branchial Water Flux

The study of branchial water permeability (and its modulation) is linked inextricably to that of the environmental physiology of osmoregulation, which I will review briefly so that its importance may be comprehended. It is considered generally that life on earth originated and became initially established in an aquatic environment (probably in the oceans). The evolution of the major animal phyla from primitive eukaryotic organisms appears to have occurred within the confines of the marine environment and indeed the distribution of the majority of extant classes of animal is still restricted to the world's oceans. Water constitutes 70 to 80% of body weight in aquatic animals. In triploblastic Metazoa, a circulating extracellular fluid isolates the mesoderm from the exterior and bathes the basolateral surfaces of ectoderm and endoderm. At the macroscopic level, this fluid is always iso-osmotic with the intracellular milieu when an animal is in a steady state with its environment. The exchanges of water across the ectoderm (in particular those across the permeable branchial epithelium) may be of considerable physiological importance as body fluid volumes and composition must be regulated with respect to any net volume osmotic flux.

The major patterns of animal osmoregulation are summarized in Figure 3. Most marine invertebrates are osmoconformers, having little or no power to regulate their extracellular fluid composition, and as a consequence, their body fluids remain virtually isotonic and iso-osmotic with the seawater which they inhabit. In general, osmoconformers tolerate only

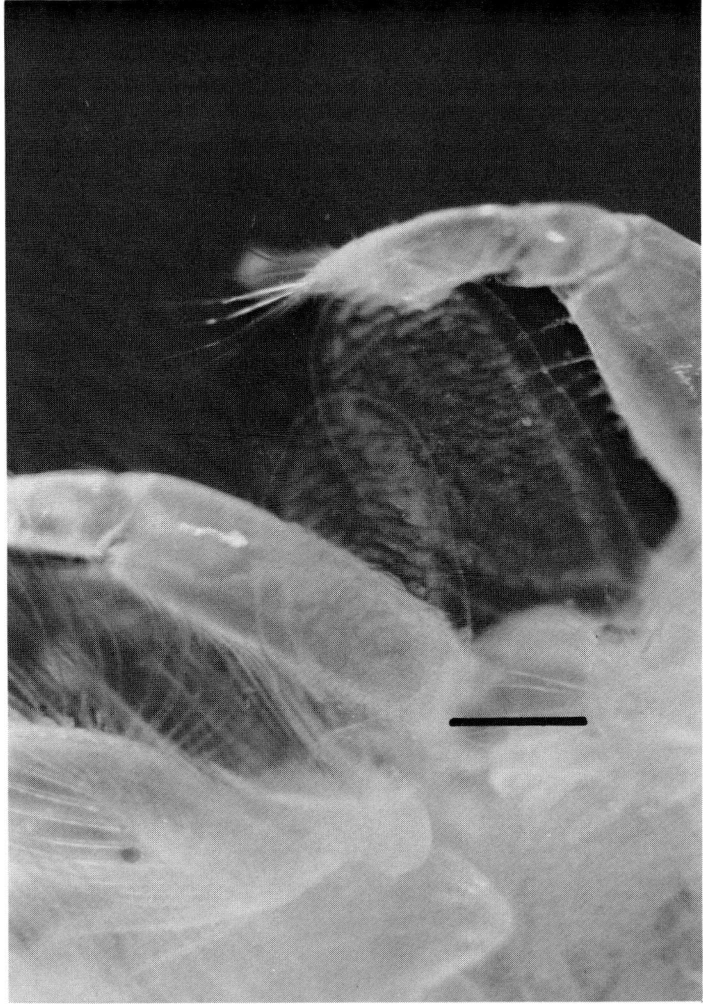

FIGURE 1. The simple lamellar gills of the amphipod crustacean *Corophium volutator*. Scale bar = 0.25 mm.

slight dilution or concentration of seawater and are referred to as being stenohaline, i.e., they have a restricted salinity tolerance. Animals capable of surviving in a wide range of salinities are termed euryhaline and include estuarine and seashore species as well as marine/freshwater migrants such as salmon and eel. Many aquatic animals are capable of maintaining their extracellular fluid at an osmotic concentration which is significantly different from that of the environment over (at least) part of their salinity tolerance range; the osmotic gradient across the body surface generates net water movement which must be compensated by appropriate effector organs. These animals are the osmoregulators, in which regulation of extracellular fluid osmolarity isolates cells from potentially deleterious environmental salinities. Osmolarity is regulated by active control of ionic (principally Na^+ and Cl^-) concentrations and fluxes, as water movements are invariably passive. Mechanisms of ionic regulation fall outside the scope of this article and the interested reader is directed to recent reviews of the subject.[8,9] In the current terminology, regulation refers to control of extracellular fluid composition, and it must be pointed out that euryhaline osmoconformers may have a considerable capacity for intracellular osmoregulation in the face of fluctuations in

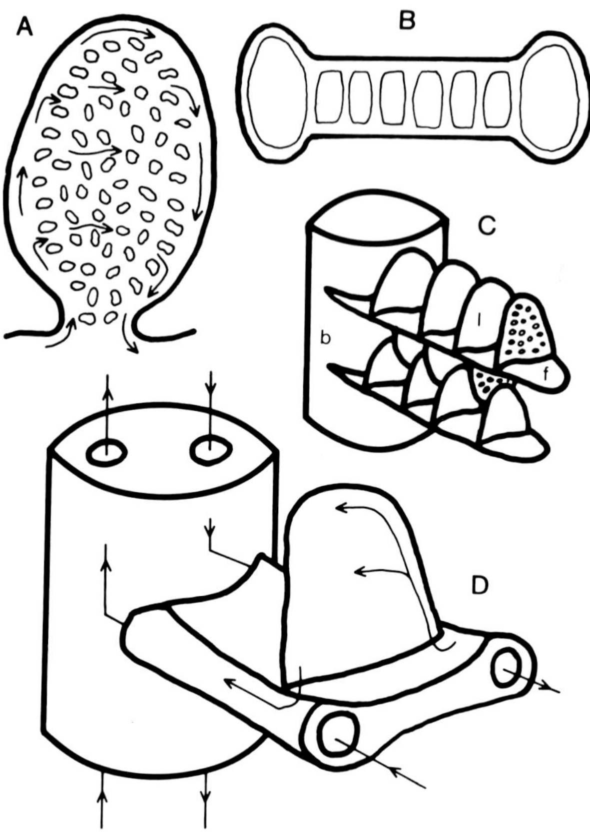

FIGURE 2. Diagrams illustrating basic branchial structure and perfusion pattern. (A) Lateral view of a representative gill lamella (the functional branchial unit) with arrows indicating circulation of extracellular fluid along the marginal canal and between trabeculae. (B) Cross section of gill lamella showing lacunae formed by trabecular elements and a widened marginal canal. (C) Tertiary structure of gills;[2] lamellae (l) may be arranged on filaments (f) borne on a branchial arch (b). (D) Unidirectional circulation through the branchial elements.

extracellular osmolarity.[10,11] Osmoregulators in general show a greater degree of euryhalinity than do osmoconformers. There are two principal categories of osmoregulator:[12]

1. Hyperosmotic regulators, which are hyperosmotic to the medium when acclimated to dilute media but are almost iso-osmotic with seawater. The salinity range over which hyperosmoticity can be maintained varies with species, as does the degree of hyperosmoticity attained. Freshwater animals form a subset of hyperosmotic regulators which rarely tolerate a rise of external osmotic concentration to a level above that normal for their extracellular fluid, i.e., they are comparatively stenohaline. Elasmobranch fish are hypoionic regulators in seawater, but by maintaining high plasma concentrations of urea and trimethylamine oxide (TMAO) they stay slightly hyperosmotic to the marine environment.
2. Hypo-osmotic regulators (or "strong" osmoregulators), which maintain their body fluids hypo-osmotic to seawater, although many are capable of hyperosmotic regulation in dilute media. The capacity for hypo-osmotic regulation is virtually restricted to

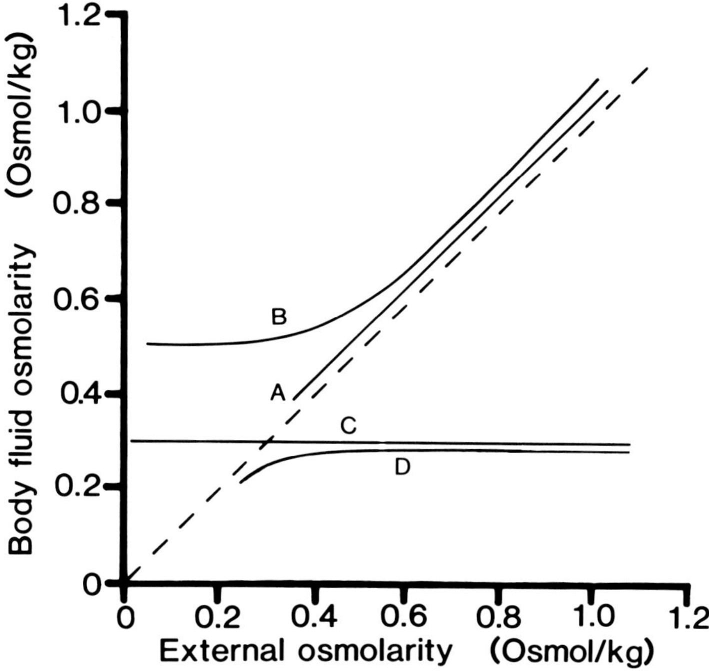

FIGURE 3. The principal patterns of osmoregulation in aquatic animals. The osmotic concentration (osmolarity) of full-strength seawater (≃34.5‰ salinity) is of the order 1.07 Osm kg^{-1} which is equivalent to an osmotic pressure of 2.5 MPa. The principal osmotic effector of environment and extracellular fluid is NaCl. (A) Osmoconformity (the majority of marine invertebrates). (B) Hyperosmotic regulation (most euryhaline invertebrates, all freshwater animals). (C) Hyper/hypo-osmotic (or strong) regulation (euryhaline fish and few Crustacea). (D) Hypo-osmotic regulation (stenohaline marine fish). Highly euryhaline species (e.g., the branchiopod crustacean *Artemia salina*) may tolerate salinities close to that of crystallizing brine (i.e., >600% SW).

euryhaline and marine teleost fish and to certain branchiopod and decapod crustaceans, and enables many of these animals to maintain a constant body fluid composition in all but the extremes of environmental salinity.

In general, hyperosmotic regulators excrete excess water as urine while hypo-osmotic regulators balance passive loss of water with absorption across the wall of the intestine. Absorption or clearance of free water requires considerable expenditure of energy to fuel solute transport mechanisms. It is therefore of energetic advantage for an osmoregulator to adopt the lowest surface permeability compatible with its respiratory requirements. Branchial water transfer linked to ion transport may be of importance in (1) volume regulation[13] and (2) the uptake of water under iso- or hypo-osmotic conditions to produce either urine for excretory purposes[14] or in the Arthropoda, expansion at molt.[15]

The study of gill water exchanges permits comparisons of branchial permeability coefficients, enabling functional classification of the epithelium to be made and affording correlation between membrane permeability and cellular structure as a function of the environment.

II. THEORETICAL CONSIDERATIONS OF WATER PERMEABILITY

A. Diffusion and Permeability

Structural adaptations facilitating dissolved-gas exchange are generally assumed to be incompatible with solute and water impermeability. Gases, water, and many solutes traverse the gill integument by diffusion only, the rate of which is directly proportional to the exchange surface area and inversely proportional to the thickness of the diffusion barrier. Passive movement of water and other nonelectrolytes across the gill is unaffected by changes in transepithelial electrical-potential difference (TEP), the only driving force being the gradient of chemical potential. In a steady state, the net diffusional flux (J^{net}; mol sec^{-1}, where a positive value indicates a net flux into the animal and a negative value a net flux out of the animal) of a nonelectrolyte moving independently across the branchial barrier may be described by the relation

$$J^{net} = -AD(C^{in} - C^{out})/\chi \tag{1}$$

where A is the exchange area (m^2), D is the barrier diffusion constant ($m^2\ sec^{-1}$), C^{in} and C^{out} are the internal and external chemical concentrations (strictly, chemical activity), respectively (mol m^{-3}), and χ is the barrier thickness (m). The ratios D/χ and $(C^{in} - C^{out})/\chi$ represent, respectively, the permeability coefficient (P; m sec^{-1}) and the gradient of potential energy across the barrier. In practice, D is not accessible to measurement and χ may vary spatially over the branchial surface so P is determined experimentally; P incorporates all the inherent factors of a barrier/traversing-nonelectrolyte system and will determine the probability of any molecule actually crossing the barrier.[1] For a homogeneous, symmetrical membrane $P = D^m K/\chi$, where D^m is the molecular diffusion coefficient within the membrane (dependent on membrane viscosity and molecular size) and K is the dimensionless local lipid-water partition coefficient. The pattern of gill microcirculation in many species shows temporal variation, however, one result being that the exchange area A cannot be regarded as a constant. The lumped parameter AD/χ (the "transfer factor" of respiratory physiology) may therefore be considered as a practical permeability, being a function of the ratio effective area/effective thickness for the branchial barrier. From Equation 1 it can be seen that this factor is equivalent to the net flux per unit concentration difference, i.e., the diffusion conductance.

It is evident that the evolution of a large, thin area of branchial integument for efficient gas exchange has resulted in a general increase in surface permeability.

B. Water Movements: Pressure and Concentration

Water molecules do not move independently in aqueous solution due to the high molar concentration, and the potential gradient for water flow is generally expressed in terms of a difference in solution pressure. The SI pressure unit, the Pascal (Pa), is used here. The osmotic pressure of a solution, a colligative property related to the mole fraction of solute, is often expressed in terms of osmotic concentration (the osmotic concentration of a 1-molal solution of ideal solute is 1 Osm kg^{-1}). It is assumed for both practicality and generality that physiological solutions are ideal and dilute; thus an osmotic concentration of 1 Os kg^{-1} is equivalent to an osmotic pressure of 2.3 MPa. Corrections for solute activity are rarely applied in studies of branchial water flux, although it should be borne in mind that this may lead to an overestimation of effective osmotic pressure. Volume flow is proportional to a difference in osmotic or hydrostatic/hydraulic pressure and, as the hydraulic and osmotic conductivities are equal to one another,

$$J^v = -A\ L_p(\Delta\rho - \Delta\pi) \tag{2}$$

where J^v is the volume flux of water (m^3 sec^{-1}), Lp is the conductivity (m sec^{-1} Pa^{-1}), $\Delta\rho$ is the hydrostatic pressure difference (Pa), and $\Delta\pi$ is the osmotic pressure difference (Pa); here

$$P = L_p \, RT/v^{\omega} \tag{3}$$

where R is the gas constant (8.3 m^3 Pa mol^{-1} K^{-1}), T is the absolute temperature (K), and V^{ω} is the partial molar volume of water ($\equiv 1.8 \times 10^{-5}$ m^3 mol^{-1}).

The branchial epithelium is not completely impermeable to solutes and an osmotic volume flux may not be independent of solute fluxes. Microscopic osmotic gradients may be developed across epithelial membranes by active transport of solute to drive transepithelial water movement under essentially iso-osmotic conditions (e.g., the standing and nonstanding gradient models of fluid transport[16-18]). Furthermore, there may be coupling of the diffusional fluxes of solute and solvent. The extent of solute-solvent interactions may be described by a dimensionless reflection coefficient (σ) such that

$$J^v = A \, L_P(\Delta\rho - \sigma\Delta\pi) \tag{4}$$

where in general $O \leq \sigma \leq 1$, the boundary conditions being $\sigma = 1$ for a nonpermeating solute and $\sigma = 0$ for a solute indistinguishable from water molecules.[19] This type of coupling implies a common pathway for solute and water and may be interpreted as a hindering of osmotic flux by frictional interaction with the reciprocal net flux of solute. Values of branchial Lp calculated from osmotic flux data may therefore be overestimates, as gills are known to be permeable to NaCl, the major solute present. The correction is rarely applied, however (i.e., σ is assumed to be unity), and indeed, there is some evidence that water and solutes traverse the gills of crustaceans and fishes by separate pathways.[20,21] Nevertheless, Lp derived from osmotic studies should be interpreted with caution.

Animals rarely maintain osmotic pressure differences balanced by hydrostatic pressures as is the case with plants; animal osmoregulation is essentially regulation of solute concentrations, and transmembrane osmotic concentration differences occur where either the membrane is impermeable to water or the solution compositions are maintained by physiological processes in the face of water movement along its potential gradient. The hydrostatic/hydraulic pressure of the circulatory fluid (fluid pressure) in aquatic animals rarely exceeds 5 kPa, while the transintegumental osmotic pressure difference in osmoregulators may attain a value of the order 1.5 MPa ($\simeq$700 mOsm kg^{-1}). In general, the effect of fluid pressure on osmotic movement across the gills is, or is assumed to be, negligible. The osmotic concentration difference can be expressed in terms of a difference in mole fraction of water or solute: for net branchial water flux, Equation 2 may be rewritten as

$$J^v = A \, P[(M^{out} - M^{in})/M^{out}] \tag{5}$$

where M^{in}, M^{out} represent the internal and external mole fractions of water, respectively; M = 55.556/(55.556 + S), S being the osmolarity of the solution (Osm kg^{-1}).

III. EXPERIMENTAL TECHNIQUES AND THEIR APPLICATIONS

A. Experimental Preparations

Water permeability of gills has been studied at several experimental levels: (1) whole animal studies,[4,5,7] (2) experiments using in vitro perfused preparations of isolated gills or whole fish heads,[22,23] (3) isolated, ligatured gills,[24,25] and (4) single branchial lamellae mounted as a "double-epithelium" preparation between Ussing-type half chambers.[26] The type of study used depends on whether whole animal osmoregulation or membrane permeability is of particular interest; each has distinct methodological problems, many of which are considered below, and the transformation of raw data on water flux into branchial permeability must be undertaken with great caution. There is considerable hormonal control of water exchanges in fish, therefore branchial permeability is usually studied using in vitro preparations. It is important to recognize that there is generally a significant net water flux across the gills even when an animal/medium system is in a steady state, the flux being balanced by that through other osmoregulatory organs (e.g., kidney).

B. The Branchial Exchange Area

It is necessary to obtain a quantitative estimate of the exchange surface area of the branchial barrier in order to determine water permeability. This area is calculated from morphometric measurements of the gills.[27,28] The branchial area bears an allometric correlation with body weight intraspecifically and can be related directly to oxygen requirement (i.e., animal activity) interspecifically, although the morphometry of gills in some species may alter in relation to oxygen or solute availability.[2] The allometric relationship between water flux and body weight in teleost fish may be correlated to that between gill area and weight, from which it has been concluded that water flux occurs principally over the gill integument.[29] Increase in lamellar area is more important than development of new filaments during animal growth, where filament elongation leads to an increase in interlamellar distance.

A useful method of determining the relative permeability of different regions of the body surface is silver staining;[30] the animal is exposed to a silver chloride solution and the subsequent staining density of a particular area is considered to be directly proportional to the (chloride ion) permeability. The comparative importance of the gills as sites of water exchange has been demonstrated acutely in crustaceans both by application of a hydrophobic coat to the branchial surface[7] and by gill ablation;[31] in both types of experiments, the rate of water exchange decreases by an order of magnitude after treatment, indicating that the gills represent over 90% of the permeable surface area.

Amphibian gills have received little attention from physiologists, but they are known to be involved in ion regulation, particularly when the transport systems of the skin are poorly developed. The external gills of the aquatic urodele *Necturus* are thought to account for between 50 and 60% of gas (and perhaps, therefore, water) exchange.[31] There are discrete vascular shunts at the gill bases in amphibians and the relative perfusion of gills and lungs may be regulated to optimize gas exchange in a bimodal system. Regulation of gill perfusion may also be important as a control mechanism for water exchanges in amphibians and fishes, as it is known that gill perfusion is reduced if blood pO_2 is high.[32]

It is now recognized that the complex structure of gills, in particular fish gills, does not allow for the treatment of water exchanges as diffusion across a homogeneous membrane. There may be functional separation of the gas and ion exchange areas in a gill. In fish, the osmoregulatory chloride cells are virtually restricted to the primary lamellae. There are functional differences between anterior and posterior gills in crabs, where ion-transporting cells (10 to 20 μm thick) are concentrated in patches on posterior gill pairs interspersed with the flattened respiratory cells (1 to 2 μm thick) common to every gill pair.[33,34] The simple lamellar gills of amphipod crustaceans are composed of a single, multifunctional type of cell.[35] The functional vascularization of the fish gill has now been described, allowing for separation of the fluxes across the respiratory and nonrespiratory epithelium,[20] i.e., the secondary lamellae, representing the greater part (≃96%) of the exchange surface and the primary lamellae (≃4% of the exchange surface), respectively. It has been established for gill tissue exhibiting regional specialization of function that water exchanges occur principally through the respiratory cells. The branchial respiratory epithelium is composed of flat cells which in most cases are linked by deep intercellular junctions, and it is suggested that the cell membranes, rather than these tight junctions, are the limiting factor of branchial water permeability (i.e., water traverses the gill by a transcellular route).[20]

In fishes, the respiratory epithelium covering the free part of the secondary lamellae has an exclusive relationship with the arterial vasculature of the gill, the primary lamellae having a close association with the venous compartment. The apparent exchange area may be subject to circulatory limitations on the basal surface in fishes and invertebrates.[2,36,37] Water fluxes are generally normalized to the external area of the gill integument, although in fact it may be an underlying membrane surface which is rate limiting for water transfer. The permeability properties of cuticle, endothelium, and overlying mucus are frequently overlooked. The chitin of arthropod cuticle is much more permeable to water than to salt;[26] calcification of crustacean integument appears to reduce surface permeability and indeed there is little calcification of the comparatively permeable gill integument.

C. Measurement of Branchial Water Exchange

Water permeability may be determined either from the rate of diffusion of tracer water or from the magnitude of an osmotic water flux. Diffusional and osmotic fluxes may be expressed directly in terms of the branchial exchange area when working with isolated preparations, but are generally normalized to a weight or water space when there is uncertainty about the effective magnitude of the branchial area.

1. Osmotic Permeability (P^{os})

a. Direct Measurements of P^{os}

The osmotic water flux across a gill may be measured directly, as a volume change in the animal or medium compartment, or indirectly, in whole animal studies as the algebraic sum of the water fluxes through the organs effecting osmoregulation in the steady state. Direct measurement of branchial osmotic flux generally necessitates a disruption of the steady state, either by an experimental change in the transintegumental gradient of osmotic concentration, or by effective isolation of the gills from other organs of osmoregulation. Net volume flux in whole animal studies is measured either as a change of body weight or as a direct volume change in the water ventilating the branchial region (in which case a steady-state flux can be measured). The initial rate of change of body weight in an animal subjected to osmotic transfer can be used to estimate water permeability if the osmotic gradient across the gills is known (Figure 4). This approach assumes that the concentration/dilution of system compartments by net water movements is negligible over the initial stages of an experiment, or alternatively, a mean osmotic gradient for this period may be calculated. It is also assumed that the gills are the sole site of water exchange and the osmoregulatory mechanisms of the animal are ineffective immediately following transfer (the excretory openings may be blocked to achieve this end[37,42]).

Kirsch and co-workers[5] restrained live eels by a membranous girdle separating the two chambers of an apparatus which afforded independent irrigation of the head and tail regions combined with precise determination of the volume of recirculated irrigation fluids. Volume changes in the fluid bathing the eel head reflect the sum of branchial osmotic flux and drinking (imbibition); simultaneous measurement of drinking rate by a radiotracer technique enables the osmotic component of volume change to be isolated and branchial permeability to be calculated. The fluid volume perfusing a gill or ventilating a branchial chamber in recirculating preparations may be monitored similarly to yield the branchial permeability. The major problem with perfused preparations when used for osmotic studies is that of perfusate leakage.

Studies with isolated, unperfused gills utilize simple osmometry; the base of the excised gill is ligatured after flushing with the appropriate internal experimental medium, the closed gill ''bag'' is immersed in a bathing medium, and any changes in internal (gill) fluid volume are determined either directly from the length of a fluid column in a micropipette tied into the gill base[6] or assessed from measurements of gill weight[24] or concentration of a nonpermeating indicator substance in the internal medium.[25]

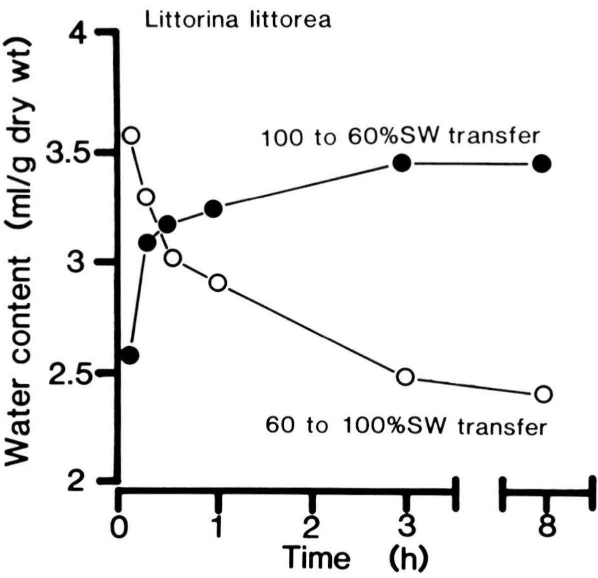

FIGURE 4. Direct measurement of osmotic water permeability (P^{os}) in the osmoconforming mesogastropod mollusk *Littorina littorea* (unpublished data of P. M. Taylor and E. B. Andrews). Animals acclimated to 100 or 60% seawater are transferred to the other salinity and the change in body water content with time is followed. Water content is normalized to body dry weight, as there are great differences in tissue hydration between animals acclimated to the two salinities. There is little osmoregulation, although passive net movement of NaCl across the integument does occur, and the body swells or shrinks following hypo-osmotic or hyperosmotic salinity shock, respectively, until a new osmotic steady state is established. The initial transintegumental osmotic concentration gradient is 0.4 Osm kg^{-1} (≡0.92 MPa)[38] and the specific net volume osmotic flux (m^3 g^{-1} dry wt sec^{-1}) may be estimated from the initial change in water content after transfer. The gills of gastropods represent the principal region of surface permeability[39] and have an area of the order 2.7×10^{-3} m^2 g^{-1} dry wt,[40] allowing branchial osmotic conductivity (Lp) to be estimated from Equation 2 and subsequently P^{os} from Equation 3. P^{os} for influx (100 to 60% SW transfer) and efflux (60 to 100% SW transfer) are 3.3×10^{-5} and 2×10^{-5} m sec^{-1}, respectively. Rectification of water permeability is described frequently in biological systems[20,41] and these results cannot be taken in isolation as evidence for reduced branchial permeability of *Littorina* with dilution of the environment.

b. Indirect Measurements of P^{os}

Indirect determination of the net osmotic flux in whole animals necessitates measurement of the rates of drinking (V^i) and urine flow (V^u).

Drinking of the external medium is usually quantified by measurement of the rate of uptake into the gut of medium marked with an inert, nonpermeating dye (e.g., amaranth)[43] or radiomolecule (e.g. [^{51}Cr] EDTA, [^{125}I] PVP):[21,44] here $V^i = M^t/(tM^m)$, where M^t is the quantity of marker imbibed (g^{-1} animal) at time t (hr) and M^m is the external marker concentration ($m\ell^{-1}$ medium); hence, V^i is expressed in units of $m\ell\ g^{-1}\ hr^{-1}$. It is assumed generally that marker associated with the animal is restricted to the gut. Certain markers may become adsorbed onto the integument, however, and this must be corrected for to avoid overestimation of drinking rate. Drinking rate may also be measured by esophageal cannulation in larger species.[5]

The rate of urine flow (mℓ g^{-1} animal hr^{-1}) may be measured directly by urine collection following the insertion of a cannula into the excretory opening(s). This method of measurement is highly invasive, however, and animal handling during cannulation may in itself affect the rate of urine flow. An alternative approach is to measure initially the rate of urine formation, determined in general as the clearance rate of an injected filtration marker (e.g., inulin, [^{51}Cr] EDTA)[45] from the extracellular fluid (ECF) of the experimental animal. The rate of clearance (V^{cl}; mℓ g^{-1} animal hr^{-1}) is calculated from either the rate of disappearance from the ECF, or the rate of appearance in the medium, of the filtration marker. The rate constant for clearance k^{cl} (hr^{-1}) may be determined as the slope of a plot of $\ln C^t$ against time (hr), where C^t is the marker concentration of the ECF at time t; $V^{cl} = k^{cl} E$, where E is the ECF volume of the animal (mℓ g^{-1} animal). The volume E is taken as the marker distribution volume; $E = C^o/(QW)$, where C^o is the concentration of marker (mℓ$^{-1}$ ECF) at the commencement of a clearance experiment, Q is the quantity of marker injected and W is the animal weight (g). The rate of urine flow may then be estimated from the relation $V^u = V^{cl} F^{in}/F^u$, where F^{in}/F^u is the steady-state concentration ratio of marker between ECF and urine (this value is not necessarily unity, as water may be added to or withdrawn from the urine postfiltration).

The resultant calculated net osmotic flux, that is $V^u - V^i$, is subject to potential error as it is derived from two independent variables on certain rather broad assumptions, i.e., the animal is in water balance, all imbibed water is absorbed, and the quantity of urine derived from metabolic water and fecal water loss are negligible. Nevertheless, it provides an extremely useful estimate of the actual flux from which an operational water permeability may be obtained. Errors may be particularly high if P^{os} is determined indirectly for osmoconformers, where the transintegumental osmotic gradient is low and fluid plus colloidal osmotic pressure effects may be considerable.[46]

2. Diffusional Permeability (P^d)

Unidirectional diffusional water flux across the gill integument is measured by introducing a small (tracer) quantity of labeled water ([^{2}H] or [^{3}H] water) into an experimental system and following its movement between the component compartments of the system. In the simplest model, the system is conveniently expressed in terms of two fluid-filled compartments, the animal body fluid and the external medium, separated by the integumental barrier. It is generally assumed that the entire water pool of each compartment is freely exchangeable with that of the other compartment and with the tracer water. This simple approach is not an authentic representation of the real system; a whole animal consists of at least three major fluid-filled compartments, the intra- and extracellular spaces and the intestinal lumen (limited by certain membranes which are likely to have different permeability characteristics to the integument). At the level of the epithelial cells, fluxes across apical and basal membranes must be considered and account taken of the intraepithelial water pool. In addition, a certain proportion of water in the body fluid pools may not be freely exchangeable as it becomes sequestered by macromolecules or organized into quasi-crystalline arrays. Nevertheless, branchial diffusion is generally rate limiting for whole body water exchange with the medium, but at least the two-compartment model suffices for whole animal studies. Modeling of a three-compartment system is considered elsewhere.[47] The tracer water concentration is low and tracer molecules can be treated as an uncharged solute. Tracer movements between body fluid and medium in the two-compartment model are described by a single exponential function (i.e., a first-order rate equation) and may be related directly to diffusional water fluxes: for an intact animal,

$$P^d = k[H_2O]^{in}/A^s \qquad (6)$$

where k (sec^{-1}) is the rate constant (alternatively transfer coefficient) for tracer water flux, representing the fractional turnover rate of body water, $[H_2O]^{in}$ is the body water content ($m^3\ g^{-1}$) and A^s is the weight-specific branchial area ($m^2\ g^{-1}$). The rate constants for influx and efflux (k^{in}, k^{out}, respectively) may be determined sequentially. To measure k^{in}, the animal is placed in a large volume of medium containing tracer water, and tracer appearance in the animal is followed until the system reaches equilibrium: the rate constant is determined from the relation

$$Q^t = Q^{eq} (1 - \exp(-k^{in}\ t)) \tag{7}$$

where Q^t and Q^{eq} are the quantities of tracer in the animal at time t and at equilibrium, respectively; a plot of $-\ln (1 - Q^t/Q^{eq})$ against t has a slope k^{in}. Tracer is then washed out of the animal in a second experiment from which k^{out} is determined according to the relation

$$Q^t = Q^{eq}\ \mathrm{esp}(k^{out}\ t) \tag{8}$$

where k^{out} is the slope of a plot $\ln(Q^t/Q^{eq})$ against t (Figure 5). A rate constant (k) is expressed alternatively as the half-time for exchange ($t^{0.5}$), where $t^{0.5} = 0.693/k$. Animal tracer content is often calculated indirectly from the proportion of experimental-system tracer in the medium as this minimizes handling effects, although rate constants for flux in terms of medium water turnover can be determined directly from changes in size of the external tracer pool.[4] Tracer movement may underestimate the actual water flux by some 5 to 10% due to isotopic disparity between the diffusion rates of $[^3H]$, $[^2H]$, and $[^1H]$ water.[48]

Diffusional water permeability of the gill in perfused preparations may be determined from branchial clearance of tracer water from the perfusate. An isolated trout head preparation affords separate study of transbranchial water fluxes in lamella and filament; the efferent arterial and venous perfusates are collected separately after passing over lamellar and filamental epithelia, respectively.[20,23] This arrangement affords separate analysis of water exchange across the respiratory and osmoregulatory epithelia using an appropriate in vitro preparation. The large respiratory surface area represents the major site of passive permeability. These studies appear to demonstrate that the basal membrane is about eight times less permeable than the apical one and constitutes the limiting barrier for water (and ion) diffusion in the gill. This arrangement has been suggested to be inconsistent with the exchange role of the gill, and an alternative interpretation of the branchial water clearance data might indicate that the apical membrane was rate limiting for transfer.[49] The intracellular water is also evident as an intermediate compartment for trans-gill water exchange, although only a small fraction of the intracellular water pool ($\simeq 0.13$) participates in the transepithelial flux.

D. Analytical Problems

The diffusional water permeability (P^d), as measured by unidirectional labeled water fluxes, and permeability calculated from net water movements in the same system when as osmotic gradient is present (P^{os}), have been found to differ, often markedly, in a wide range of epithelial tissues. Explanations to account for these differences have implied that either hydrophilic pores are present in the epithelium or that there are unstirred layers on one or both surfaces of the epithelial barrier. These theories have been developed and reviewed extensively.[41,50]

1. Water Transfer Through Pores

If aqueous pores of certain dimensions (>0.3 nm in diameter) are present in an epithelium, the net movement of water across the epithelium by bulk flow through these pores when a transepithelial osmotic pressure gradient is applied, is greater than that predicted by a purely

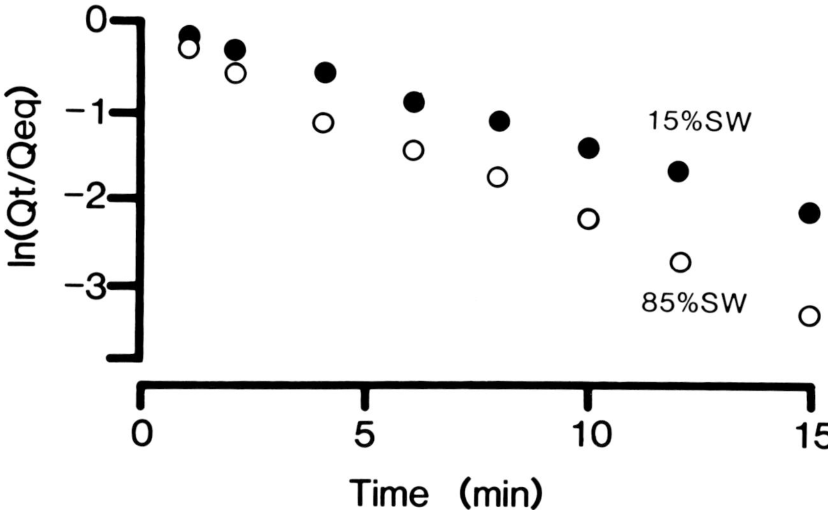

FIGURE 5. Measurement of diffusional water permeability (P^d) in the euryhaline amphipod crustacean *Corophium volutator* using the radioactive tracer ^3H_2O (unpublished data of P. M. Taylor and R. R. Harris). Results are shown for ^3H_2O washout from animals acclimated to 85 and 15% seawater; the radioactivity of animals preloaded with ^3H_2O to isotopic equilibrium is displayed as a function of time. The slope of the semilogarithmic plot represents the rate constant for efflux (k). The lamellar gills represent the principal exchange surface with an area of 9.9×10^{-4} m^2 g^{-1} wet wt[36]; body water content is 8.5×10^{-7} m^3 g^{-1} yielding from Equation 6 values for P^d of 3.6×10^{-6} m sec^{-1} and 2.4×10^{-6} m sec^{-1} in animals acclimated to 85 and 15% seawater, respectively.[21] The animals acclimated to the low (hypo-osmotic) salinity thus appear to have reduced branchial permeability to water.

diffusional process (i.e., one being a function of P^d). The bulk flow, which is limited by viscous forces as described by the Poiseuille relation, appears to originate as the result of a hydrostatic pressure gradient within a pore, which exists even when there is no such gradient across the epithelium.[51] Water molecules may pass through smaller pores ($\simeq$0.2 nm in diameter) in single file.[52]

2. Unstirred Layers

Unstirred layers persist adjacent to epithelial boundary membranes even in well-mixed systems. A significant solute concentration gradient may develop across the unstirred layer, reducing the true concentration difference across the membrane. In this situation, the diffusion of labeled water across the unstirred layer(s) at the epithelial boundary(ies) may become a limiting factor in the determination of P^d (i.e., the tracer movement underestimates the true diffusional permeability). For this reason, diffusional permeability is often referred to as "apparent water permeability" and changes in P^d are given the cautious label of "apparent permeability changes" (P^d differences may be due to alterations in the size of unstirred layers rather than true changes of permeability). The role of unstirred water layers must be considered carefully when measuring diffusional water flux although the vigorous perfusion and irrigation of gills necessary to maintain adequate levels of gas exchange in most aquatic animals (due to the poor oxygen-carrying capacity of water) will tend to minimize their thickness. An unstirred layer may be considered as a membrane with a permeability coefficient D/δ, where D is the self-diffusion coefficient of water (2×10^{-9} m^2 sec^{-1}) and δ

is the layer thickness. The effect of unstirred layers on diffusional permeability may be expressed by the relation $P^{app} = (P^d[D/\delta])/(P^d + [D/\delta])$, where δ represents the total thickness of unstirred layer across the epithelial layer. Unstirred layers are of importance only if P^d approaches D/δ in magnitude and, although this is not the general case for gills of intact animals,[20] they may affect experiments using isolated, unperfused lamellae which have a comparatively high water permeability.

A comparison of P^{os} and P^d measured in the same species under identical conditions may provide an indication as to the value of tracer flux measurements as indicators of branchial water permeability. The difficulties inherent to calculation of permeability per se have already been discussed, so to minimize the errors associated with an osmotic/diffusional permeability comparison, a variable reflecting permeability (i.e., fP) can be considered; e.g., the specific net volume flux ($m\ell\ g^{-1}\ hr^{-1}$) is readily accessible to measurement by both osmotic and diffusional techniques and is not normalized to exchange area. Net volume osmotic flux (fP^{os}) is determined as described above either from direct volume changes or the sum of osmoregulatory effector fluxes. Volume flux from tracer water exchange (fP^d) is calculated from a derivation of Equation 5 where the term AP is substituted by the equivalent $k[H_2O]^{in}$ (i.e., the transfer factor, Equation 6). The ratio fP^{os}/fP^d represents an index of P^d viability and can be used to test the authenticity of "apparent" permeability changes.

E. Activation Energy of Water Flux

The normal temperature range of aquatic environments (−2 to +40°C) is similar to that in which proteins remain stable and in which enzymatic reactions can proceed optimally. The mechanism of water exchange can be investigated by examining the temperature dependence of k, the rate constant for water flux. The Arrhenius relation states

$$k = A \exp(-E^a/RT) \tag{9}$$

where E^a is the activation energy for water transfer ($J\ mol^{-1}$), A is a constant termed the frequency factor (sec^{-1}), R is the gas constant expressed as 8.3 $J\ mol^{-1}\ K^{-1}$, and T is the temperature (K): a plot of ln k against 1/T should yield a straight line of slope E^a/R. The temperature relations of water flux are expressed conveniently in terms of an empirical temperature quotient Q^{10}, defined as $k^{(T+10)}/k^T$ (k^T, $k^{(T+10)}$ are the rate constants at temperatures T and T + 10, respectively); Q^{10} differs over different temperature ranges and the experimental temperature range is given on each citation. Biochemical processes (e.g., active transport) in general have values of Q^{10} between 2 and 3, whereas physical processes (e.g., diffusion) have lower temperature sensitivities (i.e., Q^{10} closer to unity).

IV. FACTORS INFLUENCING WATER PERMEABILITY

A. Salinity

A wide range of euryhaline crustaceans exhibit apparent reductions in P^d on dilution (Table 1). In addition, strong osmoregulators such as the brine shrimp *Artemia salina*[8] (Branchiopoda) and the decapod *Paleomonetes pugio*[61] display reduced k for water exchange with increase in hyperosmotic salinity, and *P. pugio* at least exhibits low k in dilute media to which it is hyperosmotic; i.e., it appears that permeability is highest when extracellular fluid and medium are close to iso-osmoticity. Branchial osmotic permeability of isolated gills of the euryhaline teleost *Oreochromis (Sarotherodon) mossambicus* is also at its highest in fish acclimated to iso-osmotic saline.[62] Euryhaline teleosts tend to exhibit their lowest values of k in raised salinities (Table 2). These results demonstrate the fundamental impermeability of fish gill epithelium to water.

The changes in k of Crustacea shown in Table 1 are thought to represent authentic

Table 1
DIFFUSIONAL WATER PERMEABILITY (P^d) OF CRUSTACEAN GILL INTEGUMENT IN RELATION TO ENVIRONMENTAL SALINITY

Species	Acclimation medium	k (hr^{-1})	P^d ($m\ sec^{-1} \times 10^6$)	Ref.
Amphipoda				
Gammarus duebeni	100% SW (iso)	5.9	1.53	
	70% SW (iso)	4.2		53, 54
	2% SW (hyper)	2.4	0.66	
Corophium volutator	85% SW (iso)	14.5	3.60	21
	15% SW (hyper)	9.9	2.40	
Crangonyx pseudogracilis	FW (hyper)	4.5	1.20	55
Isopoda				
Sphaeroma serratum	100% SW (iso)	5.9	1.71	56
	50% SW (hyper)	2.5	0.70	
Decapoda				
Carcinus maenas	100% SW (iso)	2.4	0.65	57
	30% SW (hyper)	1.8	0.48	
Pachygrapsus marmoratus	130% SW (hypo)	1.3	0.34	
	75% SW (iso)	1.4	0.37	58
	30% SW (hyper)	1.1	0.29	
Uca pugilator	100% SW (iso)	0.33	0.08	59
	3% SW (hyper)	0.33	0.08	
Astacus fluviatilis	FW (hyper)	0.20	0.07	60
Branchiopoda				
Artemia salina	566% SW (hypo)	0.04		
	100% SW (hypo)	0.09	0.07	8
	38% SW (iso)	0.13		

Note: Iso — hemolymph iso-osmotic to medium. Hyper — hemolymph hyperosmotic to medium. Hypo — Hemolymph hypo-osmotic to medium. SW — seawater (100% SW ≡ 34.5‰ salinity). FW — freshwater (<0.5% SW).

Table 2
BRANCHIAL WATER PERMEABILITY AND P^{os}/P^d RATIO IN FISHES

Species	Acclimation medium	P^d ($m\ sec^{-1} \times 10^{-7}$)	P^{os}/P^d	Ref.
Teleost				
Anguilla anguilla	FW	2.5	3.2	4
	SW	1.8	1.1	
Platichythys flesus	FW	2.7	2.6	4
	SW	1.7	0.8	
Salmo gairdneri	FW	0.77	3.0	63
	SW	0.44	7.0	
Carassius auratus	FW	3.4	3.3	4, 64
Serranus scriba	SW	1.2	1.3	65
Elasmobranch				
Raja erinacea	SW	0.02	4.0	66
Scyliorhinus canicula	SW	0.06	14.5	67

Note: FW — freshwater, in which all species are hyperosmotic regulators. SW — seawater, in which teleosts are hypo-osmotic regulators and elasmobranchs are iso-osmotic/hypoionic regulators.

alterations in the rate of branchial water exchange, not least because P^{os}/P^{d} (or fP^{os}/fP^{d}) ratios are low (of the order 1 to 2) and appear to remain essentially constant with respect to environmental salinity.[58,68,69] These changes are interpreted in terms of absolute changes in branchial water permeability (P), although alterations in factors such as branchial perfusion pattern and epithelial ultrastructure (which might change the effective branchial area rather than permeability per se) must still be considered as conceivable alternatives, or at least as contributary agents.[36,68,70] The magnitude of k (i.e., the proportion of body water exchanged in unit time) is perhaps the most realistic indicator available of the degree of osmotic adaptation to a particular environment by any species. The P^{os}/P^{d} ratio in freshwater fishes is often markedly greater than unity (Table 2), and this has been ascribed to the presence of water-filled channels in the branchial barrier.[20]

The reduction of integumental permeability to water in aniso-osmotic environments may be considered adaptive in that it tends to minimize the energy cost of osmoregulation.[12] It is possible that the maintenance of a comparatively high water permeability by hyperosmotic regulators at high salinities is also of adaptive significance, in that by maximizing water gain along the reduced transintegumental osmotic gradient, the availability of sufficient urine for excretory purposes is ensured.[14]

In *Artemia salina,* k appears to bear an almost linear relation to the osmotic gradient between extracellular fluid and medium, where k varies by a factor of three in the salinity range 38 to 566% seawater. This is not the case with *Gammarus duebeni*, where for animals acclimated to salinities within the range 2 to 150% seawater, the largest change in k occurs in the range 50 to 70% seawater, although the branchial osmotic gradient is small between these salinities (Table 1). Sudden changes in salinity which cause *G. duebeni* to become aniso-osmotic to the medium result in a decrease in k, where the time scale of these apparent changes in water permeability is <1 min.[53] On transfer to a hyperosmotic salinity, the reduced value of k is maintained until iso-osmoticity is reached, when a rapid increase in k occurs. These changes in k are also apparent if *G. duebeni* are maintained in a regime of cycling salinity (2 to 98% seawater), which exposes the animal to hypo- and hyperosmotic environments alternately.[68] The external osmotic concentration in itself does not determine the magnitude of k, as substitution of nonelectrolyte (e.g., mannitol, sucrose) for all or some of the ions in the medium results in a change of k to a value appropriate to the residual ionic concentration. Animals such as *G. duebeni*, which inhabit waters liable to rapid and extensive changes of salinity (e.g., rock and salt-marsh pools reached by spring tides only), might find adaptive a capacity to effect rapid reductions in surface permeability when exposed to aniso-osmotic environments. This would limit net water movement over the branchial surface while osmoregulatory effectors were mobilized.[68] Salinity shock effects appear to be absent in other euryhaline crustaceans studied, which for the most part inhabit estuaries (where salinity changes are comparatively slow and predictable), and which may in addition exhibit adaptive avoidance responses (e.g., burrowing) to unfavorable salinities.

Gradual changes in k may be classified under two headings: (1) those occurring in the initial period of salinity acclimation, where variations in P^{d} parallel adjustments of extracellular osmotic concentration toward a new steady state[71] and (2) long-term effects, possibly associated with modification of membrane lipid composition.[72,73]

Changes in external concentrations of divalent cations may have an important influence on branchial water permeability; depletion of environmental calcium increases water exchange in freshwater fishes[74,75] and certain crustaceans. Increased magnesium concentration of the medium reduces water flux in *Artemia salina*.[8]

There are differences in branchial morphology between fishes adapted to freshwater and seawater. The number of chloride cells and their associated accessory cells increases in seawater by a factor of about three and loose apical junctions appear between them. These ultrastructural modifications, which develop rapidly after salinity transfer, do not change the relative exchange surface of chloride cells, but epithelial permeability is increased by the development of leaky junctions.[20] In euryhaline Crustacea also, there are ultrastructural differences between the gill epithelium of high and low salinity acclimated animals, which may be associated with phenotypic differences in integumental transport capacity.[33,35]

B. Temperature

There is evidence that the permeability of a membrane is related to its fluidity, which is dependent on lipid composition and physical factors such as temperature.[49,76] Poikilotherms are known to increase the degree of membrane lipid unsaturation with decrease in environmental temperature so as to retain optimum fluidity (and, presumably, permeability).

The Q^{10} for water influx in the crab *Hemigrapsus nudus* is of the order 1.6 in the temperature range 10 to 20°C[77] and in isopods of the genus *Mesidotea* varies between 1.7 and 2.7 in the temperature range 0 to 10°C.[78] The Q^{10} for water flux in a variety of teleost fish is of the order 1.9 in the temperature range 10 to 20°C.[29] Q^{10} appears to decline with salinity in *Uca pugilator*[59] and with temperature in *Cyathura carinata*[8] and *Mesidotea* species.[78] In general, these results are consistent with diffusional transfer of water across the gill integument.

C. Hormones

Studies of the control of branchial water permeability in fishes have been concerned mainly with the effects of hormones released by the pituitary adenohypophysis (prolactin, adrenocorticotrophic hormone (ACTH)) and the adrenals (corticosteroids, catecholamines). The secretion of cortisol, the principal fish corticosteroid, is controlled by ACTH. Published results highlight the difficulty of isolating primary endocrine effects on permeability from indirect effects, which may be due to modulation of other physiological control mechanisms, particularly in the case of whole animal studies. Hypophysectomy of freshwater, euryhaline, or marine fishes results in a reduction of the rate of water turnover, due apparently to decreased branchial water permeability.[67,67,74] Administration of prolactin, cortisol, or ACTH may restore permeability to control levels, but in the case of prolactin there is an apparent discrepancy between effects on branchial water permeability in vivo and those in vitro, where in the latter, prolactin reduces osmotic permeability.[75] Prolactin is of particular importance to fish in freshwater, where its principal role may be to regulate sodium balance.[79] Prolactin action is related to the calcium concentration of freshwater, which in itself has an important influence on branchial water permeability; increase of environmental calcium reduces water flux across the gill integument.[74,75] Receptors for glucocorticoids (e.g., cortisol) have been identified in the gill tissue of fish,[80] providing further evidence for hormonal control of osmoregulation. Arginine vasotocin, a neurohypophysial hormone, induced large reductions in the rate of water turnover in both intact and hypophysectomized goldfish.[64] The catecholamines, adrenaline and noradrenaline, increase water permeability of fish gill epithelium, but it is not known to what extent this is as an indirect result of their hemodynamic actions, which might increase the effective branchial exchange area via lamellar recruitment. It appears at present that the greater part of the catecholamine effect represents an increase in membrane permeability.[49,81] Heavy exercise or handling stress in fishes leads to release of catecholamines with resultant increase of branchial permeability. This may be a secondary result of an adaptive increase in diffusional capacity for gas exchange,[82] where over short periods of maximal activity osmoregulation becomes subordinate to the respiratory requirement for oxygen.[49] Little is known of the mechanisms underlying hormonal control of branchial water permeability.

Studies on decapod crustaceans indicate that there is some neurohormonal control of branchial water permeability[83,84] and of fluid permeability in tissues such as the foregut of *Gecarcinus lateralis*[85] and the coelomosac wall of the excretory organ in *Carcinus maenas*.[86]

The principal neuroendocrine centres in Crustacea are the brain, eyestalks, and thoracic ganglionic mass (TGM). Fractions of brain and TGM homogenates alter P^d in *Thalamita crenata*.[84] Aqueous extracts of *C. maenas* TGM caused a decrease in ^3H_2O influx across its isolated, perfused gill, particularly when extract from animals acclimated to low salinity was tested on the perfused gill of a crab acclimated to high salinity;[83] this corresponds well with in vivo studies of water flux. There is little direct evidence that handling or other stress factors alter water permeability of crustacean gills.[59]

D. Molt

Transintegumental water flux approximately doubles at molt in crustaceans such as the amphipod *Gammarus duebeni* and the isopod *Idotea linearis*.[87] This may be obligatory for a process which involves temporary loss of the hard, comparatively impermeable cuticle covering the general body surface and does not necessarily represent a marked increase of branchial permeability per se. The increase in flux appears to be independent of the external salinity for *G. duebeni*. The flux returns to the control level within 5 days of molt. Inward sodium transport across the gill integument is also stimulated at molt (perhaps under the influence of a molting hormone), and this may effect fluid uptake for expansion as the old exoskeleton is cast off.[15]

E. Pollutants

The most severe chemical pollutants can be divided into three groups: heavy metals, pesticides and polychlorinated biphenyls, and oil and dispersants. There may be considerable disruption of osmoregulatory processes in fishes and crustaceans following laboratory exposure to potential pollutants.[88] In many cases, this disruption can be ascribed to inhibition of active transport mechanisms on the branchial epithelium. Organochlorine pesticides and heavy metals are known to inhibit the enzyme $Na^+ + K^+$ ATPase, the activity of which is often related directly to branchial ion transport capacity. Organic pollutants may alter cell membrane configuration by binding to the lipid component, which as well as impairing the activity of membrane-bound ion transporters, tends to increase passive permeability to water[70,89] and salts. Salts of heavy metals such as cadmium, copper, mercury, and zinc are also reported to cause damage to the branchial epithelium. The toxicity of pollutants to aquatic animals must be considered in relation to synergistic effects of other environmental variables, where breakdown of processes other than osmoregulation may be limiting for survival.

V. SUMMARY

The osmoregulatory adaptation of reducing surface water exchanges in aniso-osmotic salinities may be common to many euryhaline animals. However, the practical and theoretical problems of elucidating the exact nature of these reductions are yet to be fully overcome. The complexities of gill structure have hindered the development of an experimental preparation where membrane permeability can be studied independently of hemodynamic and other factors. It is unfortunate that the simpler branchial designs are, in general, confined to extremely small animals which offer no convenient in vitro study basis. The use of cultured gill cells[49,90] or plasma membrane vesicles[91] may provide alternative strategies for the study of branchial water permeability, a subject of great interest at both environmental and physiological control levels.

ACKNOWLEDGMENTS

I wish to thank Drs. R. R. Harris and E. B. Andrews for advice and discussion, and for allowing me to present previously unpublished details of our work supported by SERC grants

78307501 and GR/C/21953. I am indebted to my family for their support and considerable assistance in manuscript preparation, and to Professor A. P. M. Lockwood for my initiation into the current project. I am grateful to Mr. A. Robertson for help with the figures and Professor M. J. Rennie for continued support.

REFERENCES

1. **Kirschner, L. B.,** Physical basis of solute and water transfer across gills, in *Gills, Society for Experimental Biology Seminar,* Series 16, Houlihan, D. F., Rankin, J. C., and Shuttleworth, T. J., Eds., University Press, Cambridge, 1982, 63.
2. **Hughes, G. M.,** An introduction to the study of gills, in *Gills, Society for Experimental Biology Seminar,* Series 16, Houlihan, D. F., Rankin, J. C., and Shuttleworth, T. J., Eds., University Press, Cambridge, 1982, 1.
3. **Burggren, W. W., McMahon, B. R., and Costerton, J. W.,** Branchial water- and blood-flow patterns and the structure of the gill of the crayfish *Procambarus clarkii, Can. J. Zool.,* 52, 1511, 1974.
4. **Motais, R., Isaia, J., Rankin, J. C., and Maetz, J.,** Adaptive changes of the water permeability of the teleostan gill epithelium in relation to external salinity, *J. Exp. Biol.,* 51, 529, 1969.
5. **Kirsch, R., Guinier, D., and Meens, R.,** L'equilibre hydrique de l'anguille europeene *(Anguilla anguilla L.).* Etude du role de l'oesophage dans l'utilisation de l'eau de boisson et etude de la permeabilite osmotique branchiale, *J. Physiol. (Paris),* 70, 605, 1975.
6. **Bielawski, J.,** Chloride transport and water intake into isolated gills of crayfish, *Comp. Biochem. Physiol.,* 13, 423, 1964.
7. **Thuet, P.,** Les transferts d'eau en fonction de la salinite du milieu chez le Crustace isopode *Sphaeroma serratum* (Fabricius), *Arch. Int. Physiol. Biochem.,* 86, 1011, 1978.
8. **Lockwood, A. P. M.,** Transport and osmoregulation in Crustacea, in *Transport of Ions and Water in Animals,* Gupta, B. L., Moreton, R. B., Oschman, J. L., and Wall, B. J., Eds., Academic Press, New York, 1977, 673.
9. **Evans, D. H.,** The roles of gill permeability and transport mechanisms in euryhalinity, in *Fish Physiology,* Vol. 10, Hoar, W. S. and Randall, D. J., Eds., Academic Press, New York, 1984, 239.
10. **Rankin, J. C. and Davenport, J.,** *Animal Osmoregulation,* Blackie, Glasgow, 1981.
11. **Mantel, L. H. and Farmer, L. L.,** Osmotic and ionic regulation, in *The Biology of Crustacea,* Vol. 5, Bliss, D. E. and Mantel, L. H., Eds., Academic Press, New York, 1983, 54.
12. **Kirschner, L. B.,** Control mechanisms in crustaceans and fishes, in *Mechanisms of Osmoregulation in Animals,* Gilles, R., Ed., John Wiley & Sons, New York, 1979, 157.
13. **Lockwood, A. P. M.,** The involvement of sodium transport in the volume regulation of the amphipod crustacean *Gammarus duebeni, J. Exp. Biol.,* 53, 737, 1970.
14. **Lockwood, A. P. M. and Inman, C. B. E.,** Water uptake and loss in relation to the salinity of the medium in the amphipod crustacean *Gammarus duebeni, J. Exp. Biol.,* 58, 149, 1973.
15. **Lockwood, A. P. M. and Andrews, W. R. H.,** Active transport and sodium fluxes at moult in the amphipod *Gammarus duebeni, J. Exp. Biol.,* 51, 591, 1969.
16. **Diamond, J.,** Standing-gradient model of fluid transport in epithelia, *Fed. Proc. Fed. Am. Soc. Exp. Biol.,* 30, 6, 1971.
17. **Sackin, H. and Boulpaep, E. L,** Models for coupling of salt and water transport, *J. Gen. Physiol.,* 66, 671, 1975.
18. **Gupta, B. L., Hall, T. A., and Naftalin, R. J.,** Microprobe measurements of Na, K and Cl concentration profiles in epithelial cells and intercellular spaces of rabbit ileum, *Nature (London),* 272, 70, 1978.
19. **Janacek, K.,** Relations between solutes and water: analysis of solute transport, in *Mechanisms of Osmoregulation in Animals,* Gilles, R., Ed., John Wiley & Sons, New York, 1979, 47.
20. **Isaia, J.,** Water and nonelectrolyte permeation, in *Fish Physiology,* Vol. 10, Hoar, W. S. and Randall, D. J., Eds., Academic Press, New York, 1984, 1.
21. **Taylor, P. M.,** Water balance in the estuarine crustacean *Corophium volutator* Pallas (Amphipoda), *J. Exp. Mar. Biol. Ecol.,* 88, 21, 1985.
22. **Cantelmo, A. C.,** Water permeability of isolated tissues from decapod crustaceans. I. Effect of osmotic conditions, *Comp. Biochem. Physiol.,* 58A, 343, 1977.
23. **Isaia, J.,** Effects of environmental salinity on branchial permeability of rainbow trout, *Salmo gairdneri, J. Physiol. (London),* 326, 297, 1982.

24. **Isaia, J. and Hirano, T.,** Effect of environmental salinity change on osmotic permeability of the isolated gill of the eel, *Anguilla anguilla L., J. Physiol. (Paris),* 70, 737, 1975.
25. **Bergmiler, E. and Bielawski, J.,** Role of the gills in osmotic regulation in the crayfish *Astacus leptodactylus* Esch., *Comp. Biochem. Physiol.,* 37, 85, 1970.
26. **Dunson, W. A.,** Permeability of the integument of the horseshoe crab *Limulus polyphemus* to water, sodium and bromide, *J. Exp. Zool.,* 230, 495, 1984.
27. **Gray, I. E.,** A comparative study of the gill area of crabs, *Biol. Bull. (Woods Hole, Mass.),* 112, 34, 1957.
28. **Hughes, G. M., Perry, S. F., and Piiper, J.,** Morphometry of the gills of the elasmobranch *Scyliorhinus stellaris* in relation to body size, *J. Exp. Biol.,* 121, 27, 1986.
29. **Evans, D. H.,** Studies on the permeability to water of selected marine, freshwater and euryhaline teleosts, *J. Exp. Biol.,* 50, 689, 1969.
30. **Croghan, P. C.,** The mechanism of osmotic regulation in *Artemia salina:* the physiology of the branchiae, *J. Exp. Biol.,* 35, 235, 1958.
31. **Dawson, M. E.,** personal communication, 1984.
32. **Johansen, K.,** Respiratory gas exchange of vertebrate gills, in *Gills, Society for Experimental Biology Seminar,* Series 16, Houlihan, D. F., Rankin, J. C., and Shuttleworth, T. J., Eds., University Press, Cambridge, 1982, 99.
33. **Copeland, D. E. and Fitzjarrell, A. T.,** Salt-absorbing cells in the gills of the blue crab *Callinectes sapidus* with notes on modified mitochondria, *Z. Zellforsch. Mikrosk. Anat.,* 92, 1, 1968.
34. **Dehnel, P. A.,** Gill tissue respiration in the crab *Eriochier sinensis, Can. J. Zool.,* 52, 923, 1974.
35. **Milne, D. J. and Ellis, R. A.,** The effects of salinity acclimation on the ultrastructure of the gills of *Gammarus oceanicus* (Segerstrale, 1947) (Crustacea; Amphipoda), *Z. Zellforsch. Mikrosk. Anat.,* 139, 311, 1973.
36. **Taylor, P. M.,** The pattern of gill perfusion in two species of *Corophium* (Crustacea: Amphipoda) and its relation to environmental salinity, *J. Zool.,* 205, 29, 1985.
37. **Cornell, J. C.,** Salt and water balance in two marine spider crabs, *Libinia emarginata* and *Pugettia producta.* II. Apparent water permeability, *Biol. Bull. (Woods Hole, Mass),* 157, 422, 1979.
38. **Todd, M. E.,** Osmotic balance in *Littorina littorea, L. littoralis* and *L. saxatilis* (Littorinidae), *Physiol. Zool.,* 37, 33, 1964.
39. **Mangum, C. P. and Polites, G.,** Oxygen uptake and transport in the prosobranch mollusc Busycon canaliculatum. I. Gas exchange and the response to hypoxia, *Biol. Bull. (Woods Hole, Mass.),* 158, 77, 1980.
40. **Ghiretti, F.,** Respiration, in *Physiology of Mollusca,* Vol. 1, Wilbur, K. M. and Yonge, C. M., Eds., Academic Press, New York, 1966, 175.
41. **Dick, D. A. T.,** Structure and properties of water in the cell, in *Mechanisms of Osmoregulation in Animals,* Gilles, R., Ed., John Wiley & Sons, New York, 1979, 3.
42. **Davenport, J.,** Effects of size upon salinity tolerance and volume regulation in the hermit crab *Pagurus bernhardus, Mar. Biol.,* 17, 222, 1972.
43. **Hickman, C. P.,** Ingestion, intestinal absorption and elimination of seawater and salts in the southern flounder, *Paralichthys lethostigma, Can. J. Zool.,* 46, 457, 1968.
44. **Dall, W. and Smith, D. M.,** Measurement of water drinking rates in marine Crustacea, *J. Exp. Mar. Biol. Ecol.,* 30, 199, 1977.
45. **Taylor, P. M. and Harris, R. R.,** Osmoregulation in *Corophium curvispinum* (Crustacea: Amphipoda), a recent coloniser of freshwater. II. Water balance and the functional anatomy of the antennary organ, *J. Comp. Physiol. B,* 156, 331, 1986.
46. **Mangum, C. P. and Johansen, K.,** The colloidal osmotic pressure of invertebrate body fluids, *J. Exp. Biol.,* 63, 661, 1975.
47. **Payan, P., Girard, J. P., and Mayer-Gostan, N.,** Branchial ion movements in teleosts: the roles of respiratory and chloride cells, in *Fish Physiology,* Vol. 10, Hoar, W.S. and Randall, D. J., Eds., Academic Press, New York, 1984, 39.
48. **Smith, R. I.,** Apparent water permeability variation and water exchange in crustaceans and annelids, in *Perspectives in Experimental Biology,* Vol. 1, Spencer-Davies, P., Ed., Pergamon Press, Oxford, 1976, 17.
49. **Rankin, J. C. and Bolis, L.,** Hormonal control of water movement across the gills, in *Fish Physiology,* Vol. 10, Hoar, W. S. and Randall, D. J., Eds., Academic Press, New York, 1984, 177.
50. **House, C. R.,** *Water Transport in Cells and Tissues,* Edward Arnold, London, 1974.
51. **Dainty, J.,** Osmotic flow, *Symp. Soc. Exp. Biol.,* 19, 75, 1965.
52. **Hebert, S.C. and Andreoli, T. E.,** Water permeability of biological membranes. Lessons from antidiuretic hormone-responsive epithelia, *Biochim. Biophys. Acta,* 650, 267, 1982.

53. **Lockwood, A. P. M., Inman, C. B. E., and Courtenay, T. H.,** The influence of environmental salinity on the water fluxes of the amphipod *Gammarus duebeni, J. Exp. Biol.*, 58, 137, 1973.
54. **Lockwood, A. P. M., Croghan, P. C., and Sutcliffe, D. W.,** Sodium regulation and adaptation to dilute media in Crustacea as exemplified by the isopod *Mesidotea entomon* and the amphipod *Gammarus duebeni,* in *Perspectives in Experimental Biology,* Vol. 1, Spencer-Davies, P., Ed., Pergamon Press, Oxford, 1976, 93.
55. **Horton, S.,** Aspects of Branchial Osmoregulation in Amphipod Crustacea: a Morpho-Physiological Study, Ph.D. thesis, University of Leicester, U. K., 1985.
56. **Thuet, P.,** Etude des flux de diffusion de l'eau en fonction de la concentration du milieu exterieur chez l'isopode *Sphaeroma serratum* (Fabricius), *Arch. Int. Physiol. Biochem.*, 86, 289, 1978.
57. **Smith, R. I.,** The apparent water permeability of *Carcinus maenas* (Crustacea; Brachyura, Portunidae) as a function of salinity, *Biol. Bull. (Woods Hole, Mass.),* 139, 351, 1970.
58. **Thuet, P.,** L'ionoregulation et l'osmoregulation chez les Crustaces. II. Aspects physiologiques, *Oceanus,* 5, 769, 1980.
59. **Hannan, J. and Evans, D. H.,** Water permeability in some euryhaline decapods and *Limulus polyphemus, Comp. Biochem. Physiol.*, 44A, 1199, 1973.
60. **Rudy, P. P.,** Water permeability in selected decapod Crustacea, *Comp. Biochem. Physiol.*, 22, 581, 1967.
61. **Roesijadi, G., Anderson, J. W., Petrocelli, S. R., and Giam, C. S.,** Osmoregulation of the grass shrimp *Palaeomonetes pugio.* I. Effects on chloride and osmotic concentrations and chloride- and water-exchange kinetics, *Mar. Biol.*, 38, 343, 1976.
62. **Wendelaar Bonga, S. E. and Van der Meij, J. C. A.,** Effect of ambient osmolarity and calcium on prolactin cell activity and osmotic permeability of the gills in the teleost *Sarotherodon mossambicus, Gen. Comp. Endocrinol.*, 43, 432, 1981.
63. **Isaia, J., Payan, P., and Girard, J. P.,** A study of water permeability of trout gills *(Salmo gairdneri)* adapted to freshwater and to seawater. Mode of action of epinephrine, *Physiol. Zool.*, 52, 269, 1979.
64. **Lahlou, B. and Giordan, A.,** Controle hormonal des echanges et de la balance de l'eau chez le teleosteen d'eau douce *Carrasius auratus,* intact et hypophysectomise, *Gen. Comp. Endocrinol.*, 14, 491, 1970.
65. **Isaia, J.,** Comparative effects of temperature on the sodium and water permeabilities of the gills of a stenohaline freshwater fish (*Carassius auratus*) and a stenohaline marine fish *(Serranus scriba, Serranus cabrilla), J. Exp. Biol.*, 57, 359, 1972.
66. **Payan, P., Goldstein, L., and Forster, R. P.,** Gills and kidneys in ureosmotic regulation in euryhaline skates, *Am. J. Physiol.*, 224, 367, 1973.
67. **Payan, P. and Maetz, J.,** Balance hydrique chez les elasmobranches: Arguments en faveur d'un controle endocrinien, *Gen. Comp. Endocrinol.*, 16, 535, 1971.
68. **Lockwood, A. P. M., Bolt, S. R. L., and Dawson, M. E.,** Water exchange across crustacean gills, in *Gills, Society for Experimental Biology Seminar,* Series 16, Houlihan, D. F., Rankin, J. C., and Shuttleworth, T. J., Eds., University Press, Cambridge, 1982, 129.
69. **Bolt, S. R. L.,** Urine clearance rates and apparent permeability of *Gammarus duebeni* exposed to varying conditions, *J. Exp. Biol.*, 114, 673, 1985.
70. **Dawson, M. E.,** Aspects of Osmoregulation in the Amphipod. *Gammarus duebeni:* the Effects of Changing Salinity and Some Potential Pollutants, Ph.D. thesis, University of Southampton, U.K., 1982.
71. **Capen, R. L.,** Studies of water uptake in the euryhaline crab, *Rhithropanopeus harrisi, J. Exp. Zool,* 182, 307, 1972.
72. **Morris, R. J., Lockwood, A. P. M., and Dawson, M. E.,** An effect of acclimation salinity on the fatty acid composition of the gill phospholipids and water flux of the amphipod crustacean *Gammarus duebeni, Comp. Biochem. Physiol.*, 72A, 497, 1982.
73. **Shaner, S. W., Crowe, J. H., and Knight, A. W.,** Long-term adaptation to low salinities in the euryhaline shrimp *Crangon franciscorum* (Stimpson), *J. Exp. Zool.*, 235, 315, 1985.
74. **Potts, W. T. W. and Fleming, W. R.,** The effects of prolactin and divalent ions on the permeability to water of *Fundulus kansae, J. Exp. Biol.*, 53, 317, 1970.
75. **Ogasawara, T. and Hirano, T.,** Effects of prolactin and environmental calcium on osmotic water permeability of the gills in the eel *Anguilla japonica, Gen. Comp. Endocrinol.*, 53, 315, 1984.
76. **Hazel, J. R. and Prosser, C. L.,** Molecular mechanisms of temperature compensation in poikilotherms, *Physiol. Rev.*, 54, 620, 1974.
77. **Smith, R. I. and Rudy, P. P.,** Water exchange in the crab *Hemigrapsus nudus* measured by use of deuterium and tritium oxides as tracers, *Biol. Bull. (Woods Hole, Mass.),* 143, 234, 1972.
78. **Percy, J. A.,** Temperature tolerance, salinity tolerance, osmoregulation and water permeability of arctic marine isopods of the *Mesidotea (Saduria)* complex, *Can. J. Zool.*, 63, 28, 1985.
79. **Loretz, C. A. and Bern, H. A.,** Prolactin and osmoregulation in vertebrates, *Neuroendocrinology,* 35, 292, 1982.
80. **Sandor, T., Dibattista, J. A., and Mehdi, A. Z.,** Glucocorticoid receptors in the gill tissue of fish, *Gen. Comp. Endocrinol.*, 53, 353, 1984.

81. **Haywood, G. P., Isaia, J., and Maetz, J.,** Epinephrine effects on branchial water and urea flux in rainbow trout, *Am. J. Physiol.*, 232, R110, 1977.
82. **Daxboeck, C., Davie, P. S., Perry, S. F., and Randall, D. J.,** Oxygen uptake in a spontaneously ventilating, blood-perfused trout preparation, *J. Exp. Biol.*, 101, 35, 1982.
83. **Berlind, A. and Kamemoto, F. I.,** Rapid water permeability changes in eyestalkless euryhaline crabs and in isolated perfused gills, *Comp. Biochem. Physiol.*, 58A, 383, 1977.
84. **Tullis, R. E. and Kamemoto, F. I.,** Separation and biological effects of C.N.S. factors affecting water balance in the decapod *Thalamita crenata, Gen. Comp. Endocrinol.*, 23, 19, 1974.
85. **Mantel, L. H.,** The foregut of *Gecarcinus lateralis* as an organ of salt and water balance, *Am. Zool.*, 8, 433, 1968.
86. **Norfolk, J. R. W. and Craik, J. C. A.,** Investigation of the control of urine production in the shore crab, *Carcinus maenas, Comp. Biochem. Physiol.*, 67A, 141, 1980.
87. **Lockwood, A. P. M. and Inman, C. B. E.,** Changes in the apparent permeability to water at moult in the amphipod *Gammarus duebeni* and the isopod *Idotea linearis, Comp. Biochem. Physiol.*, 44A, 943, 1973.
88. **Bouquegneau, J. M. and Gilles, R.,** Osmoregulation and pollution of the aquatic medium, in *Mechanisms of Osmoregulation in Animals,* Gilles, R., Ed., John Wiley & Sons, New York, 1979, 563.
89. **Lockwood, A. P. M. and Inman, C. B. E.,** Diuresis in the amphipod *Gammarus duebeni* induced by methylmercury, DDT, Lindane and Fenithrothion, *Comp. Biochem. Physiol.*, 52C, 75, 1975.
90. **Naito, N. and Ishikawa, H.,** Reconstruction of the gill from single-cell suspensions of the eel, *Anguilla japonica, Am. J. Physiol.*, 238, R165, 1980.
91. **Towle, D. W.,** Regulatory functions of Na + K ATPase in marine and estuarine animals, in *Lecture Notes on Coastal and Estuarine Studies,* Vol. 9, Pequeux, A., Gilles, R., and Bolis, L., Eds., Springer-Verlag, Berlin 1984 157.

Chapter 12

INVERTEBRATE PERMEABILITY WITH WHOLE ORGANISMS

Stephen R. L. Bolt

TABLE OF CONTENTS

I. INTRODUCTION

Control of the body water budget is an important aspect of the physiology of all organisms but is particularly important in forms inhabiting fluctuating salinities. All animals must respire, and this necessitates the presence of an area of surface that is permeable to gaseous exchange; such areas are generally also permeable to water so that when an osmotic gradient is present water passes across the body surface. Corrective measures must then be taken by the animal in order to maintain the correct water budget. Aquatic forms experience passive fluxes into and out of the body across the permeable surface. These fluxes will be unequal if there is an osmotic gradient across the body wall resulting in a net flow of water in one direction. Many euryhaline invertebrates maintain an osmotic gradient between their hemolymph and the external medium. For instance, the euryhaline isopod *Sphaeroma rugicauda*[1] is hypotonic at high salinities and hypertonic at low salinities. If the hemolymph is more concentrated than the external medium, then the net flow will be into the animal and vice versa. These net fluxes can threaten the homeostasis of the hemolymph and hence the cells if the animals are unable to compensate. Measurement of the osmotic gradient across the surface together with the gross fluxes enable an assessment to be made of the net water movements and hence further interpretations of the various adaptive mechanisms of organisms to a variety of habitats.

Animals experiencing salinity changes would benefit from the ability to restrict the passage of water and ions when large osmotic gradients are present betweeen the body fluids and the external medium. This would necessitate a mechanism controlling the permeability of the body surface in relation to the concentration gradient between the hemolymph and external medium. Several species of euryhaline Crustacea have been demonstrated to change their apparent permeability to water when the external medium concentration is altered. The euryhaline crab *Rhithropanopeus harrisi, Carcinus maenas*, and the amphipods *Corophium volutator, Gammarus setosus*, and *G. duebeni*[2-6] all display varying degrees of apparent permeability when acclimated to different salinities.

In contrast, the degrees of permeability to water in terrestrial groups are often linked to water conservation and resistance to desiccation. In this case, the animal is in an environment with little or no water present externally and the water fluxes are almost entirely out of the animal. Water loss must be replaced by behavioral mechanisms such as drinking and from food. Control of water permeability is often limited to restricting the water loss in arid circumstances and work has thus concentrated on adaptive structures such as the structural adaptions of molluscs in reducing water loss. Machin[7] concludes that adaptions of the shell and epiphragm are important factors in the colonization of arid environments. He argues that the rates of water loss are due to the physical structure and not the physiology of the snails.

II. MEASUREMENT OF WATER PERMEABILITY

In order to determine water permeability across the body surface of an animal, it is necessary to measure at least some of the components of the water fluxes. In aquatic animals, the passive fluxes are large compared to the net flow so that measurement of the influx of a marker tends not to significantly differ from the outflux, even if there is an osmotic gradient between the internal fluids and the external medium. Indeed, the exchange rates of water in either direction are at least an order of magnitude larger than the net unidirectional fluxes. Thus, measurement of either the influx or the outflux of a marker can be used to assess the permeability to water.

A. Tritiated Water as a Marker

The most commonly used marker for measurement of the permeability of whole organisms is tritiated water (THO). This can be used to monitor the uptake or loss of water to or from the external medium, and has been used widely by different workers.[8-12] The THO can be sampled indirectly after allowing prelabeled animals to "unload" in nonradioactive medium or directly by sampling the hemolymph.

1. Indirect Sampling

a. Outflux

Prior to the experiment, animals are immersed in tritiated water long enough for sufficient THO to be taken up, preferably so that they are loaded to a steady state. They are then thoroughly rinsed to remove superficial tracer and transferred to a volume of unloading medium which is large relative to the volume of the organism. It is important during the experimental period to keep the animal chamber capped to avoid exchange of tracer with water vapor in the air. The time is noted, and samples of the medium are then taken at known time intervals, placed in a suitable liquid scintillation cocktail, and counted. The last sample is removed when the THO is tending towards a steady state in the external medium. This last reading, taken as being the C_∞ reading, should be at least ten times longer than the expected half time of exchange for all the body water. If the log values of $C_t - C_\infty$ are plotted against time, and if the permeability to water of the animal is constant throughout the experiment, then the result is a straight line using the equation $y = mx + c$ where x and y are the axes, m is the gradient, and c is the y intercept. From this equation, the time taken for half the body water to be exchanged with the outside medium can be found by taking the x value of the y point corresponding to $\log (C_\infty - C_t) - \log 2$ (Figure 1). It has been found by Bolt that the correlation coefficient of the line obtained as described is sufficiently good as to suggest that it is valid for only a single C_t value to be taken at a time approximating to the time corresponding to the $t^1/_2$. However, this presupposes that the $t^1/_2$ is known approximately and therefore such a shortcut cannot be used when the order of magnitude of the answer is not known. If this method is used then the $t^1/_2$ can be calculated simply as:

$$k = 1/t \times \ln(C_\infty/C_t - C_\infty)$$

and

$$t1/2 = \ln 2/k$$

where k is the rate constant, C_t is the count at time t, and C_∞ is the count at equilibrium. Water flux as a percentage of total body water exchanged per unit time (R) is given by $R = 100$ k. The absolute water flux can be calculated from $J = RW$, where J = water flux and W = water content. "Absolute permeability" can now be calculated from the expression, $P = J/A$ where P is the absolute permeability and A is the area where exchange occurs. A significant problem in calculating the absolute permeability in these terms is in assessing the area of the site of permeable exchange in the whole animal. Although it is thought that the most permeable surface of the animals is the gill area, it is entirely possible that other structures may also be significant sites for water exchange. Thus, measuring permeability in terms of any presupposed area of exchange can be misleading. Furthermore, the respiratory surface may be convoluted and difficult to measure or may even vary in area under different conditions. It is thus concluded that measuring the permeability of the whole animal as a half-time of exchange rather than attempting to calculate an absolute permeability gives the most meaningful interpretation of the data for comparative purposes.

This technique can be used to investigate the rate of change of water permeability in animals which vary their $t^1/_2$ over a period of time. For example, when *Gammarus duebeni*

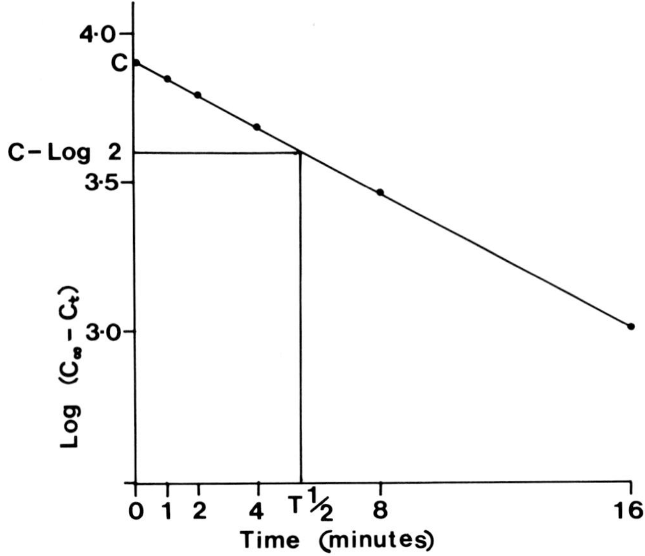

FIGURE 1. A complete determination of $t^1/_2$ by outflux of THO for an individual *Gammarus duebeni* acclimated to 100% seawater.

is transferred from 2 to 100% seawater, then the $t^1/_2$ remains at approximately 22 min for 16 hr until the animal's hemolymph reaches isotonicity with the external medium.[9] At this point, the $t^1/_2$ appears to change rapidly to around 6 min. Using the outflux method outlined above, it is possible to investigate the rate of change by plotting loss of tritiated water over the critical time, 16 hr after the transfer (Figure 2). The gradient of the line clearly shows a very rapid change indicating an extremely rapid switch in water exchange rates in this amphipod. The conclusion from this experiment is that the change in permeability cannot be controlled by any mechanism that would need more than a few minutes to operate.

b. Influx

In experiments involving animals subjected to gradually changing salinity, it is not always possible to preload the animals with THO prior to the experiment. In this case, the technique may be modified as follows: the animal is taken from the medium along with a sample of the medium and the medium ''spiked'' with THO. After a short but precisely measured time (again approximating to the $t^1/_2$ of the animal) the animal is removed, rinsed, and placed in the same, unspiked medium. It is then left until a steady state has been reached. A sample is taken and counted, giving a value equivalent to the C_t reading in the outflux technique. The animal is then reloaded in the spiked medium, until equilibrium is reached, rinsed, and transferred back to the ''clean'' medium for a final unload. A sample (C_∞) is taken when fully unloaded for a second time and the $t^1/_2$ is again calculated using the same technique as above. It is necessary to unload the animals in a volume of medium sufficiently large to make the quantity of THO in the unloading medium sufficiently dilute compared to the THO within the animal so that the reuptake of tritiated water from the unloading medium is insignificant. General criticisms of the use of THO as a marker will be discussed later.

Both these techniques are effectively comparing the quantity of tritiated water in the animals at time 0 (C_∞) to that left in the animal after time t. It is thus important to note that the permeability is only measured during the first t minutes of the technique and that after that reading, the animals are always loaded and unloaded to saturation, which is not dependent on the permeability of the animal. Since these techniques always compare samples of identical

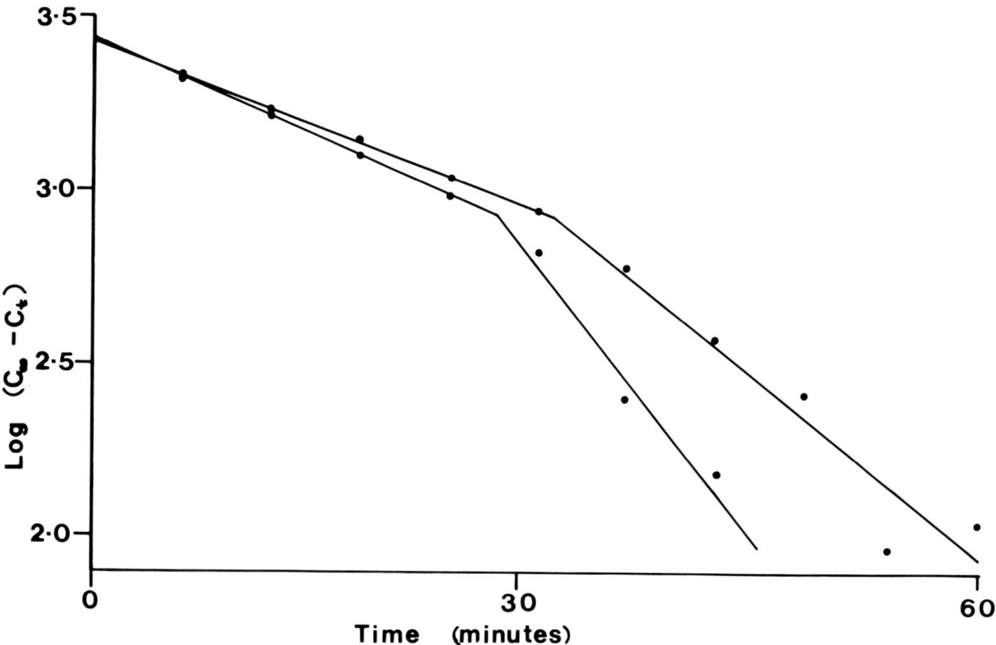

FIGURE 2. Outflux of THO from two *G. duebeni* exhibiting a rapid change in water permeability.

salinity, it is not necessary to know the efficiency of the counter used. It is assumed that the quench for all the samples will be identical. Thus, the counts per minute as measured by the counter can be used directly. For experiments in which samples vary in salinity, quench curves are required.

These techniques were used extensively by Lockwood,[6] Bolt,[9] Taylor,[4] and Dawson[10] with a variety of euryhaline amphipods in order to demonstrate the ecophysiological importance of water permeability as an osmoregulatory mechanism. Lockwood et al.[13] and Bolt[5] compared the $t^1/_2$ of several U.K. species and an Arctic species of amphipod, comparing the $t^1/_2$ of the animals with reference to the hemolymph/medium gradient in different salinity regimes. Brackish water populations of *Gammarus duebeni* clearly showed large variations in water permeability, with a $t^1/_2$ of approximately 22 min in fresh water and a $t^1/_2$ of 6 to 8 min after acclimation to full strength seawater at 15°C. Comparisons with Southern Irish freshwater populations of the subspecies *G. duebeni celticus* demonstrated that the freshwater animals did not increase their permeability after acclimation to seawater. This result concurred with the observation that the freshwater population was osmotically embarrassed when transferred directly from freshwater to seawater, although they appeared healthy if their medium was increased from freshwater to seawater in daily stages of 25% seawater. This example is particularly significant as it shows two subspecies in identical experimental conditions with the same osmotic gradient between the hemolymph and the medium yet differing in their physiological response. Indeed, although Stock and Pinkster[14] recognized these populations as subspecies on the basis of morphological differences, they required a sample of at least 30 adult males to show a significant difference. The physiological difference appears to give a much more positive separation of the two races. The less euryhaline species, *Chaetogammarus marinus* and *G. locusta*, seem unable to vary their permeability to water to the same extent as *G. duebeni*. This is apparent when the animals are exposed to the salinity cycle (Figures 6 to 8), where *C. marinus* maintains a permeability varying from 6 to 9 min and *G. locusta* from 3 to 5 min. These permeability values are reflected in the animals' ability to survive the salinity cycle. *G. duebeni* will survive indefinitely, *C.*

marinus has low mortality, while *G. setosus* suffered approximately 25% mortality during the first five salinity cycles. The Arctic species *G. setosus* is unique among the amphipods studied in its location. This species is found under the sea ice during winter and spring. During the winter months there is no fresh water present, however, in spring the melted water percolates through and forms a layer of fresh water under the ice. The amphipods are found within this layer feeding on the micro algae which blooms on the under surface of the ice during the spring. The animals are observed to move freely across the fresh/seawater interface and swim in the saline midwater. Thus, they are in a unique situation where there is an easily accessible wide range of salinity. They are therefore able to "choose" the salinity of their environment. In these animals, water permeability responses differ from those of *G. duebeni* and are such as to allow rapid uptake of salts when the animals drop into the seawater. After a rapid change in external medium from 2 to 100% seawater, *G. duebeni* has an initial $t^{1}/_{2}$ of approximately 22 min while *G. setosus* has a $t^{1}/_{2}$ of 10 min (Figure 3a). The hypothesis developed from these results is that the animals have evolved mechanisms by which they can quickly "top up" their hemolymph concentration by swimming into the seawater periodically. Experiments have shown that the hemolymph of *G. setosus* reaches equilibrium with seawater on transfer from freshwater in 3 to 4 hr while *G. duebeni* would require 16 hr (Figure 3b). Figure 4 indicates three different strategies of water permeability management in three forms: *G. duebeni duebeni, G. duebeni celticus,* and *G. setosus* exposed to a sudden increase in external medium concentration. This comparison emphasises the different "tactics" evolved by varied ecological pressures.

c. *Intrusive Sampling*

The alternative to passively allowing the THO to be taken up by the animal and then unloaded into clean medium is to place the animal into the tritiated medium and take hemolymph samples every 5 to 15 min thus measuring the influx of water. This technique was adopted by Shaner et al.[15] on the euryhaline shrimp *Crangon franciscorum*. THO was added to the medium and hemolymph samples taken between 5 and 15 min and placed in a liquid scintillation cocktail. These samples were counted and the quench taken into account. The rate constant was then calculated using the equation:

$$k = \frac{1}{t} \cdot \frac{C_t}{C_\infty - C_t}$$

where k = total body water exchanged per hour, C_t = counts at time t, and C_∞ = counts at equilibrium. The $t^{1}/_{2}$ was not calculated by these authors.

B. Water Loss as a Measurement of Permeability

It may not be possible or desirable to place terrestrial animals in water in order to examine whole body permeability. In these circumstances, permeability may be measured by examining the loss of water to the atmosphere. Cannings[16] measured the weight loss of *Cenocorixa bifida*, an hemipteran corixid insect, in dry CO_2 and used this as a measure of permeability, expressing the permeability as mg of water cm^{-1} hr^{-1}. The CO_2 acted as a narcotic on the insects. Using this technique, Cannings[16] showed that the permeability of the animal varied with the salinity of the lake water the animals were collected from (Figure 5). The permeability varied from 3 to 6 mg water cm^{-1} hr^{-1} in animals collected from water varying from 0 to 15 μmho $cm^{-1} \times 10^3$. This can be converted to approximately 0 to 8‰ salinity. Cannings[16] concludes that although the biological reasons for the variation in cuticular permeability are not clear, the permeability increases as problems of ion regulation diminish. Thus, at very low salinities and relatively high salinities, the animals are less permeable to water. Jarial and Scudder[17] were able to show that this change in permeability

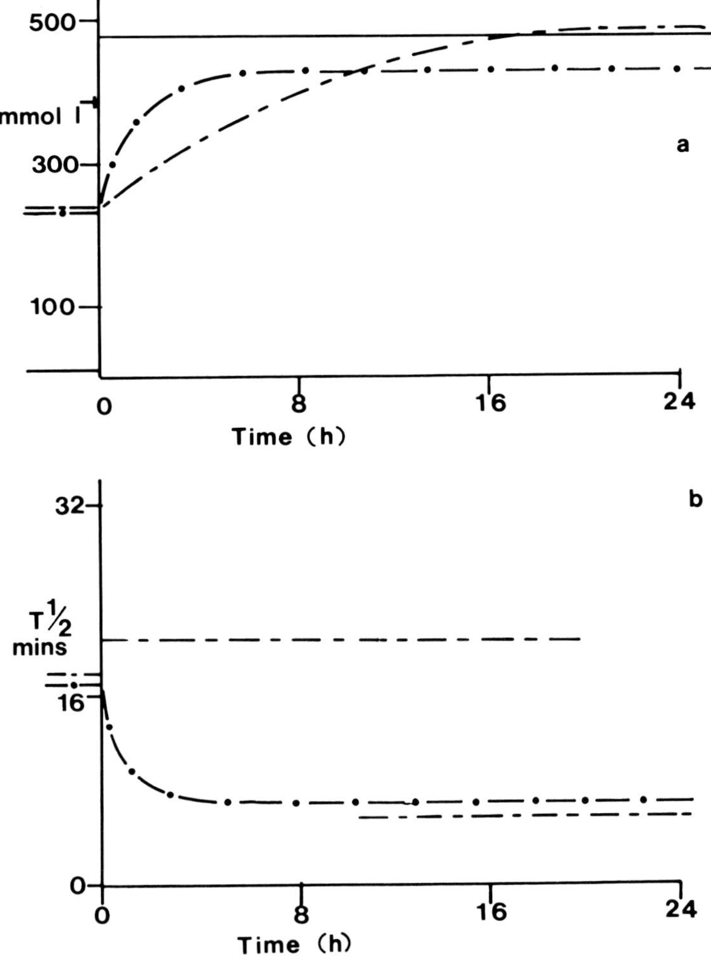

FIGURE 3. Hemolymph concentration (a) and half time of exchange (b) of *G. duebeni* (— - —) and *G. setosus* (— · —) exposed to a sudden increase in medium concentration (—).

of the cuticle of *C. bifida* can be linked to neurosecretory changes. They found that the type A neurosecretory cells of the pars intercerebralis have axons that pass directly to the region of the dorsal aorta. When the insects are in their normal environment, the protocerebral region stains deeply, the cells being loaded with neurosecretory material. The depletion of this material under potential dehydration conditions strongly suggests that the water balance and possibly the water permeability of these animals is under neurosecretory control.

III. NET FLOW OF WATER IN WHOLE ORGANISMS

An important aspect of water relations in whole organisms closely correlated to the water permeability of the organism is the net flow out of or into the animal. As already stated, if there is a dynamic exchange of water and an osmotic gradient across the body wall then there must be an overall unidirectional movement of water. This flow can be calculated from the half-time of exchange ($t^1/_2$) and the osmotic concentrations of the hemolymph and the medium using the following equations:

$$M\text{m} \quad \text{or} \quad M\text{a} = 55.56/(55.56 + \text{L})$$

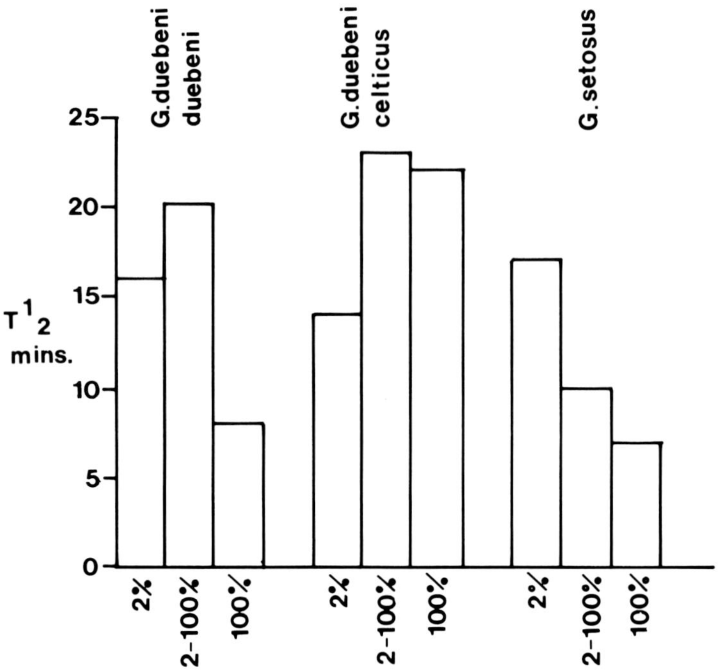

FIGURE 4. Half-times of exchange in three forms of amphipod before, immediately after, and 24 hr after a sudden increase in medium concentration.

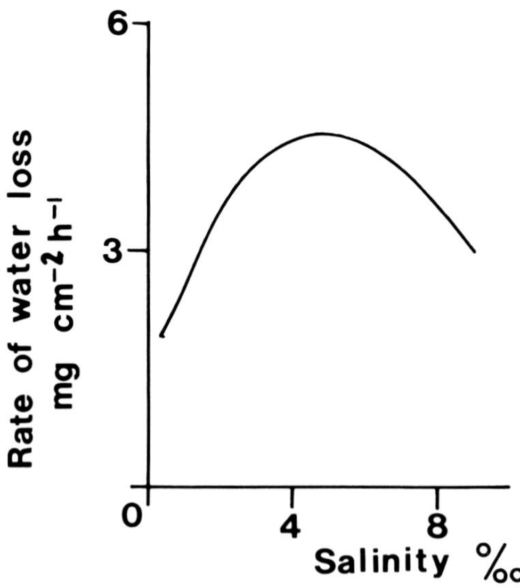

FIGURE 5. Rate of water loss as a measure of water permeability in the corixid *Cenocorixa bifida* collected over a range in salinities.[10]

where *M*m is the molar concentration of the medium, *M*a is the molar concentration of the hemolymph, and L is the osmotic concentration of the medium or hemolymph, and

$$\left(\frac{M_m - M_a}{M_m}\right) \times \frac{100 \ln 2}{t1/2} = Os$$

where $t^1/_2$ is the half-time of exchange in minutes and Os is the % blood volume loss per minute.

Calculating this net flow for the different species of amphipod allows some insight into the benefit of changing the permeability in varying osmotic conditions. If the amphipods are subjected to a cycling salinity regime ranging from 5 to 100% seawater, then *Gammarus duebeni* exhibits large changes in permeability while the other species do not (Figure 6). Calculations of the expected net water flux throughout the cycle clearly show that *G. duebeni* is potentially able to restrict the net fluxes throughout the cycle (Figure 7) more effectively than the other species. As a possible reflection of this, there is higher mortality in the cycle of the less euryhaline species. The difference between these three species is further emphasized by calculations of the percentage water uptake over a period of one cycle (= 12.25 hr). This is achieved by integrating under the curve on Figure 7 for each animal. *G. duebeni* has a % body water uptake of 15% during one cycle, *C. marinus* +31.9%, and *G. locusta* +60.6% body water per cycle. It is assumed that the urine flow matches the influx of water and the volume of the animals remains constant. These relative values of water flux are reflected in the level of hemolymph control these species have throughout the cycle (Figure 8). It is noticeable that *G. duebeni* is able to maintain a relatively constant hemolymph concentration throughout the cycle, while that of the other two species fluctuates.

Net water flux into and out of the animals may be an important factor in limiting the range of conditions which an animal is capable of colonizing. If flux can be limited by changes in $t^1/_2$, then osmotically embarrassing situations may be averted. Perhaps the most spectacular example is the brine shrimp *Artemia salina*, which maintains its hemolymph strongly hypotonic to the medium at high salinities. In an extreme case with the animal in 566% seawater, the hemolymph concentration is 590 mmol ℓ^{-1} (Croghan[18]) and the medium 6920 mmol ℓ^{-1}. There is therefore a medium to hemolymph gradient of 6330 mmol ℓ^{-1}! If the animals had a half-time of exchange in the same order as other crustaceans of a similar size (30 min as an example), then the net flux should be approximately 380% body water out of the animal per day. However, the measured $t^1/_2$ of 17.84 hr (Stewart[19]) reduces this value to a more reasonable value of 10% body water per day. Knowledge of the means by which this species is able to maintain a $t^1/_2$ at least 40 times larger than that of most crustaceans could give an important insight into the mechanisms used in permeability control.

IV. CRITICISMS OF USING THO TO MEASURE WATER PERMEABILITY

There are three possible interpretations of the observed changes in water fluxes using THO as a marker:

1. The differences are artifacts of the experimental technique and do not represent changes in epithelium permeability which affect the net passage of water. Such an explanation of flux differences would imply that the results obtained were of no biological significance.
2. Flux variations are caused by variation in surface area over which water movement can occur due to circulatory variations. No change in hydraulic permeability per unit area is postulated. This interpretation could represent biological responses and confer advantages to the species.
3. Real variations occur in the hydraulic permeability reflecting the $T^1/_2$ for water exchange. This option alone involves changes at the intracellular level and represents real changes in water permeability.

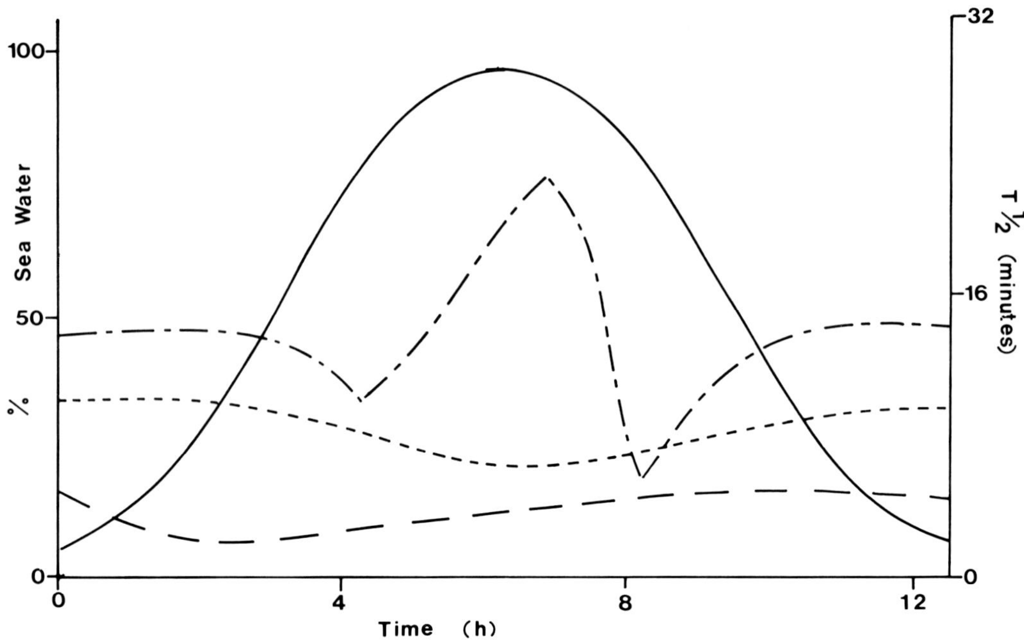

FIGURE 6. Half-time of exchange in three species of amphipod exposed to a 12.25-hr salinity cycle from 2 to 30%. *G. duebeni* = — - —, *C. marinus* = - - - -, and *G. locusta* = — —.

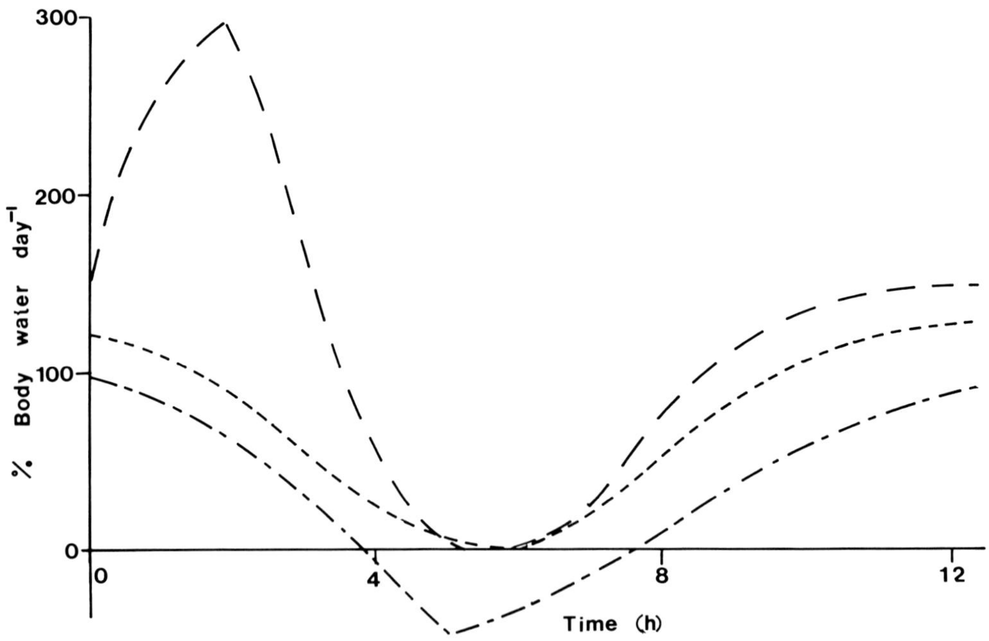

FIGURE 7. Net water flux in three species of amphipod exposed to a 12.25-hr salinity cycle from 2 to 30%. *G. duebeni* = — - —, *C. marinus* = - - - -, and *G. locusta* = — —.

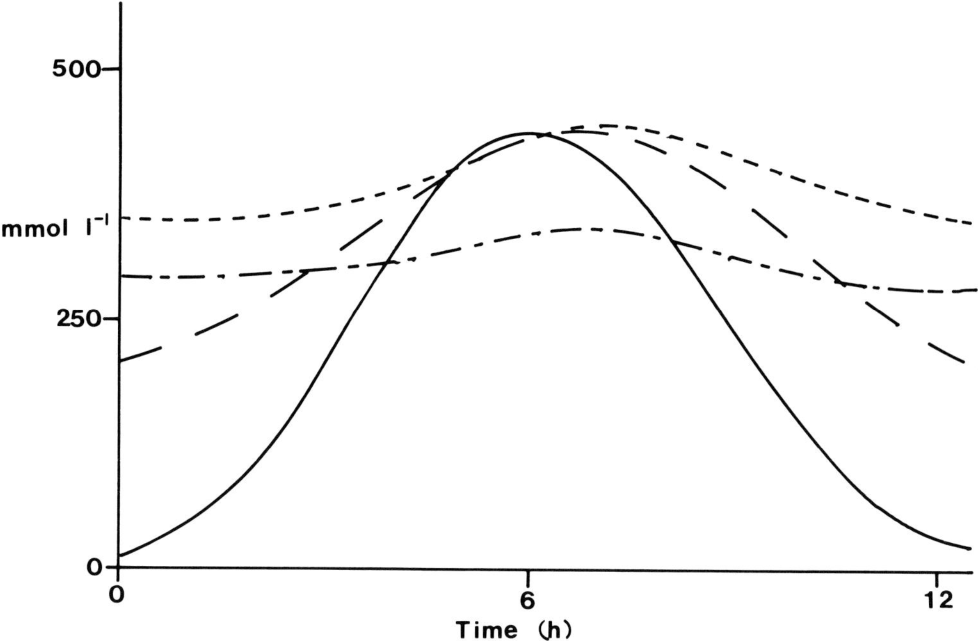

FIGURE 8. Hemolymph and medium concentration of three species of amphipod exposed to a 12.25-hr salinity cycle from 2 to 30%. *G. duebeni* = — - —, *C. marinus* = - - - -, *G. locusta* = — —, and Medium = —.

A. Possible Artifacts in Measurement

Dainty and House[20] pointed out that variation in the thickness of unstirred layers on either side of the membrane will be likely to result in changes in the apparent permeability to water. Similarly, the passage of water through narrow pores in a membrane might be expected to result in distortions in the theoretical diffusional exchange from one side to the other across the membrane.

The ratio of the osmotic permeability (p_{os}, as determined from the net transfer of water) and the diffusional permeability (p_{dif}, determined from the flux of THO) can differ markedly, especially when isolated tissues without adequate irrigation are used. For instance, permeability measurements based on flux measurements on frog skin can substantially underestimate the net passage of water down an osmotic gradient. Thus, the p_{os}/p_{dif} ratio for amphibian skin can exceed 10, *Bufo regularis*, 27.2; *Xenopus laevis*, 9.3; and *Rana esculenta*, 10.1[22] By contrast, the $p_{os}dip_{dif}$ ratio of irrigated tissues such as gills generally gives lower p_{os}/p_{dif} ratios. The teleost fish Anguilla and *Platichthys* have ratios of 3.16 and 2.59 in freshwater and 1.05 and 0.83 in seawater.[23] Similarly, the crabs *Libinia emarginata* and *Carcinus maenas* give p_{os}/p_{dif} ratios ranging between 1 and 2.5.[11] Presumably the p_{os}/p_{dif} ratios observed in gills can be ascribed to the irrigation of the outside by respiratory currents and of the inside by the blood, so that unstirred layers are diminished. Bolt[27] noted that the $t^1/_2$ in *Gammarus duebeni* rose sharply in animals after heart failure from 6 to approximately 60 min immediately after death. This emphasizes the importance of respiratory and circulatory currents in minimizing the unstirred layers.

In some species, the decrease in apparent permeability commonly observed in transfer of an individual from a high salinity to a lower one may be attributed to a change in heart rate. Cornell[12] reports that transfer of the crab *Libinia* from 100 to 80% seawater is followed by a reduction in heart rate which seems to correlate with the observed change in water flux. However, in more euryhaline forms it seems unlikely that the changes observed in apparent water permeability can be attributed to changes in heart rate. *G. duebeni* shows only small changes in heart rate when transferred between salinities,[24] and those that do occur are such that the rate is faster in dilute medium when the animal is less permeable. Similarly, the heart rate of *Carcinus maenas* increases on acclimation to dilute medium.[25]

Comparative studies of two subspecies of *G. duebeni* (*duebeni* and *celticus*) show that they differ markedly in their water fluxes in identical conditions. The two stocks have rather similar $t^1/_2$s when in fresh water at 15 and 14 min respectively. However, *G. duebeni duebeni*, the brackish water form, maintains a $t^1/_2$ of 6 min while acclimated to seawater, while *celticus*, the freshwater form, has a $t^1/_2$ of 22 min (Figure 4) in seawater. The hemolymph of both forms is slightly hypertonic to the external medium. It seems unlikely that in such similar forms the difference in apparent permeability can be attributed to an artifact in the experimental technique.

In order to consolidate further the THO measurements of water permeability, THO flux measurements were compared with calculated permeabilities using urine flow rates measured using Cr EDTA as a marker.[26] In interpreting the results, it was assumed that the urine flow out of the animal matched the net flow into the animal. With this value and the osmotic gradient between the hemolymph and external medium known, the theoretical $t^1/_2$ could be calculated. Unfortunately, this method only allows permeability readings to be taken when the animal is urinating, which limits experiments in practice to periods when the animal is hypertonic to the external medium. However, the results show a reasonable agreement between the measured $t^1/_2$ using THO and the calculated $t^1/_2$ using the urine measurements, thus providing some confirmation of the validity of the THO technique in assessment of net flux. The general conclusion to be drawn from the heart rate and comparative and urine measurements is that it is highly unlikely for unstirred layers or other artifacts to be responsible for the often large changes in apparent permeability observed on transfer between different salinities in the species studied. Thus, the observed flux changes must be either due to changes in the effective surface area where exchange can occur or a real change in hydraulic permeability. In view of the lack of positive evidence that passive factors could offer an explanation of the changes in apparent permeability as large as those observed, it would appear that these changes are reflecting real changes in water permeability of the animal.

V. CONCLUSIONS

The measurement of water permeability in various organisms has proved an extremely useful tool in interpreting the "whole animal" physiology, especially as related to the ecology of the species in question. Techniques have been broadly split into two categories: the first dealing with water loss in terrestrial or semiterrestrial forms and the second dealing with aquatic and often euryhaline organisms.

Techniques concerning the loss of water from the surface of animals have often been concerned with the survival of animals in adverse conditions where desiccation is of prime importance, while water permeability in aquatic animals has often centered on the ability of some species to survive changes in the external environment. Thus, the first method has concentrated on insects and mollusks, the second on crustaceans, especially on the intertidal and estuarine forms.

Measurements of the water permeability when used in conjunction with other investigations allows the water budget of an animal to be studied. The techniques involve assumptions and are open to some criticism, however it is felt that for invertebrate organisms the methods offer an important methodology which should not be ignored. Furthermore, many of the criticisms have now been answered, and it is felt that the changes in water permeability measured reflect a real effect. These measurements are especially useful as a comparative measurement between different species or between populations within a species. Finally, it must be emphasized that water permeability is only a small aspect of the biology of the ecophysiological mechanism and should be used in conjunction with other important facets of this field.

REFERENCES

1. **Harris, R. R.,** Aspects of Ionic and Osmotic Regulation of two Species of *Sphaeroma* (Isopoda), Ph.D. thesis, University of Southampton, Southampton, U.K.,
2. **Smith, R. I.,** Osmotic regulation and adaptive reduction of water permeability in a brackish water crab, *Rhithropanopeus harrisi* (Brachyura, Xanthidae), *Biol. Bull. (Woods Hole, Mass.),* 144, 643, 1967.
3. **Smith, R. I.,** The apparent permeability of Carcinus maenas (Crustacean, Brachyura, Portunidae) as a function of salinity, *Biol. Bull. (Woods Hole, Mass.),* 139, 351, 1970.
4. **Taylor, P. M.,** Water balance in the estuarine crustacean *Corophium volutator* (Pallas) (Amphipoda), *J. Exp. Mar. Biol. Ecol.,* 88, 21, 1985.
5. **Bolt, S. R. L.,** Ecological, behavioural and physiological observations on under ice populations of arctic amphipods associated with salinity anomalies, *Prog. Underwater Sci.,* 11, 1986, 127.
6. **Lockwood, A. P. M. and Inman C. B. E.,** Water uptake and loss in relation to the salinity of the medium in the amphipod crustacean *Gammarus duebeni, J. Exp. Biol.,* 58, 149, 1973.
7. **Machin, J.,** Structural adoptions for reducing water-loss in three species of terrestrial snail, *J. Zool. (London),* 152, 55. 1967.
8. **Bolt, S. R. L.,** Ecophysiological Responses to Salinity Changes in Selected Euryhaline Ampipods with Special Reference to *Gammarus duebeni,* Ph.D. thesis, Southampton University, Southampton, U.K., 1982.
9. **Bolt, S. R. L.,** Haemolymph concentrations and apparent permeability in varying salinity conditions of *Gammarus duebeni, Chaetogammarus marinus* and *Gammarus locusta, J. Exp. Biol.,* 107, 129, 1983.
10. **Dawson, M. E.,** Aspects of Osmoregulation in the Amphipod *Gammarus duebeni:* the Effects of Changing Salinity and Some Potential Pollutants, Ph.D. thesis, Southampton University, Southampton, U.K., 1982.
11. **Rudy, P. P.,** Water permeability in selected decapod crustacea, *Comp. Biochem. Physiol.,* 22, 581. 1967.
12. **Cornell, J. C.,** Salt and water balance in two marine spider crabs, *Libinia emarginata* and *Pugettia producta.* II. Apparent water permeability, *Biol. Bull. (Woods Hole, Mass.),* 157, 221, 1979.
13. **Lockwood, A. P. M., Bolt, S. R. L., and Dawson, M. E.,** Water exchange across crustacean gills, in *Gills (Society for Experimental Biology Seminar),* 16th Series 16, Houlihan, D. F., Rankin, J. C., and Shuttleworth, T. J., Eds., University Press, Cambridge, 1982, 129.
14. **Stock, J. H. and Pinkster, S.,** Irish and French freshwater populations of *Gammarus duebeni* subspecifically different from brackish water populations, *Nature (London),* 228, 874. 1970.
15. **Shaner S. W., Crowe, J. H., and Knight, A. W.,** Long term adaption to low salinities in the euryhaline shrimp *Crangon franciscorum, J. Exp. Zool,* 235, 3, 1985.
16. **Cannings, S. G.,** The influence of salinity on the cuticular permeability of *Cenocorixa bifida* Hungerfordi Hemiptera Corixidae, *Can. J. Zool.,* 59(8), 1505, 1981.
17. **Jarial, M. and Scudder G. G. E.,** Neurosecretion and water balance in *Cenocorixa bifida* (Hung.) (Hemiptera, Corixidae), *Can. J. Zool.,* 49, 1369, 1971.
18. **Croghan, P. C.,** The osmotic and ionic regulation of *Artemia salina* (L.), *J. Exp. Biol.,* 35, 219, 1958.
19. **Stewart, A. J.,** The influence of environmental salinity, temperature, and ionic composition on the water fluxes of *Artemia salina* (L.), Master's dissertation, Southampton University, Southampton, U.K., 1974.
20. **Dainty, J. and House, C. R.,** Unstirred layers in frog skin, *J. Physiol. (London),* 182, 66, 1966.
21. **Koefoed-Johnson, V. and Ussing, H. H.,** The contribution of diffusion and flow in the passage of D_20 through living membranes. Effect of hypophyseal hormone on anuran skins, *Acta. Physiol. Scand.,* 28, 60, 1953.
22. **Maetz, J.,** Salt and water metabolism, in *Perspectives in Endocrinology. Hormones in the Lives of Lower Vertebrates,* Barrington, E. J. W. and Jorgenson, C. B., Eds., Academic Press, New York, 1968, 47.
23. **Motais, R., Isaia, J., Rankin, J. L., and Maetz, J.,** Adaptive changes of water permeability of the teleostean gill epithelium in relation to external salinity, *J. Exp. Biol.,* 51, 529, 1969.
24. **Bolt, S. R. L., Dawson, M. E., Inman, C. B. E., and Lockwood, A. P. M.,** Variation of apparent permeability to water and sodium transport in *Gammarus duebeni* exposed to fluctuating salinities, *Comp. Biochem. Physiol.,* 67B, 465, 1980.
25. **Hume, R. I. and Berlind, A.,** Heart and scaphognathite rate changes in a euryhaline crab, *Carcinus maenas* exposed to a dilute environmental medium, *Biol. Bull. (Woods Hole, Mass.),* 150, 241, 1976.
26. **Bolt, S. R. L.,** Urine clearance rates and apparent permeability of Gammarus duebeni exposed to varying conditions, *J. Exp. Biol.,* 673, 1985.
27. **Bolt, S. R. L.,** unpublished observations.

Chapter 13

TRANSPIRATION OF WATER THROUGH THE INSECT CUTICLE

Vincent B. Wigglesworth

TABLE OF CONTENTS

I. INTRODUCTION

The great majority of insects are air-breathing terrestrial animals. They are of small size, with a consequent large surface-to-volume ratio, and therefore an urgent need for a waterproof surface structure. The detailed structure of the insect integument is extremely diverse, but the principles involved in waterproofing can be demonstrated by considering some typical examples, together with some characteristic specializations.

II. THE CUTICLE

The visible integument is the secreted product of the continuous monolayer of epidermal cells over the entire surface of the body. This product, termed the cuticle, is a lifeless secretion penetrated to a variable extent by fine channels. It may have a thickness of only 1 or 2 μm or up to 12 μm or more.

The cuticle may be hard, stiff, and horny, or flexible and extensible, but its permeability to water is not necessarily related to its hardness. Some of the most pliable cuticles (for example, the soft white cuticle of clothes-moth larvae) are the most impermeable.

The most familiar constituent of the cuticle is the nitrogenous polysaccharide chitin; but this rarely forms more than 50% of the substance of the cuticle. Chitin is a polymer of acetylglucosamine and closely related in its properties to cellulose. Like cellulose, it exists in the form of submicroscopic crystallites or micellae. In the inner parts of the cuticle (endocuticle), these tiny rodlets are aligned to form fibrils and these fibrils tend to be oriented, all in one direction, at one level in the cuticle and then take up a new preferred orientation in the succeeding lamina, at some definite angle to the first.[1] It is this changing orientation of the fibrils which is responsible for the apparent lamination of the cuticle as seen in vertical section.

The fibrils of chitin are bound together by a protein matrix after the manner of fiberglass. Indeed, the two may perhaps be chemically united to form a mucopolysaccharide. The proteins are of many kinds; more than a score have been recognized: some are enzymes or carriers of enzyme substrates and others are purely structural proteins associated with chitin.

Where the cuticle is hard and stiff it is the outer part (exocuticle) which is hardened by a horny component termed ''sclerotin''. Sclerotin resembles vertebrate horn or keratin, but whereas keratin is described as ''vulcanized'' protein, a substance in which the adjacent protein chains are chemically bound together by means of sulfur linkages, sclerotin is described as ''tanned'' protein.[2] It is produced by the action of quinones formed in the cuticle from the oxidation of various diphenols. The quinones react with adjacent protein chains and bind them firmly together, converting a soft, white, extensible material into a hard and horny substance that may vary in color from pale amber to deep brown. Gelatin tanned with benzoquinone gives rise to somewhat similar material.

Figure 1 summarizes the classic conception of sclerotin formation.[3] But other milder forms of polymerization, some involving the inclusion of lipids, occur in the cuticular proteins. For example, there is evidence that so-called β-sclerotin can be formed by the combination of the side chains in acetyl dopamine with proteins without involving quinone production.[4]

III. THE EPICUTICLE

The *endocuticle* is responsible for the extensibility of the integument, and for combining toughness with flexibility. The *exocuticle* provides the rigidity in the hard parts, such as the head capsule, the segments of the limbs, and so forth; it provides an external skeleton to which the muscles are attached. The *epicuticle* is responsible for the impermeability of the cuticle and particularly its power of preventing loss of water by transpiration.

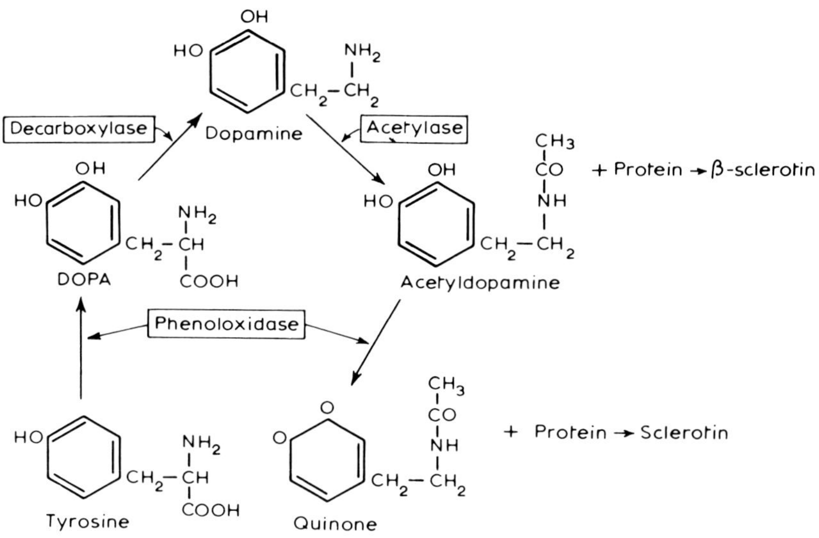

FIGURE 1. The biochemistry of sclerotin formation in the larvae of *Calliphora*[3] and β-sclerotin.[4]

When the epicuticle of the fully developed integument is examined microscopically, it appears as a refractile amber-colored layer, usually not more than 1 μm in thickness. It is inextensible; but where the cuticle is liable to stretch it is deeply folded, as in recently molted caterpillars. The epicuticle is a complex structure; the only way in which it has been possible to learn something of its composition has been to observe the stages in its development when a new cuticle is being formed before molting.[5]

The electron microscope has shown that the cuticle is an extracellular structure. The surface of each epidermal cell, when it becomes separated from the old cuticle before molting, shows tongue-like projections of microvilli about 0.5 μm long. The tips of the microvilli contain a lipid-staining deposit (Figure 2a). Then a cap of clear material forms over the tip of each microvillus (Figure 2b). Gradually, the caps fuse with their neighbors to form a continuous glassy membrane about 17 nm thick (Figure 2c). This is the *outer epicuticle*, which is constant in all insects. It takes up lipid stains and is perhaps composed of some unknown polymers of lipophilic materials.

The epidermal cells now lay down the *inner epicuticle* which is composed of protein with lipid and reaches a thickness of 0.5 to 1 μm (about 25 to 50 times that of the outer epicuticle (Figure 2d).[5]

IV. PORE CANALS AND TUBULAR FILAMENTS

During the next few days the laminated endocuticle is laid down. Figure 2e represents this stage. Three open channels (the "pore canals") can be seen running vertically from the epidermis through the newly formed endocuticle to the inner epicuticle. The pore canals arise from pits between the microvilli at the surface of an epidermal cell.

Each pore canal contains a bundle of fine "tubular filaments" which arise from the plasma membrane of the epidermis. These tubules, which have a lumen of about 5 to 10 nm in diameter, contain lipid-staining material (presumably wax precursors) and argentaffin material (presumably sclerotin precursors). On reaching the base of the inner epicuticle, Figure 2e shows the tubules discharging into a single "epicuticular channel" of about 25 nm in diameter;[6] but it is more usual for the tubular filaments to separate and form branching

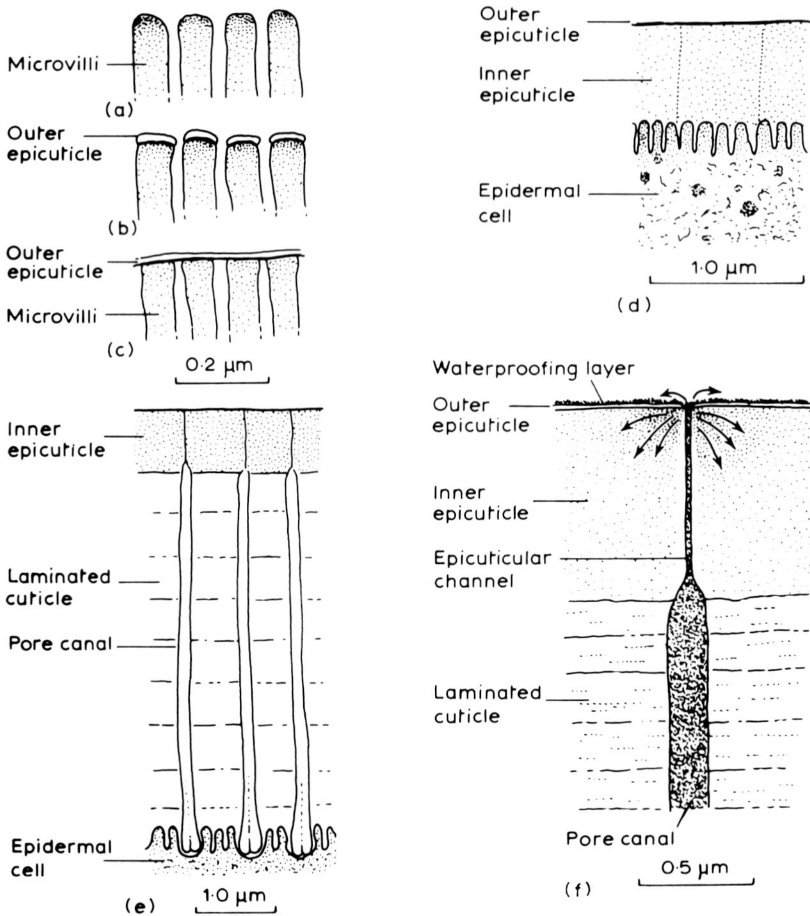

FIGURE 2. Stages in formation of epicuticle and waterproof layer; (a), (b), and (c) deposition of outer epicuticle; (d) deposition of inner epicuticle; (e) three pore canals ending in epicuticular channels; and (f) discharge of sclerotin and wax precursors from pore canal.

channels of 5 to 10 nm internal diameter to run through the inner epicuticle. In either case, the tubular filaments pass through the outer epicuticle and discharge their contents to form a very thin extracuticular layer on the surface. At the same time, the contents of tubular filaments are also discharged into the substance of the inner epicuticle (see arrows in Figure 2f).[6,7]

By the time that this development is complete, the enzymes involved in sclerotization will have been activated: all the argentaffin material will harden, enclosing with it the secreted waxes which will no longer be accessible to lipid staining. The cuticle is hardened, darkened, and waterproof.

It should be noted that tubular filaments carrying wax precursors and sclerotin precursors run not only to the substance and surface of the epicuticle, but also to the substance of all those areas of the main cuticle which are destined to harden and form the exocuticle.[7,8]

IV. LOCATION OF WATER

The water in the cuticle comes very close to the surface. The laminated cuticle of chitin and protein doubtless contains water throughout, and the pore canals and tubular filaments

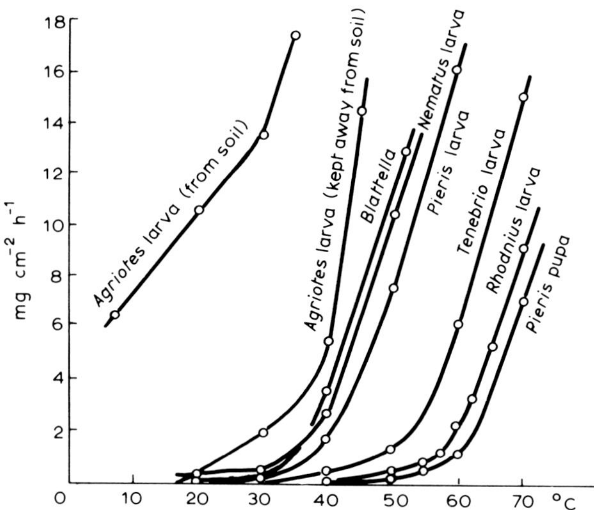

FIGURE 3. Rate of transpiration from insects, with spiracles sealed, in relation to temperature.[9]

also have water in the lumen. The tubular filaments pass through the outer epicuticle in the epicuticular channels so that water is separated from the atmosphere only by the thin covering of the sclerotin and wax waterproofing layer.

Most cuticles (excepting only those which are heavily sclerotized throughout) become permeable to water if the surface is lightly rubbed with an abrasive dust (such as Linde powder: synthetic sapphire dust). For this reason, insects burrowing in the soil are no longer waterproof. But this abrasion is rapidly repaired in the living insect by renewed secretion.[9]

If a small insect is immersed in a lipid solvent (olive oil or liquid paraffin) for perhaps half an hour, minute droplets of water appear on the surface of the cuticle. When the wax is dissolved, the long-chain molecules crowd into the water surface and cause it to be drawn out in the form of droplets. This effect is rendered more dramatic if an insect is immersed in a mixture of equal parts of light liquid paraffin and butanol. As soon as the surface wax is extracted, so that water is exposed, the paraffin and butanol separate and begin to effervesce: minute droplets appear to be streaming out of the cuticle. The time required before the effervescence begins is a measure of the strength of the waterproofing.[10]

A rise in temperature by itself commonly reduces the waterproofing and allows water to escape by transpiration. Different insects have waterproofing waxes with different melting points and this is correlated with differences in the temperature at which transpiration increases sharply (Figure 3).

Doubt has been cast on the significance of the apparently abrupt increase in transpiration at certain temperatures.[11] But it has recently been shown that permeability can be gauged in another way: by observing the temperature at which ammoniacal silver gains access to the argentaffin component in the wax/sclerotin waterproofing layer. In the various insects studied, this temperature has been approximately the same as that obtained for the more or less abrupt increase of transpiration (Figure 3).[10]

VI. WAX LAYERS

It had long been thought that the insect cuticle was waterproofed by a comparatively thick layer of wax over the surface. In the above description, the waterproofing layer is depicted as a very thin deposit (in the 10 to 20 nm range) of a mixture of wax and sclerotin. This description would apply to the cuticle of the locust *Schistocerca*, to the mealworm beetle *Tenebrio*, and to the bloodsucking bug *Rhodnius*. Some insects such as the cockroach and caterpillar, possess a thin waterproofing layer of the kind described, but this is overlaid by a substantial layer of free wax 0.5 to 1 μm thick.[7]

It is often supposed that the epidermal cells below the cuticle may actively control the water content of the cuticle. That seems very probable. But there is no evidence that such changes regulate transpiration which appears to be controlled locally by injury and repair of the waterproofing layer.

REFERENCES

1. **Neville, A. C.,** Chitin lamellogenesis in locust cuticle, *Q. J. Microsc. Sci.*, 106, 269, 1965.
2. **Pryor, M. G. M.,** On the hardening of the cuticle of insects, *Proc. R. Soc. London Ser. B.*, 128, 393, 1940.
3. **Karlson, P. and Sekeris, C. E.,** N-Acetyl dopamine as sclerotizing agent of the insect cuticle, *Nature (London)*, 195, 183, 1962.
4. **Andersen, S. O.,** Biochemistry of insect cuticle, *Annu. Rev. Entomol.*, 24, 29, 1970.
5. **Locke, M.,** Pore canals and related structures in insect cuticle, *J. Biophys. Biochem. Cytol.*, 10, 589, 1961.
6. **Wigglesworth, V. B.,** Incorporation of lipid into the epicuticle of *Rhodnius* (Hemiptera), *J. Cell Sci.*, 19, 459, 1975.
7. **Wigglesworth, V. B.,** The transfer of lipid in insects from the epidermal cells to the cuticle, *Tissue Cell*, 17, 249, 1985.
8. **Wigglesworth, V. B.,** Sclerotin and lipid in the waterproofing of the insect cuticle, *Tissue Cell*, 17, 227, 1985.
9. **Wigglesworth, V. B.,** Transpiration through the cuticle of insects, *J. Exp. Biol.*, 21, 97, 1945.
10. **Wigglesworth, V. B.,** Temperature and the transpiration of water through the insect cuticle, *Tissue Cell*, 18, 99, 1986.
11. **Gilby, A. R.,** Transpiration, temperature and lipids in insect cuticle, *Adv. Insect Physiol.*, 15, 1, 1980.

Chapter 14

TRANSPIRATION, VAPOR ABSORPTION, AND CUTICULAR PERMEABILITY IN WOODLICE (*ONISCIDEA, ISOPODA, CRUSTACEA*)

Dieter Coenen-Stass

TABLE OF CONTENTS

I. INTRODUCTION

Woodlice are terrestrial isopods of the suborder Oniscidea (formerly *Oniscoidea*). They are the only crustacea which permanently inhabit terrestrial habitats.

In general, woodlice show in vapor-unsaturated air a higher water loss than most insects (literature reviewed by Edney,[1] Cloudsley-Thompson,[2] Sutton,[3] and Wieser[4]). However, the degree of adaptation to terrestrial life is very different among woodlice. Thus, the genus *Ligia* inhabits the littoral zone, members of the genus *Oniscus, Porcellio*, and *Armadillidium* prefer mesic habitats, and species like *Venezillo arizonicus, Armadillo albomarginatus*, and *Hemilepistus reaumuri* live in deserts or semidesert regions.[5-11] For this reason, the water permeability of their integument is of considerable biological interest.

The present review considers only water relations in normal hydrated woodlice. Hyperhydration and especially dehydration cause a distinct change of the osmolality in the different body water compartments (cells, hemolymph, gut, hepatopancreas), a theme which is too voluminous to be included in this review. However, investigations on *H. reaumuri, A. officinalis*, and *Armadillidium pallasii* revealed that their cuticular permeability is not much influenced by the hydrating status of these woodlice.[12]

II. METHODS

Generally, water exchange in terrestrial arthropods may result in a gain of body water by drinking free water, ingesting wet food, producing metabolic water, absorbing water vapor from the surrounding air, or may lead to a loss of body water by secreting or excreting, and by evaporating water from cuticular and respiratory surfaces.

With regard to the water balance, the absorption of water vapor and the transpiration through the cuticula are important since they characterize the cuticular permeability, which is, apart from behavioral adaptation, the most important factor in the animal's ability to withstand terrestrial conditions.

It depends on the investigation method as to how exactly the cuticular permeability can be assessed from the obtained results about transpiration and vapor absorption rates. Until recently, for technical reasons, only weight measurements were used to determine water exchange rates. In order to investigate the cuticular permeability by weight change or to assess reported results obtained by this method respectively, the following considerations have to be taken into account.

Transpiration and vapor absorption through the cuticle can be described in a simplified manner by the diffusion of water across a homogeneous membrane in each direction. The flux rate is determined by the Fick's diffusion law:[1,13]

$$F = D \cdot \Delta c \cdot \Delta x \quad (mg \cdot cm^{-2} \cdot sec^{-1}) \tag{1}$$

where: F is the rate of flux (mass · surface unit^{-2} · time^{-1}); D is the diffusion coefficient (length of diffusion channel^2 · time^{-1}, cm^2 · s^{-1}); Δc is the difference in water concentrations between the outside and inside of the animal (see the following explanation); and Δx is the thickness of the cuticle (cm).

Water concentrations can be expressed by water activities. For liquid water, i.e., inside the animal's body, the acticity is equal to 1. In the vapor phase, its value is % relative humidity divided by 100. Osmotic changes of the body water within the range normally occurring in arthropods do not have much effect on the water activity of body water (see calculation in Arlian and Veselica[14]).

As shown in Equation 1, the water flux rate is, apart from the cuticular permeability or diffusion coefficient, respectively, also influenced by the air humidity. To eliminate the effect of different environmental temperatures and humidities on the diffusion rate, the transpiration rates should be divided by the corresponding vapor pressures. However, as will be discussed later, this is only true in the transpiration rates obtained in dry air.

The cuticular water exchange is usually expressed either as a proportion of the total body weight or as a mass rate per unit surface. The first term does not take into account that with a decrease in body size the surface area-to-volume ratio increases. For a reliable comparison of water exchange rates between larger and smaller arthropods this term is not suitable. The second expression bears the problem that the exact surface area is difficult to obtain. They are usually derived from the animal's weight by the empirical surface-weight relationship $S = k \cdot w^{0.67}$, where S is the surface area in cm^2, k is an animal specific constant, and w is the animal weight in grams. Edney[15] and Cloudsley-Thompson[16] found for various woodlice an average k of about 12. In the literature, both terms have been used. Unfortunately, the body weight is often not mentioned. Thus, some results published by different authors are difficult to transform into comparative units.

An exact method to measure transpiration rate and cuticular permeability per a definite unit surface in dry air has been developed by Hadley et al.[17] Cuticle segments are taken from pereonal tergites, cleaned of all adherent tissue, and then exposed in a temperature-regulated chamber. Dry air is circulated over the cuticle where it acquires the moisture diffusing through the cuticle. The humidified air then passes over an aluminum oxide water sensor which registers its water content.

Both the method of Hadley et al.[17] and the traditional weight measurements provide only information about the net water exchange. When experiments are performed in dry air, only transpiration occurs. In this case, weight measurements do represent the cuticular permeability and they are identical with the actual transpiration rate. However, with rising air humidity, water vapor absorption may also take place. Then only the difference between transpiration and absorption of water vapor can be recorded. Thus, if woodlice are exposed to humid air, weight measurements are not an adequate method to determine transpiration and absorption rate or cuticular permeability.

To discriminate between transpiration and absorption rate and to obtain adequate results about cuticular permeability in different air humidities, a combination of weight measurements and tracing of tritiated water is necessary. A suitable method was originally developed by Wharton and Devine[18] and since then has been used several times (for latest literature see Coenen-Stass[19]). The first two authors could demonstrate that the rate of cuticular transpiration in arthropods is related to the amount of exchangeable body water by a first-order or a multiple-order rate, according to how many separate body water compartments in the animals exhibit different water exchange rates.

A constant percentage of the exchangeable water is transpired from each water compartment, which is expressed by the so-called transpiration rate constant k_T. The actual transpiration rate is given by the multiplication of the transpiration rate constant and the amount of exchangeable water left inside each water compartment (see Equation 6). The transpiration rate therefore changes continuously in time as long as the amount of body water is either reduced by transpiration or increased by vapor absorption. In other words, in environmental conditions where both transpiration and vapor absorption occur, the transpiration rate constant and not the weight change is the adequate measure for the cuticular permeability. Only in experiments performed in dry air in which the body weight is not changing much compared to the original value, the weight change expressed as percent of original body weight is nearly equivalent to the transpiration rate constant.

Details about the experimental procedure are described by Wharton and Devine[18] and Devine and Wharton,[20] while a complete calculation example with all necessary experimental data is given by Coenen-Stass.[19] Calculations of the water exchange in animals with more than one body water compartment are described by Wharton and Richards[21] and Arlian and Veselica.[13]

A short description is given below about the way to calculate the exchange rates of an animal with a single water compartment. The transpiration rate constant, k_T, is identical to the rate constant for the loss of tritiated water from an animal exposed to a nontritiated atmosphere:[18]

$$-k_T \cdot t = \ln T_t/T_o \tag{2}$$

where: t is the time in hours spent in the test humidity, ln is the logarithmus naturalis, T_o is the initial tritium content (counts per minute), and T_t is the final tritium content after the test time, t.

The transpiration rate constant, k_T, is also given by the equation:[20]

$$-k_T = \ln[(m_t - m_\infty)/m_o - m_\infty)] \tag{3}$$

where: m_o is the initial body water mass after standardization, m_t is the final body water mass after spending the time, t is the investigated air humidity, and m_∞ is the body water mass when the organism is in equilibrium with its ambient air humidity.

The equilibrium body water mass m_∞ can be obtained at equilibrium conditions when a balance is achieved between transpiration and absorption rate and the animal does not show weight change during further exposure to the tested air humidities. In unsaturated air, m_∞ may not be obtainable by weight measurements since the animal may die before the balance is reached. In this case, m_∞ can be calculated by rearranging Equation 3:[14,22]

$$m_\infty = (m_t - m_o\, e^{-k_T \cdot t})/1 - e^{-k_T \cdot t} \ [\text{mg}] \tag{4}$$

m_o and m_t can be determined by weight measurements before and after the exposure to corresponding air humidities for the time t and k_T can be obtained from Equation 1 by the measured change in specific radioactivity of the body water during exposure time.

After the determination of m_∞, the absorption rate m_S can be calculated by the equation:[18]

$$m_S = m_\infty \cdot (-k_T) \quad [\text{mg} \cdot \text{h}^{-1}] \tag{5}$$

It has been shown for all arthropods so far investigated that this passive absorption, as will be discussed later, at any given ambient humidity, follows a zero-order rate relationship (reviewed by Wharton and Richards[21]). This means that this kind of vapor absorption is independent of the hydration condition or the amount of actual body water mass (m_t) of the animal and, as long as there is no change in cuticular permeability, its rate is only dependent on the surrounding air humidity.

In contrast to that, the transpiration rate is dependent on m_t. It is given by the equation:[20]

$$m_T = m_t \cdot (-k_T) \quad [\text{mg} \cdot \text{h}^{-1}] \tag{6}$$

Vapor absorption measurements at a constant temperature on the red wood ant[19] and on woodlice (see the following chapter) have shown that with rising air humidity, water vapor absorption increases gradually. Since in 100% relative humidity (r.h.) for the red wood ant a weight loss and for the woodlice only a small weight gain is recorded, the transpiration rate constant cannot be reduced to nil as it should be if transpiration where related only to the drying power of the surrounding air. Therefore, the higher the humidity of the air, the lesser the transpiration rate is related to the saturation deficit. When the effect of temperature on cuticular permeability has to be investigated, only measurements in dry air allow the elimination of a variable drying power of the air by dividing the transpiration rate through the corresponding saturation deficit.

III. RESULTS AND DISCUSSION

A. Absorption of Water Vapor From the Atmosphere

It is well established for many woodlice that after desiccating they regain weight during exposure to water-saturated air. Thus, previously desiccated woodlice of the genus *Oniscus, Porcellio, Philoscia, Cylisticus, Ligia,* and *Armadillidium* could compensate at least some of their weight loss in 100% air humidity, but not in 95% r.h. (Edney[15]). *Porcellio scaber*[23] and also desiccated *Oniscus asellus*,[24] *Armadillidium vulgare, Hemilepistus reaumuri*, and *H. aphganicus*[25] showed a weight increase until a constant weight was regained. The increase was regarded by the quoted authors as evidence that water vapor was absorbed in water-saturated air.

With the method described by Wharton and Devine,[18] it was possible to distinguish between the transpiration and absorption rate of *H. reaumuri, A. officinalis*, and *A. pallasii* and to demonstrate that absorption occurs also in water-unsaturated air but is masked by a higher transpiration rate (Figure 1). At a constant temperature of 26°C, the vapor absorption rates increase continuously with rising air humidities. According to Fick's law (Equation 1), the diffusion rate of water molecules is, apart from the cuticular permeability, also dependent on the vapor pressure of the surrounding air. Since with increasing air humidity the vapor pressure rises too, the vapor absorption increases correspondingly, although the cuticular permeability remains constant. Because in the three investigated woodlice, water vapor is absorbed disproportionately in very high humidities, their cuticular permeability increases with rising humidities. This is also established by transpiration results which will be discussed later.

For normally hydrated *H. reaumuri, A. officinalis,* and *A. pallasii*, the absorption and transpiration rates are equal in about 97% r.h. No weight change can be observed. At higher humidities, the weight increases up to about 105% of the normal body weight. At lower ambient humidities, the transpiration rates are higher than absorption rates and the woodlice lose weight (Figure 1).

According to Wharton and Richards,[21] an absorption rate which increases continuously with rising air humidities, is caused by diffusion, i.e., by a passive transport mechanism. It must be distinguished from so-called water vapor absorption shown by some insects and mites (literature reviewed by Machin et al.[26] and by Rudolph and Knülle[27]). These animals possess, near their mouth or in their rectum, special fluid compartments with high electrolyte concentrations and, therefore, with reduced water activity. If the water activity of these compartments is lower than that of the environmental air water, vapor molecules are absorbed from the atmosphere and eventually transferred to the body fluid. For woodlice, rectal or mouth water vapor absorption can be excluded since occlusion of both parts does not prevent the weight increase in saturated air.[12,25]

In woodlice, a possible site of water vapor absorption outside the excluded mouth and rectal region might be the pleoventral chamber, the space between the pleoventral body wall and the exopodites of the pleopods. In this chamber are located the endopodites which are permanently covered with a water film, apparently in relation to their respiratory function (for literature see Kümmel[28]). It was found by Kümmel[28,29] that in all species of *Oniscidea* investigated (*Oniscus, Porcellio, Hemilepistus*, and *Armadillidium*) the epithelial cells of the pleopodial endopodites show the ultrastructural organization of transporting epithelia (literature quoted by Kümmel[28]). Especially apical and basal plasma membranes are deeply folded with most of the mitochondria located in the region of the basal infoldings.

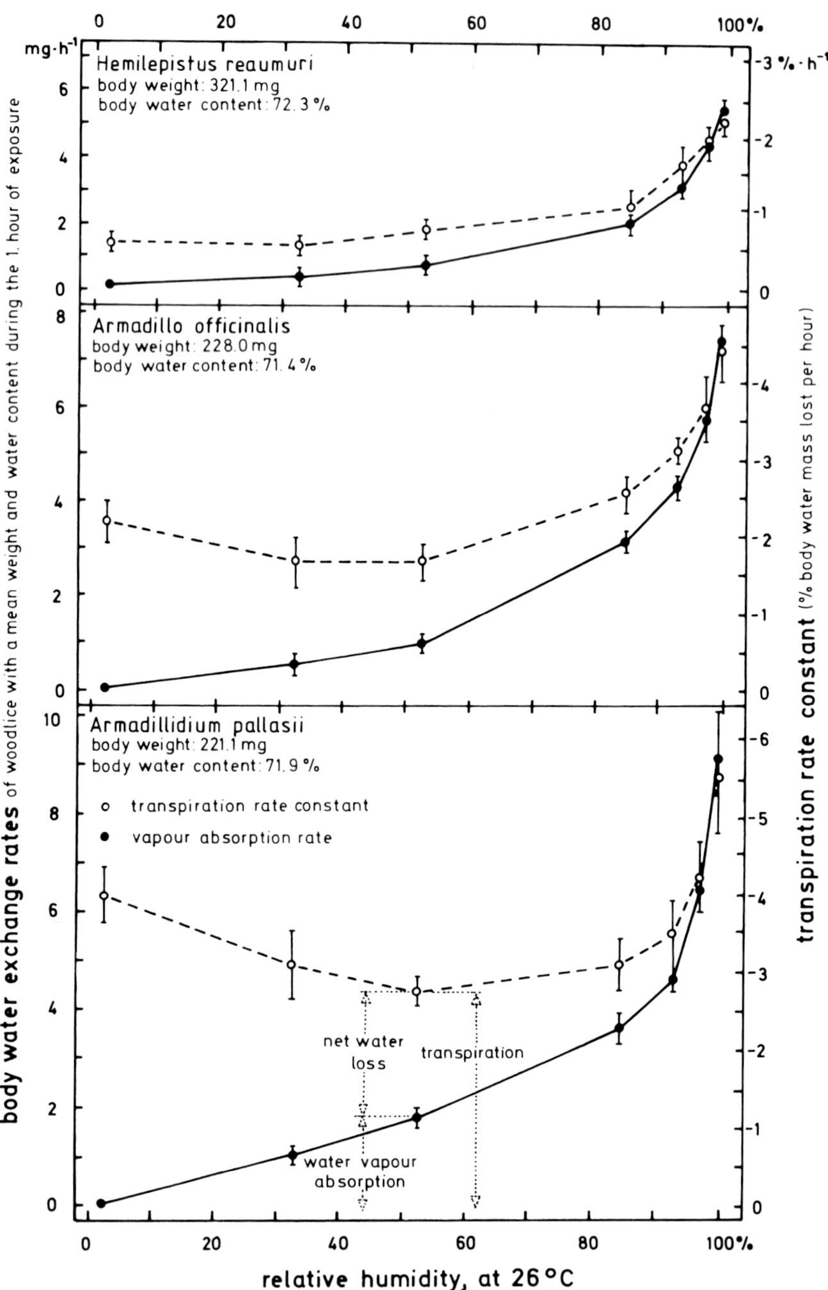

FIGURE 1. Water exchange rates of three woodlice with an average weight and an average water content during the first hour of their exposure to different humidities at 26°C ($\bar{X}$, $\pm$SD). Left ordinate: absorption and transpiration rates; right ordinate: transpiration rate constant.[12]

It might be assumed that water vapor from the surrounding atmosphere could condense in the pleoventral chamber and would be osmotically removed by the mentioned epithelial cells of the endopodites. Capillary spaces may facilitate vapor condensation. The smaller the dimensions of spaces, the lower are the air humidities at which vapor will condense. However, if capillar condensation should not cease in vapor-unsaturated air the condensed water must be absorbed from the spaces continuously.

A good model of capillary condensation has been described by O'Donnell[30] for the desert cockroach *Arenivaga*, which absorbs atmospheric water condensed on the paired pharyngeal bladders. These are densely covered by extremely fine hairs arranged parallel to one another. The cuticle of the hairs is extremely wettable so that water readily spreads along the triangular grooves formed between neighboring hairs. It is suggested that the cyclical additions of fluid secreted by the so-called frontal body alters the water affinity of the bladder cuticle so that the condensed water is released and then swallowed through the mouth.

At first sight, there might be a striking analogy to what happens in woodlice. According to Lindqvist,[31,32] *Armadillidium vulgare* and *Porcellio scaber* discharge fluid periodically from the mouth, which then should be transported via the water-conducting system towards the pleoventral chamber.[33] Both Lindqvist[31,32] and Hoese[33] believe that the secreted fluid is finally resorbed via the rectum. It is tempting to assume that similar to the O'Donnell model, the secreted fluid might support condensation and removal of water vapor from the suspected condensing surface structures by lowering their water affinity. However, in *Hemilepistus reaumuri*, the water vapor uptake was not interrupted by occlusion of the mouth and anus.[12,25] While a complete sealing of the mouth could not be guaranteed because the woodlice tried hard to remove the wax plug, the blockade of the anus was easily achieved.

What is left is the question of whether a condensation of the water vapor alone or accompanied by absorption occurs in the pleoventral chamber, even without the help of a specially secreted condensing fluid as found in *Arenivaga*. In any case, in some aquatic isopods and in all euryhaline marine isopods investigated, regions of the pleopods have been found which indicate osmoregulatory functions (literature quoted by Kümmel[28]). Since probably all land isopods have been descended from marine ancestors,[6] it is likely that the assumed osmoregulatory function of the euryhaline ancesters still remains in their terrestrial descendants.[28]

The question remains whether the epithelial cells of the pleopodal endopodites are actually involved in the water vapor absorption. In the red wood ant, *Formica polyctena*, whose water balance was investigated by the same method as used to study *H. reaumuri*, the vapor absorption also rose gradually with increasing air humidity. But it occurred at a much lower level and in water-saturated air was still distinctly lower than transpiration.[19] Thus, the ability of woodlice to absorb water is not dependent entirely on the structure of its pleoventral chamber and is related in some way to the permeability of the total cuticle. But the absorption is presumably important only in those species which regain weight in water-saturated air. They may possess additional facilities like those mentioned for insects and mites or perhaps the osmotic structures inside the pleoventral chamber of woodlice.

B. Respiratory Water Loss

An important morphological adaptation of *Isopoda* to terrestrial life is the development of an air-breathing respiratory system. In *Oniscidea*, respiratory organs are located on the exopodites of the pleopods. According to Hoese,[34] who reinvestigated the results of Verhoeff,[35] all intermediate steps from the predominantly gill-breathing genus *Ligia* to the nearly exclusively lung-breathing genus *Hemilepistus* can be found. The line of development starts with a simply folded, uncovered lung in *Oniscus* to a much more folded lung in *Trachelipus* and ending in the highly developed tubulous lung in the species *Armadillo* and *Hemilepistus* which opens into an atrium. The lungs are located on the inner side of exopodites facing the pleoventral chamber. However, their hydrophobic atrium is located outside the chamber near the insertion of the exopodites. Woodlice living in drier habitat are able to close the entrance of the lungs' cavity by vasodilation of the neighboring blood sinus.

Until now, only Edney[15] investigated the respiratory water loss in woodlice. From the difference in weight loss between uncovered woodlice and those with covered dorsal surface or with covered pleopodal area, the respiratory water loss was calculated. Freshly killed *A.*

vulgare lost respiratory water to 25% of their total water loss. Corresponding figures for *P. scaber* 42% and *L. oceanica* are surprisingly only 25%.

C. Cuticular Water Loss and Cuticular Permeability

1. Evaporation Rate as a Function of Time

In order to compare published results about water loss, the duration of exposure must be taken into account. Some woodlice show a decline in their transpiration rate during exposure to desiccating atmospheres: *O. asellus*,[8,31,32,36] *A. vulgare*,[8,36,37] *P. scaber*,[8,15,36] *P. laevis*, and *Porcellionides pruinosus*.[38,39] However, Holdich and Mayes[40] could not find a distinct initial high water loss in *O. asellus*. No initial peak of water loss was found in the xeric woodlouse *V. arizonicus*.[7] For the desert woodlouse, *H. reaumuri*, conflicting results are reported. While I could not find, by hourly weighing up to 24 hr, any change in the weight loss rate[25] and in the cuticular permeability,[12] Warburg[41] reported for the first 3 hr a somewhat higher weight loss compared to the first 24 hr of exposure in dry air.

Different opinions are stated in the literature on how to explain the initial high water loss. Edney[15] and Bursell[36] believed that this effect on *A. vulgare* and on *O. asellus* may be caused by drying off the outer layers of the integument, which thereby should become less permeable. Lindqvist[31,32] and Lindqvist et al.[42] stated that in *P. scaber* and *A. vulgare* the cuticular permeability was not changed during desiccation because the initial peak in weight loss was completely eliminated after occlusion of mouth and anus. They believed that both species periodically discharge water from the gut via mouth and anus, probably in order to moisten respiratory surfaces in the pleoventral chamber. Hoese[33] proposed that only the maxillar-nephrids located in the mouth region secrete a fluid into the water conduction system. This consists of low lateral capillary channels on the ventral side of the body extending from the mouth part to the anus.[33,35] Hoese[33] found some ammonia in the secreted fluid which may vaporize before the remaining fluid should be resorbed through the anus. It has been established that woodlice excrete nitrogen mainly as ammonia.[43]

It remains to be seen whether the initial high weight loss is actually caused by the evaporation of water secreted and probably stored in the pleoventral chamber, as Warburg[7] has also proposed.

2. Effects of Air Humidity

As mentioned above, the effect of different air humidities on the cuticular permeability of woodlice cannot be investigated by following the weight change alone, since transpiration and absorption of water vapor occur simultaneously. Using the method described by Wharton and Devine,[18] the influence of air humidities on the cuticular permeability was studied at 26°C by investigating the transpiration rate constants of three woodlice species (Figure 1).

With rising air humidity, the rate constant of *H. reaumuri* increased only slightly at lower humidities, but more distinctly at higher humidities. The effects on the other two species were similar but more marked. Therefore, contrary to what has been concluded from weight measurements (literature reviewed by Edney[1]), the cuticular permeability increases with rising humidities. This is also confirmed by the increase in vapor absorption above the corresponding rise in vapor pressure already mentioned.

To compare the transpiration constant with the absorption rate, the rate constants were used to calculate the transpiration rate for the three woodlice, with an averaged body weight and an averaged body water content, which occurs during their first hour of exposure in the investigated air humidity (see Equation 6). As seen in Figure 1, the desert woodlouse exhibits the lowest water exchange and thereby the lowest cuticular permeability in comparison with the Mediterranean species.

Since with rising air humidity the rates of vapor absorption increase faster than the transpiration rates, the weight loss of the three species is reduced as has been observed in all woodlice so far investigated (for the latest literature see Warburg[41]). At about 97% r.h., both opposed rates are equally high and no weight change occurs. Above this so-called critical equilibrium humidity, the vapor absorption rate is higher than the transpiration rate. The body weight increases until a new balance is established at about 105% of the original (standardized) body weight.[12]

3. Effects of Temperature and Cuticular Structure

As mentioned in Section II, the effect of rising temperature on cuticular permeability can only be obtained when the animal is exposed to dry air, i.e., when no absorption occurs. The effects of a variable drying power of the air can then be eliminated by dividing the transpiration or weight change through the corresponding vapor saturation deficit.

Wigglesworth[44] and Beament[45] were the first who found that with rising temperature the transpiration of insects in dry air increased rapidly above a critical, so-called transition temperature. Both authors assumed that the permeability of insect cuticle was mainly determined by a superficial lipid layer in the epicuticle and that at the transition temperature the ordered state of the lipid molecules became disturbed, causing an increased cuticular permeability. The transition phenomena was taken as an evidence for the presence of a monolayer lipid barrier. However, further investigations showed that the breaks in the transpiration-temperature curves were less expressed when the rates were divided by the corresponding vapor pressure of the surrounding air. The discussion about the existence of a lipid monolayer and the explanation of the transition phenomena is still not settled (see Chapter 13).

Published results about transpiration-temperature curves in woodlice exposed in dry air are summarized in Figure 2. On the left side, reported transpiration rates are shown (except the results of Hadley and Quinlan,[38] which are originally related to the water saturation deficit as shown in k). On the right side their values are divided by the corresponding water saturation deficit. Most of these values do not show a distinct increase at a certain temperature. This was taken as evidence for the nonexistence of a lipid barrier in woodlice. But a few exceptions are reported. Bursell[36] found in *Oniscus asellus*, during exposure to 40% r.h. at 28 and 35°C, and in *Porcellio dilatatus* at 28°C, transition points. He proposed that the lipids responsible lie deep in the cuticle since superficial abrasion had little effect. A reexamination of his results only confirms one small break for *O. asellus* at 28°C (Figure 2c, d). In addition, Mead-Briggs[46] could find no break in the transpiration temperature curve of *O. asellus* when its transpiration rate was related to the coresponding air saturation deficit. Warburg[37,41] observed transition points in *Venezillo arizonicus* at 38°C and in *Hemilepistus reaumuri* at 40°C (Figure 2e, f). In my investigation on *H. reaumuri*,[25] I found no such an effect for a temperature range up to 39°C (Figure 2e, f). Furthermore, species of the genus *Buddelundia* from Australia showed, at about 35°C, a sharp increase in weight loss.[7] To sum up, the transition phenomena seem to occur only in xeric-adapted woodlice, exhibiting a very low transpiration rate, like the genus *Buddelundia, V. arizonicus,* and probably *H. reaumuri*. Thus, only these woodlice may possess a distinctly formed lipid layer in their cuticle.

Recently, Hadley and Quinlan[38] were able to extract the cuticular lipids from the integument of *P. laevis* and studied their effect on cuticular permeability. Chromatographic analysis of the lipid extract indicated that basically the same lipid classes which characterize the cuticles of terrestrial insects and arachnids are present in this woodlouse, but to a much lesser extent than in other terrestrial arthropods. Since extraction of epicuticular lipids did not cause an increase in water loss, the authors concluded that the amount of lipids did not provide the effective barrier to water flux as found in most other terrestrial insects and arachnids.

However, in more xeric-adapted woodlice, different results emerged. Thus, the surface

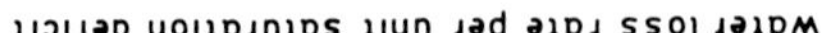

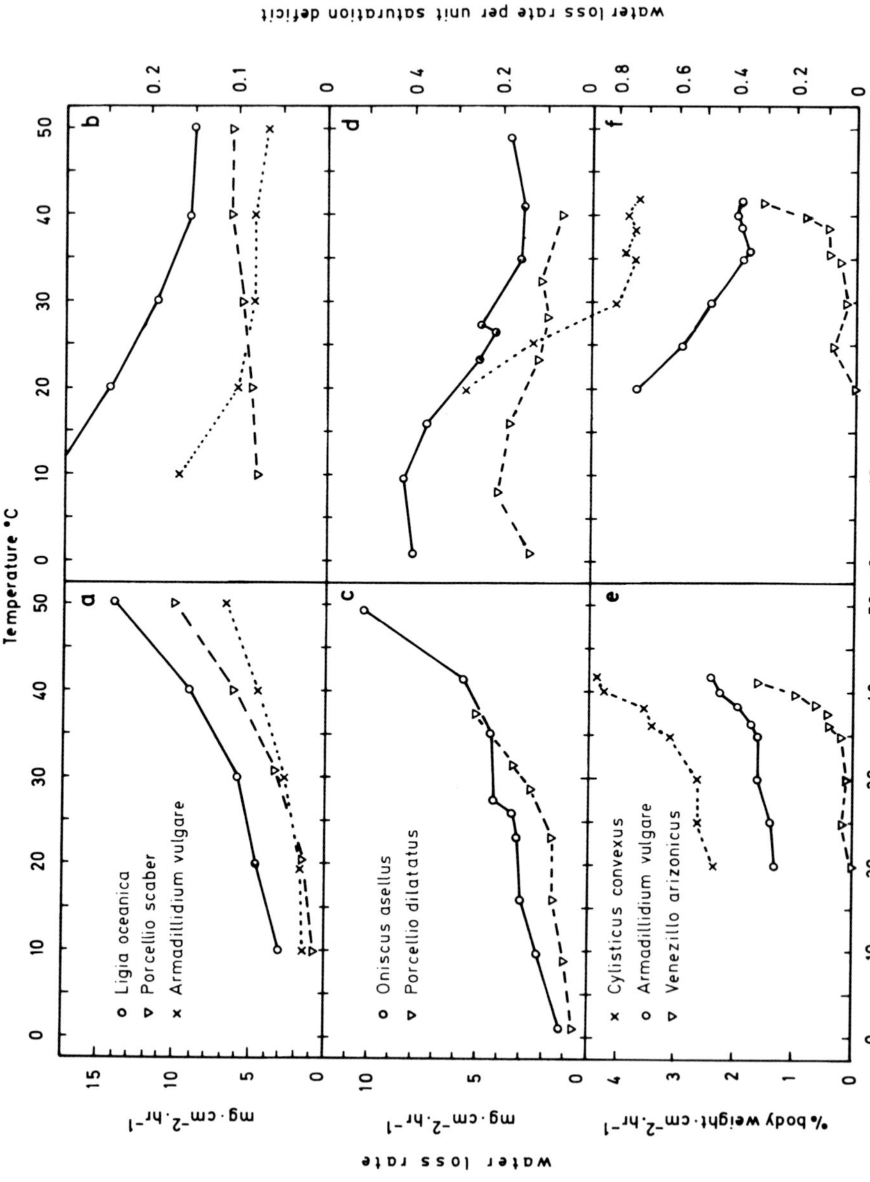
Temperature °C
water loss rate
mg·cm⁻²·hr⁻¹
%body weight·cm⁻²·hr⁻¹
water loss rate per unit saturation deficit
a
b
c
d
e
f
o Ligia oceanica
▽ Porcellio scaber
× Armadillidium vulgare
o Oniscus asellus
▽ Porcellio dilatatus
× Cylisticus convexus
o Armadillidium vulgare
▽ Venezillo arizonicus

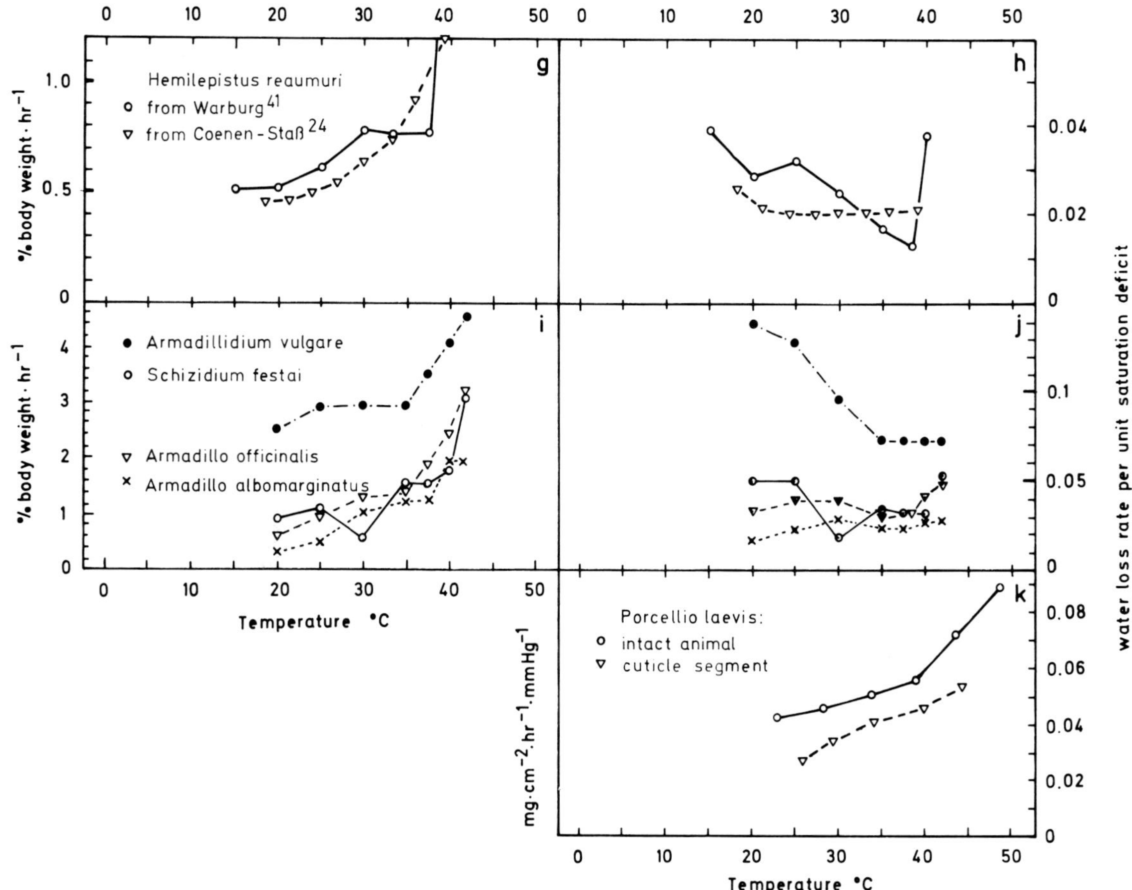

FIGURE 2. Effects of temperature on the water loss of woodlice in dry air. (a) From Edney,[15] (c) from Bursell,[36] (e) from Warburg,[7] (g) from Warburg[41] and Coenen-Stass,[19] (i) from Warburg,[41] and (k) from Hadley and Quinlan.[38]

density of cuticular lipids (hydrocarbons) for *H. reaumuri* is comparable to values reported for the desert scorpion *Hadrurus arizonicus* (see the literature quoted by Hadley and Warburg[47]). It was found that in contrast to *A. albomarginatus* and *A. officinalis*, the cuticle of the desert woodlouse *Hemilepistus reaumuri* contains many more long-chain branched compounds of hydrocarbons. Their composition is very similar to the patterns exhibited by xeric-adapted arthropods. Both authors could not say whether the differences in the molecular composition of the respective hydrocarbon fraction were responsible for the differences in water loss among the three species, but they found strong evidence that in all three species cuticular lipids contribute to a reduced cuticular permeability, contrary to the findings for *P. laevis.*[38]

Many terrestrial isopods do possess microstructures like tricorns, spherical particles in form of balls of various diameter, and small granules.[48] Hadley and Hendricks[49] investigated the ball structures of *Porcellionides pruinosus* and found that they do not primarily contain lipids. Nevertheless, it was established that their presence reduced cuticular transpiration in those species, probably by increasing the boundary layer on the animals' surface and (or) lengthening the diffusion pathway.

A further analysis of the graphs presented in Figure 2 (right side) reveals that in woodlice with a high water loss rate, the proportion "water loss rate/saturation deficit" is reduced with rising temperatures. This could lead to the rather unrealistic conclusion that their cuticular permeability is reduced with rising temperatures. However, these woodlice do also show a high initial water loss. Its effect on the transpiration rate should be more marked at lower temperatures because the transpiration rates are also low.

In general, it can be assumed that the rising of temperature has little effect on the cuticular permeability of woodlice showing a high water loss. In woodlice, with a moderate water loss, it may cause only a gradual increase in the permeability as found in *P. laevies* (Hadley and Quinlan,[38] Figure 2k). A few xeric-adapted woodlice with very low water loss may even show some kind of transition phenomena.

4. *Hormonal Control of Water Loss*

Currently no evidence is known about hormonal influences on the cuticular permeability. In regard to a hormonal effect on the transpiration rate of woodlice, possible hormonal regulations of the body water content must also be taken into account. A first report is now available suggesting that antidiuretic hormones may be present in woodlice, influencing the water content similarly to what is known from vertebrates. Takeda et al.[50] detected in the central nervous system and partly in the alimentary system, including the hepatopancreas of *Ligia exotica, P. scaber*, and *A. vulgare*, cells which are immunoreactive to antisera of antidiuretic peptides (vasopressin, vasotocin) and to the drinking inducing peptide angiotensin II. Injection of vasopressin inhibited the water loss and elongated the lethal time in desiccating conditions.

D. Water Balance and Habitat

For a direct comparison of the transpiration rates of different woodlice, they should be identically pretreated and investigated by the same method. Since different methods have been used, a compromise has to be made. As discussed in the section on methods of investigation, a great number of the results on water exchange have been expressed in weight change per initial body weight or squared centimeter. Only when the animals were exposed to completely dry air is their weight change actually related to the transpiration rate and cuticular permeability.

In order to evaluate the differences in cuticular permeability, reported data about transpiration rates in dry air at about 30°C have been compiled in Table 1. They are expressed, partly transformed from other units, both in milligrams per cuticular surface unit and in relation to the corresponding air humidity saturation deficit, regardless of the different

Table 1
COMPARATIVE RATES OF WATER LOSS OF WOODLICE IN DRY AIR[15]

Species	Air Temp. (°C)	Water loss rate mg/cm/hr	Water loss rate mg/cm/hr/mmHg	Duration (hr)	Ref.
Ligia oceanica	30	7.0	0.22	1	15
Cylisticus convexus	30	7.0	0.22	1	15
C. convexus	30	3.98	0.125	1	7
Philoscia muscorum	30	5.72	0.18	1	15
Oniscus asellus	30	4.2	0.132	1	15
Porcellio scaber	30	3.5	0.11	1	15
P. dilatatus	29	3.07	0.104	0.5—1	36
P. laevis	30	3.08	0.097	?	39
Armadillidium vulgare	30	2.7	0.085	1	15
P. pruinosus	30	2.2	0.068	?	39
A. pallasii	26	1.44	0.057	3-6	12*
Periscyphis jannonei	30	1.3	0.041	1	16
Venezillo arizonicus	30	1.01	0.032	1	7
Armadillo officinalis	30	0.65	0.02	3	4
A. officinalis	26	0.79	0.03	6—24	12*
A. albomarginatus	30	0.35	0.011	3	41
Hemilepistus reaumuri	30	0.33	0.01	8	19*
H. reaumuri ♂	30	0.4	0.013	3	47
H. reaumuri ♀	30	0.49	0.015	3	47
H. reaumuri ♂	30	0.33	0.01	24	41
H. reaumuri	26	0.25	0.01	6—24	12*

Note: *—original unit of published data changed.

investigation methods used. To avoid a possible influence of the initial high water loss, an exposure time of at least 1 hr was considered. In the case of the low transpirating desert woodlouse, results of longer lasting exposures are mentioned too.

Although this comparison is not without simplification, it should allow some conclusions to be drawn about a general correlation between the permeability of the cuticle and the local distribution of woodlice on land. As seen in the Table 1, the succession of the woodlice arranged according to their cuticular permeability does correspond to the dryness of their habitat mentioned in the introduction, i.e., woodlice with high cuticular permeability are found in humid habitats and vice versa, low cuticular permeability is shown by woodlice living in dry habitats. Of course, morphological adaptations like the evolution of the lungs as described by Hoese[34] are also important.

A general discussion about the effect of climate on the distribution and abundance of woodlice is given by Warburg et al.[51] It must be restated that woodlice retreat into spaces with high air humidities where they even may regain body water by vapor absorption. Therefore, it is more likely that the cuticular permeability may be related more closely to the time the woodlice can spend away from saturated air.[2,5] For a detailed assessment about the relations between the water balance and the dryness of the habitat, the behavior activity of the woodlice has also been considered. As mentioned by Edney,[5] this aspect may explain why, for instance, species of *Oniscus, Porcellio,* and *Armadillidium* are found sometimes in the same places although their transpiration rates are quite different.

REFERENCES

1. **Edney, E. B.,** *Water Balance in Arthropod,* Springer-Verlag, Berlin, 1977.
2. **Cloudsley-Thompson, J. L.,** *The Water and Temperature Relation of Woodlice,* Meadowfield Press, Shildon, U.K., 1977.
3. **Sutton, S. L.,** *Woodlice,* Pergamon Press, Oxford, 1980.
4. **Wieser, W.,** Ecophysiological adaptations of terrestrial isopods: a brief review, *Symp. Zool. Soc. London,* 53, 247, 1984.
5. **Edney, E. B.,** Woodlice and land habitat, *Biol. Rev.,* 29, 185, 1954.
6. **Edney, E. B.,** Transition from water to land in isopod crustaceans, *Am. Zool.,* 8, 309, 1968.
7. **Warburg, M. R.,** Water relation and internal body temperature of isopods from mesic and xeric habitats, *Physiol. Zool.,* 38, 99, 1965.
8. **Lindqvist, O. V.,** Water regulation in terrestrial isopods, with comments on their behavior in a stimulus gradient, *Ann. Zool. Fenn.,* 5, 279, 1968.
9. **Linsenmair, K. E. and Linsenmair, C.,** Paarbildung und Paarzusammenhalt bei der monogamen Wüstenassel *Hemilepistus reaumuri (Crustacea, Isopoda, Oniscoidea), Z. Tierpsychol.,* 29, 134, 1971.
10. **Shachak, M.,** Energy allocation and life history strategy of the desert isopod *H. reaumuri, Oecologia,* 45, 404, 1980.
11. **Coenen-Stass, D.,** Observations on the distribution of the desert woodlice *Hemilepistus reaumuri (Oniscidea, Isopoda, Crustacea)* in North Africa, *Symp. Zool. Soc. London,* 53, 369, 1984.
12. **Coenen-Stass, D.,** *Okophysiologische Untersuchungen an der Wüstenassel Hemilepistus reaumuri, mit einigen vergleichenden Untersuchungen an den beiden Mediterranen Arten Armadillo officinalis und Armadillidium pallasii (Oniscidea, Isopoda, Crustacea),* Habilitationsschrift, Universität Karlsruhe, Karlsruhe, W. Germany, 1987.
13. **Gilby, A. R.,** Transpiration, temperature and lipids in insect cuticle, in *Advances in Insect Physiology,* Vol. 15, Berridge, M. J., Treherne, J. E., and Wigglesworth, V. B., Eds., Academic Press, New York, 1981, 1.
14. **Arlian, L. G. and Veselica M. M.,** Water balance in insects and mites, *Comp. Biochem. Physiol.,* 64A, 191, 1979.
15. **Edney, E. B.,** The evaporation of water from woodlice and the millipede *Glomeris, J. Exp. Biol.,* 28, 91, 1951.
16. **Cloudsley-Thompson, J. L.,** Acclimation, water and temperature relations of the woodlice *Metoponorthus pruinosus* and *Periscyphis jannonei* in the Sudan, *J. Zool. London,* 158, 267, 1969.
17. **Hadley, N. F., Stuart, J. L., and Quinlan, M.,** An air-flow system for measuring total transpiration and cuticular permeability in arthropods: studies on the centipede *Scolopendra polymorpha, Physiol. Zool.,* 55, 393, 1982.
18. **Wharton, G. W. and Devine, T. L.,** Exchange of water between a mite, *Laelaps echidnina,* and the surrounding air under equilibrium conditions, *J. Insect Physiol.,* 14, 1303, 1968.
19. **Coenen-Stass, D.,** Investigations about the water balance of the red wood ant, *Formica polytena (Hymenoptera, Formicidae):* workers, their larvae und pupae, *Comp. Biochem. Physiol.,* 83A, 141, 1986.
20. **Devine, T. L. and Wharton, G. W.,** Kinetics of water exchange between a mite *Laelaps echidnina* and the surrounding air, *J. Insect Physiol.,* 19, 243, 1973.
21. **Wharton, G. W. and Richards, A. G.,** Water vapor exchange kinetics in insects and acarines, *Annu. Rev. Entomol.,* 23, 309, 1978.
22. **Knülle, W. and Devine, T. L.,** Evidence for active and passive components of sorption of atmospheric water vapour by larvae of the tick *Dermacentor variabilis, J. Insect Physiol.,* 18, 1635, 1972.
23. **Boer Den, P. J.,** The ecological significance of activity pattern in the woodlouse *Porcellio scaber* Latr. *(Isopoda), Arch. Neerl. Zool.,* 14, 283, 1961.
24. **Mayes, K. R. and Holdich, D. M.,** Water exchange between woodlice and moist environments, with particular reference to *Oniscus asellus, Comp. Biochem. Physiol.,* 51A, 295, 1975.
25. **Coenen-Stass, D.,** Some aspects of the waterbalance of two desert woodlice, *Hemilepistus apghanicus* and *Hemilepistus reaumuri (Oniscoidea, Isopoda, Crustacea), Comp. Biochem. Physiol.,* 70, 405, 1981.
26. **Machin, J., O'Donnel, M. J., and Coutchie, P. A.,** Mechanisms of water vapor absorption in insects, *J. Exp. Zool.,* 222, 309, 1982.
27. **Rudolph, D. and Knülle, W.,** Novel uptake systems for atmospheric water vapor among insects, *J. Exp. Zool.,* 222, 321, 1982.
28. **Kümmel, K.,** Fine-structural investigations of the pleopodal endopods of terrestrial isopods with some remarks on their function, *Symp. Zool. Soc. London,* 53, 77, 1984.
29. **Kümmel, G.,** Fine structural indications of an osmoregulatory function of the "gills" in terrestrial isopods *(Crustacea, Oniscoidea), Cell Tissue Res.,* 214, 663, 1981.

30. **O'Donnell, M. J.,** Hydrophilic cuticle — the basis for water vapour absorption by the desert burrowing cockroach, *Arenivaga investigata, J. Exp. Biol.*, 99, 43, 1982.
31. **Lindqvist, O. V.,** Evaporation in terrestrial isopods is determined by oral and anal discharge, *Experientia*, 27, 1496, 1971.
32. **Lindqvist, O. V.,** Components of water loss in terrestrial isopods, *Physiol. Zool.*, 45, 316, 1972.
33. **Hoese, B.,** Morphologie und Funktion des Wasserleitungssystems der terrestrischen Isopoden *Crustacea, Isopoda, Oniscoidea), Zoomorphology*, 98, 135, 1982.
34. **Hoese, B.,** Morphologie und Evolution der Lungen bei terrestrischen Isopoden *(Crustacea, Isopoda, Oniscoidea). Zool. Jahrb. Anat. Abt. Ontog. Tiere*, 107, 396, 1982.
35. **Verhoeff, K. W.,** Über die Atmung der Landasseln, zugleich ein Beitrag zur Kenntnis der Entstehung der Landtiere, *Z. Wiss. Zool. Abt. A*, 118, 365, 1920.
36. **Bursell, E.,** The transpiration of terrestrial isopods, *J. Exp. Biol.*, 32, 238, 1955.
37. **Warburg, M. R.,** The evaporative water loss of three isopods from South Australia, *Crustaceana (Leiden)*, 9, 302, 1965.
38. **Hadley, N. F. and Quinlan, M.C.,** Cuticular transpiration in the isopod *Porcellio laevis:* chemical and morphological factors involved in its control, *Symp. Zool. Soc. London*, 53, 97, 1984.
39. **Quinlan, M. C. and Hadley, N. F.,** Water relations of the terrestrial isopods *Porcellio laevis* and *Porcellionides pruinosus (Crustacea, Oniscoidea), J. Comp. Physiol.*, 151, 155, 1983.
40. **Holdich, D. M. and Mayes, K. R.,** Blood volume and total water content of the woodlouse, *Oniscus asellus*, in conditions of hydration and desiccation, *J. Insect Physiol.*, 22, 547, 1976.
41. **Warburg, M. R.,** Haemolymph osmolality, ion concentration and the distribution of water in body compartments of terrestrial isopods under different ambient conditions, *Comp. Biochem. Physiol.*, 86A, 433, 1987.
42. **Lindqvist, O. V., Salminen, J., and Winston, P. W.,** Water content and water activity in the cuticle of terrestrial isopods, *J. Exp. Biol.*, 56, 49, 1972.
43. **Wieser, W. and Schweizer, G.,** A re-examination of the excretion of nitrogen by terrestrial isopods, *J. Exp. Biol.*, 52, 267, 1970.
44. **Wigglesworth, V. B.,** Transpiration through the cuticle of insects, *J. Exp. Biol.*, 21, 97, 1945.
45. **Beament, J. W. L.,** The cuticular lipoids of insects, *J. Exp. Biol.*, 21, 115, 1945.
46. **Mead-Briggs, A. R.,** The effect of temperature upon the permability to water of arthropod cuticles, *J. Exp. Biol.*, 33, 737, 1956.
47. **Hadley N. F. and Warburg M. R.,** Water loss in three species of xeric-adapted isopods: correlations with cuticular lipids, *Comp. Biochem. Physiol.*, 85A, 669, 1986.
48. **Schmalfuss, H.,** Morphology and function of cuticular microscales and corresponding structures in terrestrial isopods *(Crust., Isop., Oniscoidea), Zoomorphologie*, 91, 263, 1978.
49. **Hadley, N. F. and Hendricks, G. M.,** Cuticular microstructures and their relationship to structural color and transpiration in the terrestrial isopod *Procellionides pruinosus, Can., J. Zool.*, 63, 649, 1985.
50. **Takeda N., Mizuno J., and Fujii, K.,** Land adaptation and neuropeptides in terrestrial isopods, *Monit. Zool. Ital.*, in press.
51. **Warburg, M. R., Linsenmair, K. E., and Bercovitz, K.,** The effect of climate on the distribution and abundance of isopods. *Symp. Zool. Soc. London*, 53, 339, 1984.

Chapter 15

ELECTRO-OSMOSIS SURMOUNTS HIGH WATER POTENTIAL DIFFERENCES IN INSECTS

J. Küppers and U. Thurm

TABLE OF CONTENTS

I. INTRODUCTION

Several species of insects and mites present a particular challenge in the field of water transport: they are able to maintain the high water activity of their body fluids ($a_w = 0.99$) by active resorption of water vapor from substantially subsaturated atmospheres. Up to now, a lot of information has been accumulated on phylogenetic and ecological aspects of this water transport, on its kinetics and regulation, as well as on the structure of the organs involved in this process. The "critical equilibrium activity" (CEA) for a species designates the water activity of the air at which water gain by active resorption just compensates for water loss by transpiration. For some mites, the CEA can be as low as 0.8 to 0.75.[1]

Among insects, we meet water-vapor absorbers, whose ability to gain water reaches down to a CEA of 0.45. That means these animals transfer water against an osmotic gradient of about 70 Osm/kg or an osmotic pressure of at least $1.1 \cdot 10^8$ Pa (= 1100 bar). The capability to absorb water vapor occurs in several species scattered throughout various orders of insects.[2-5]

The water-vapor uptake is performed by distinct but rather different organs at either end of the alimentary canal. For *Psocoptera* and *Mallophaga*,[5] and for the cockroach *Arenivaga*,[6] exposable structures at the mouth play an essential part. In *Tenebrionidae*,[7] *Thysanura*,[8] and flea larvae,[5] we find an anal route of water-vapor uptake. Generally, one function of the last part of the insect gut is the dehydration of feces besides the reabsorption of usable substances.

Anatomy and ultrastructure of the recta differ considerably between various groups, and are often very intricate (cf. to the comprehensive review of Edney[2]). In Tenebrionidae, the "cryptonephridial complex", an association of rectal tissue and the distal parts of the Malpighian tubules, dehydrates the feces and absorbs water vapor from the air. In *Thysanura* and flea larvae, water-vapor resorption is obviously the main function, if not the only one, of the anal sac or rectal sac respectively, because feces pass these segments quickly so that their lumen is usually filled with air.

One common aspect of all known cases of water-vapor uptake is a hygroscopic phase which mediates between the air and the body fluids, transforming water vapor to fluid water of low activity by passive condensation.[9]

For the *Psocoptera* and *Mallophaga*, Rudolph and Knülle[5] propose a more or less continuous secretion of a hygroscopic fluid which is ingested after equilibration with the humidity of the air. We emphasize that in such a case concentrating a soluble substance to a water activity below that of the body fluids increases the water activity at the inside and represents, therefore, the very process of (relative) water gain and the indispensably energy-requiring step. Further energy input is optional for the transfer of the equilibrated fluid from the gut into the hemolymph, since the transfer (leading to net water gain) might be effected by simple dilution with the body fluids.

The hygroscopic phase is stationary in the cockroach *Arenivaga* and in most "anal absorbers"; energy must be spent to transfer the condensed water against the difference of its chemical potential into the hemolymph.

Therefore, the principles of water vapor absorption may be tentatively subdivided into two categories: (1) those using the energy to secrete a concentrated solute to the outside and (2) those using energy to transfer water to the inside by resorption from low activity.

These two modes do not necessarily represent mutually exclusive alternatives (as, for instance (2) may follow (1)), but these categories may help to develop feasible hypotheses on the nature of the active processes which do the osmotic work and which are hardly understood. Some physicochemical principles will not work at physiological conditions. Thus, we need not seriously consider, for example, that hydrostatic pressure might be of any significance for water vapor absorption. To generate the required pressure differences of 100 to more than 1000 bars, both the forces and the firmness of the organs involved, would have to be prohibitively high.

Category (1): the excretion of a soluble substance in a (nearly) dry state may be possible. In this case, the input of metabolic energy has to occur at the site of excretion and structural equivalents of energy turnover should be found there. A problem inherent in (but not specific for) this mode is to prevent the re-equilibration of the excreted matter with the body fluids.

The salt glands of marine birds and reptiles excrete concentrated NaCl solutions.[10] In this context, however, they hardly represent a suitable model because the osmolality of the excreted solutions (0.4 to 0.6 mol/ℓ) is far below that which would be needed to account for water-vapor uptake of insects, and the salts of the main electrolytes K^+, Na^+, and Cl^- are precipitated at water activities below 0.75 so that for lower CEAs the excretion of an organic substance is assumed.

An external hygroscopic phase may possibly be created another way, by which energy input need not occur at the site or the time of excretion: a poorly soluble polymeric substance might be exuded which becomes hygroscopic at the outside by decomposition to soluble monomeres. These could, for instance, be swallowed after equilibration with the air. The energy-requiring osmotic process is then the recomposition of the monomeres to the polymere anywhere within the body, in the course of which the condensed water becomes available.

In the present paper, we shall advance a model appropriate to withdraw water from low activities (category (2)), and we shall report corroborating experimental evidence. This model has been promoted by the fascinatingly simple structure of the anal sac of *Thermobia domestica* described by Noirot and Noirot-Timothee.[11] The anal sac had been identified by Noble-Nesbitt[8,12] to be the organ which enables the animal to gain water from the atmosphere at relative humidities down to 45%.[13]

II. BASIC MORPHOLOGY OF THE WATER VAPOR-ABSORBING ORGAN

We selected *Lepisma saccharina* for our experiments, having ascertained by weighing experiments that this species is capable of a similarly potent water-vapor resorption as the related species *Thermobia*. The morphology of the anal sac of *Lepisma* (studied in parallel with the physiological measurements) is very similar to that of *Thermobia*.[14] This study yielded also specific quantitative data on *Lepisma*. The development of the hypothesis and the interpretation of our experimental results depend heavily on the anatomy and ultrastructure which shall be described first: the anal sac, or posterior rectum, of *Thysanura* is the last part of the gut separated from the rectum by a tightly closing sphincter (S in Figure 1). The narrow lumen (L) is lined by a thin intima (C) and opens via a ventral cleft to the atmosphere. Between the intima and the folded monolayered epithelium (E) extends a subcuticular space (SC) filled with an electron translucent material which proved to be mutually soluble with water.[15] Tight contact of special "sheath cells"[16] with the intima closes the subcuticular space anteriorly and posteriorly. Similar attachments between cells and cuticle occur in epidermal sensilla of insects preventing the migration of lanthanum ions.[17]

The epithelium underneath the subcuticular space consists of one cell type only (Figure 2). The most conspicuous structural features of these cells are the deep, regular infoldings of the apical (i.e., directed toward the lumen) membrane and its intimate association with mitochondria. In a medium-sized animal (length = 10 mm), the area of this membrane is about 2 cm^2. A section in the plane of the epithelium reveals that the packing of mitochondria and the membrane is of the highest possible density (Figure 3B). The membrane of this complex is of a type we regularly meet in various insect organs. It is coated on the cytoplasmic side with particles of about 12 nm in diameter. This membrane coat has been found in Malpighian tubules, recta, salivary glands, midgut goblet cells, and in the tormogen cell of epidermal sensilla. These organs and cells are known to transport potassium to the apical

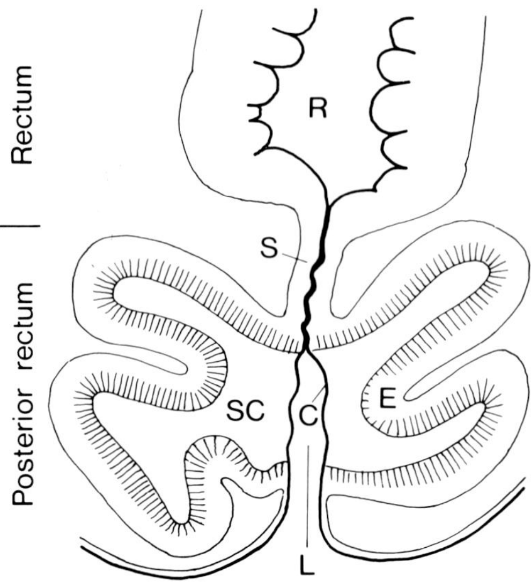

FIGURE 1. The terminal part of the gut of Lepisma. R: rectum, S: sphincter, C: cuticle, SC: subcuticular space, E: monolayered epithelium, L: lumen. (From Küppers, J. and Thurm, U., *Insect Biology in the Future*, Locke, M. and Smith, Eds., Academic Press, New York, 1980, 125. With permission.)

extracellular space.[18] The particle-studded membrane is always the apical membrane of the epithelia. Electrophysiological evidence indicates that this type of cell membrane transports potassium electrogenically from the cytoplasm to the extracellular space in goblet cells[20] and in tormogen cells.[21]

Other salient structures of the anal sac are meandering septate junctions near the apical pole of the cells which close the intercellular clefts (Figure 3A), thus separating the subcuticular from the hemolymph space. The basolateral membrane of the cells is inconspicuous and not associated with mitochondria. There are hardly any folds or invaginations (Figure 2).

A similar cell architecture is found in the coxal organs of centipedes,[22] to which a participation in water-vapor uptake from saturated atmospheres is attributed,[23] and in the rectal sac of vapor-absorbing flea larvae.[24] In the latter organ, however, the cells of its ventral half have the opposite structural polarity as membrane folds and mitochondria are found at the basal pole of the cells.

III. BACKGROUND AND HYPOTHESIS

The approach to the active step of water-vapor resorption in *Thysanura* is facilitated by the following facts. (1) Morphology and uniformity of the cells of the anal sac allow the active process unequivocally to be ascribed to a defined structure, and even to a single cell membrane and (2) there is no structural substratum to drive water flow by chemical osmosis. This is inferred from the polarity of the cells, the direction of water flow and huge gradient surmounted (CEA for *Thermobia* = 0.45), because otherwise, metabolism would have to occur within the apical folds at a water activity below 0.45. In opposition to Diamond,[25] we do not see any possibility to extend the standing gradient hypothesis to this paradigm.

To summarize: only a process that is capable of raising the water potential across the apical membrane by more than $1.1 \cdot 10^8$ Pa will square with the known structure and function.

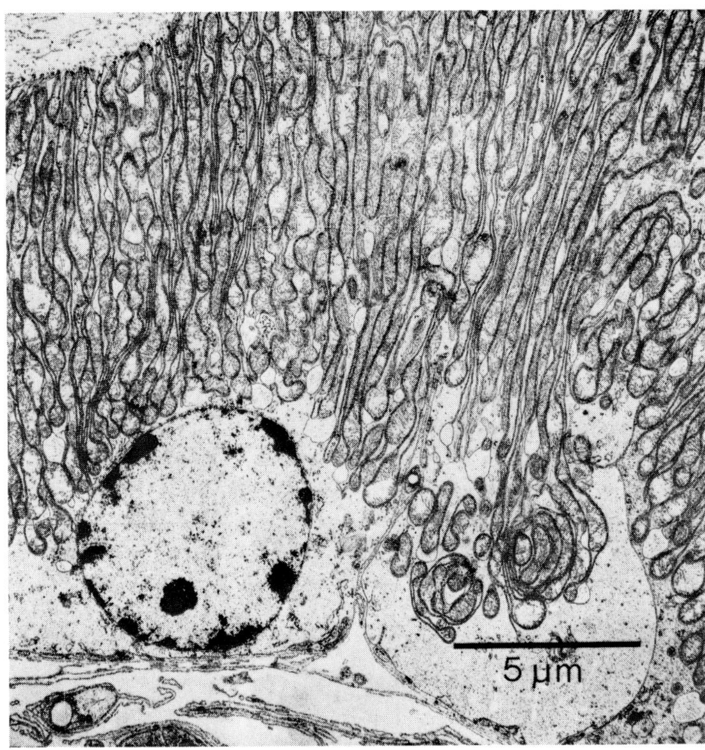

FIGURE 2. Cross section through the epithelium of the anal sac. Bottom: hemolymph space, top: subcuticular space. (From Küppers, J., Plagemann A., and Thurm, U., *J. Membr. Biol.*, 91, 107, 1986. With permission.)

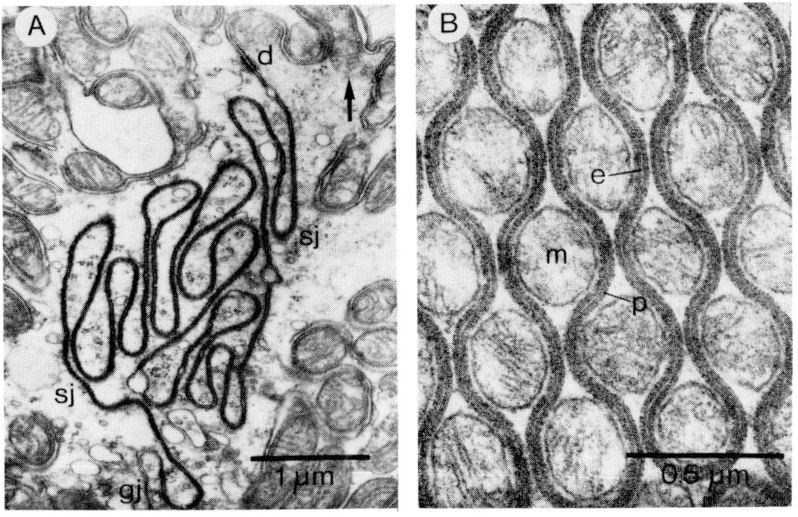

FIGURE 3. (A) Connections between adjacent epithelial cells (arrow points to the lumen); d: desmosome, sj: septate junction, gj: gap junction. (B) Aspect of the apical membrane folds (section in the epithelial plane). e: extracellular space, m: mitochondrium, p: particle coat. (From Küppers, J., Plagemann, A., and Thurm, U., *J. Membr. Biol.*, 91, 107, 1986. With permission.)

Our model, which is summarized in Figure 4, presumes the electrogenic potassium transport to be located at the apical membrane, as the characteristic particle coat suggests. Since the subcuticular space is a limited compartment with respect to solutes, any maintained transport into or out of it will lead, in the steady state, to the circulation of the transported matter. The circulation will be confined to the transporting membrane itself to the degree to which the anterior and posterior sealings and the septate junctions are tight. Having realized this, electro-osmosis comes to mind as the principle which transfers water from the low activity of the subcuticular space to the high activity of the cytoplasm: the voltage generated by an electrogenic outward transport of cations drives a corresponding inward current through ion channels of the apical membrane. The ions drive water molecules, which enter the narrow hydrophilic channels, into the cell. The net result is the transport of water, since the charge carrier cycles in the circuit.

The coupling ratio between water and ions has been determined for some ion-selective membrane channels. This ratio is 2 to 3 H_2O/K^+ for K^+ channels of the mammalian sarcoplasmic recticulum.[26] As determined by electrokinetic measurements, 6 to 7 H_2O accompany one ion through gramicidin A channels in artificial bilayers.[27] This coupling ratio varies between 7 H_2O/K^+ at low K^+ activities and 5 H_2O/K^+ at 2.5 molal activity.[28] 31 H_2O move with one ion across algal cell membranes.[29]

If electro-osmosis is expressed in terms of the general phenomenological flux equations of irreversible thermodynamics, we obtain for the volume flow J and the current I:

$$J_v = L_p \cdot P + L_{pe} \cdot E \qquad (1)$$

and

$$I = L_{pe} \cdot P + g \cdot E \qquad (2)$$

with L_p = hydraulic permeability, L_{pe} = electro-osmotic coefficient, g = conductivity, E = driving voltage, P = osmotic pressure difference calculated as $P = (RT/\bar{V}_w) \ln (a_{w1}a_{w2})$ with $\bar{V}_w$ = molar volume of water, and a_{w1}, a_{w2} = water activities in the two compartments. From Equations 1 and 2 we get

$$P_{I=0} = -E(g/L_{pe}) \qquad (3)$$

and

$$(I/J_v)_{p=0} = g/L_{pe} \qquad (4)$$

so that

$$P_{I=0} = -E(I/J_v)_{p=0} \qquad (5)$$

If there are no shunts (optimum efficiency) then $P_{I=0} = P_{Jv=0}$, so that the maximum pressure achievable is determined by the driving voltage multiplied by the charge concentration of the moving fluid at zero pressure. (For more detailed treatment of electrokinetic and electro-osmotic phenomena at ion channels cf. Rosenberg and Finkelstein.[27])

The estimated charge concentrations within the channels are in the range of $2 \cdot 10^8$ to $2 \cdot 10^9$ As/m³ (2 to 20 M/ℓ) for the paradigms cited above. It follows that a pressure of 10^8

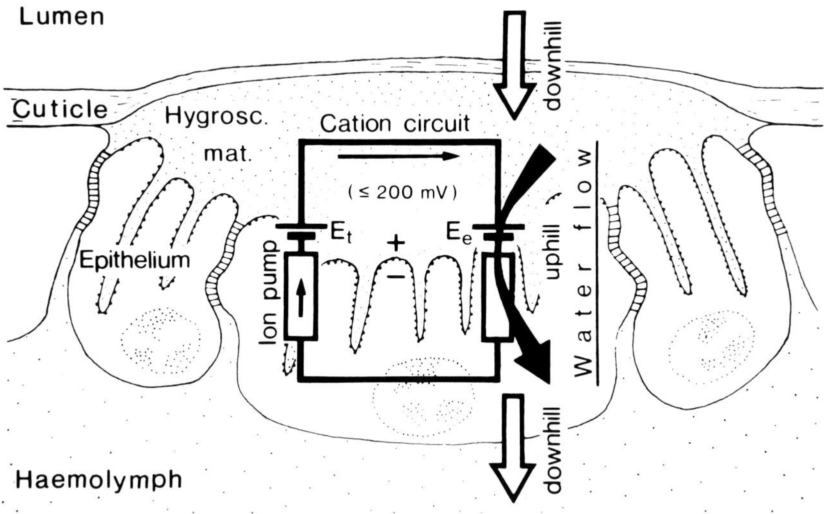

FIGURE 4. Illustration of the electro-osmotic model proposed. E_t: voltage generated by the transport, E_e: electrokinetic voltage. (From Küppers, J., Plagemann, A., and Thurm, U., *J. Membr. Biol.*, 91, 107, 1986. With permission.)

Pa might be generated by voltages of 500 to 50 mV. The above estimation of the maximum electro-osmotic pressure disregards the hydraulic conductivity of lipid bilayers which is usually of considerable magnitude[30] and which will reduce the efficiency of the transport. The problem of hydraulic shunts, however, is not specific for an electro-osmotic mechanism. The fact that net water resorption against huge gradients happens, tells us that this problem has been solved sufficiently. Possible means to reduce the hydraulic shunt permeability are suggested below.

IV. RESULTS CORROBORATING THE HYPOTHESIS

Our experiments focused on two questions to examine the value of the model. (1) Does the high energy turnover, shown by the unique ultrastructure which is apparently used for water transport only, manifest itself as a voltage (lumen positive) and an outward directed short circuit current? (2) Can ion channels be detected, the properties of which square with the postulated process? For a comprehensive and detailed representation of the methods and the experimental data see Reference 19. Here we shall summarize the results in short.

A. Voltage and Current

A transepithelial voltage (TEV) can be regularly found across the epithelium of the anal sac (between the lumen and the hemocoel). It renders the lumen positive; when the anus is unsealed and air has free access to the lumen, the TEV can reach 200 mV. A higher voltage has, to our knowledge, never been recorded at any animal epithelium. The TEV reversibly decays to zero when the animal is deprived of oxygen (Figure 5). If the voltage was clamped to zero a current of 4 μÅ on average (maximum 9 μÅ) has been derived.

These results show the existence of an ion transport at the epithelium of the anal sac. The steep reincrease of the TEV after inhibition of the metabolism points to an electrogenic transport rather than to a voltage based on diffusion potentials.

These experiments yield, of course, no evidence on the site or the specificity of the

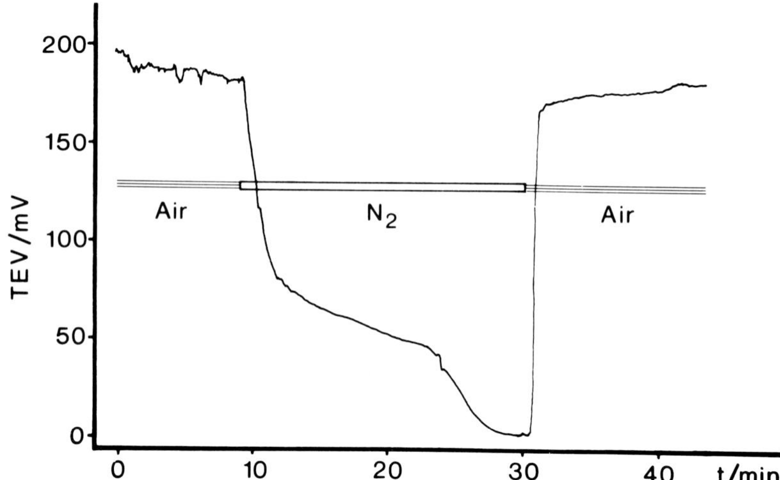

FIGURE 5. Typical change of the transepithelial voltage (TEV) during transient anoxia (anus open). (From Küppers, J., Plagemann, A., and Thurm, U., *J. Membr. Biol.*, 91, 107, 1986. With permission.)

transport. Taking into account the morphology, however, there can be little doubt, that it occurs at the apical membrane. Our assumption that we are dealing with the known electrogenic potassium transport, is based on the observation of the coat of "portasomes"[31] at this membrane. Low shunt conductivities, i.e., tight sealings of the subcuticular space and tight septate junctions must be inferred from the exceptionally high voltage.

The current density at zero voltage seems low compared to other potassium transporting insect epithelia where the membrane-mitochondria complex is less elaborated. This may indicate considerable resistances in series to the apical membrane due to the cuticle and/or the basal membrane.

B. Volume Flow

Volume flow was continuously recorded by observing the movement of an air bubble within a capillary that was connected to the sealed lumen of the anal sac. A spontaneous volume flow was observed. It depends, like the TEV, on metabolic activity (Figure 6). The average uptake rate of previously dehydrated animals was 55 nℓ/hr ($= 1.5 \cdot 10^{-14}$ m^3/sec), which corresponds to the maximum uptake rate in our weighing experiments. The electrolyte within the electrodes was slightly hypertonic to hemolymph. The KCl solution within the lumen will be dehydrated to saturation within the first few minutes of the experiment, if, as in vivo, pure water is absorbed. It may then be estimated that the water flow observed occurs against a gradient of 2.7 10^8 Pa.

The inward volume flow can be increased by an exogenous inward current and reduced by an outward current (Figure 7). Metabolic activity can be substituted during anoxia by an inward current of about 10 μÅ which restores the regular rate of volume flow. Osmotic gradients which may build up in unstirred layers from unequal transport numbers[32] cannot explain (by "pseudo-electro-osmosis") the volume-to-current coupling since the hydraulic permeability of the anal sac is too low.[19] We therefore interpret the observed coupling to be true electro-osmosis produced by the flow of cations.

The slope of the function $J_v(I)$ was on average 5.5 ± 1 nℓ/hr μÅ or = $(1.5 \pm 0.3) \cdot 10^{-9}$ m^3/Å (n = 4). This corresponds to the movement of about 7 to 8 H_2O per positive net charge or to the flow of a fluid with a charge concentration of $6.5 \cdot 10^8$ Å/m^3 (6.5 M/l). The significance of these values for in vivo performance is limited because the ionic composition of an exogenous current may differ from that of the endogenous current which is merely determined by the specificity of the active transport. Despite these reservations, the measured

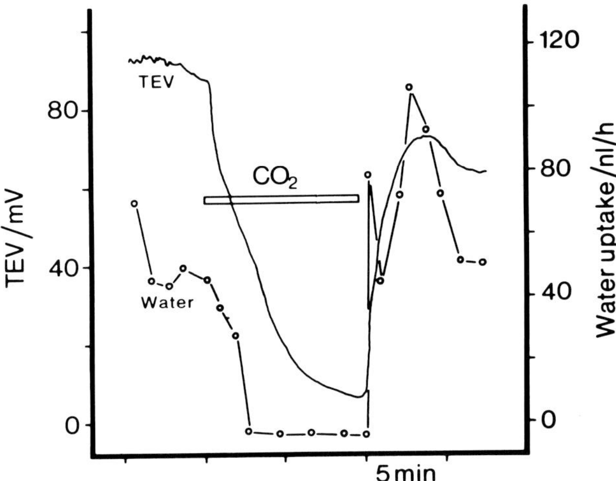

FIGURE 6. Dependence of the TEV and the water uptake on oxidative metabolism. (From Küppers, J. and Thurm, U., *Insect Biology in the Future*, Locke, M. and Smith, Eds., Academic Press, New York, 1980, 125. With permission.)

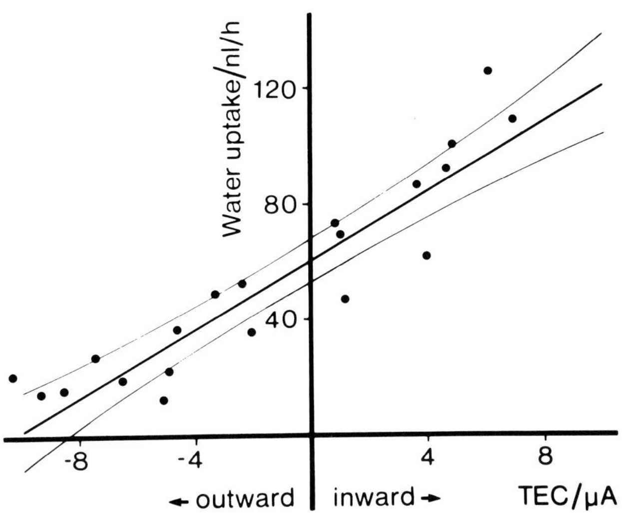

FIGURE 7. Influence of an exogenous transepithelial current (TEC) on volume flow. The 95% confidence interval for the regression line is given, data from one animal. (From Küppers, J., Plagemann, A., and Thurm, U., *J. Membr. Biol.*, 91, 107, 1986. With permission.)

coupling ratio, which is well within the range found for other ion channels (see above), may be used to estimate the possible electro-osmotic pressure: with a voltage of 200 mV, we arrive (according to Equation 3) at $1.3 \cdot 10^8$ Pa which corresponds to the water potential difference between the animal and an atmosphere of 40% relative humidity.

An inward current of about 10 μÅ has to be assumed to account for the maximum flow rate observed in vivo. From the membrane area (2 cm^2) and from the density of the short-circuit current at comparable membranes (20 μÅ/cm^2;[33]), we infer that there will be some additional transport capacity in reserve to compensate for passive water flow out of the cells and back to the subcuticular space.

V. CONCLUSION AND PROSPECTS

To conclude: the experimental results corroborate the electro-osmotic model for the uphill transport of water, though some conclusions are still circumstantial, i.e., on the kind of ion transport and on the localization of the current-water coupling channels. The quantitative data account for the water vapor resorption by *Thysanura* and reveal that this extraordinary paradigm of water transport can be explained by common elements of membrane function.

An important problem of general significance for hypo-osmotic water transport is the hydraulic shunt permeability of the organs involved. This permeability decides on the efficiency of water uptake against large gradients. "Oral absorbers" with a mobile hygroscopic phase might reduce the problem if they would follow our proposal outlined above. Because of the remote energy input, the secretion of the polymere may require only a small area of cell membrane which has to be exposed to the humidity of the air; decomposition to hygroscopic monomeres and their equilibration with the ambient humidity might occur over waterproof areas of the integument.

Rectal absorbers which withdraw water from a stationary phase should likewise be interested to keep the membrane area small, which faces the humidity of the ambient air. To visualize the magnitude of the problem: in the anal sac of *Lepisma* the area of the apical membrane is ca. 2 cm^2. We infer from the unique morphology of this membrane that its area cannot be reduced without affecting the power output. The lowest hydraulic conductance reported for a biological membrane is about $7 \cdot 10^{-17}$ m/Pa sec.[34] If the lipid proper of the apical membrane is similarly tight, and its whole area is exposed to a water activity of 0.5, the animal should lose water by transpiration at a rate of $1.5 \cdot 10^{-12}$ m^3/sec or 5000 ℓ/hr solely through this membrane (or 1400 nℓ/hr at $a_w = 0.84$; compare the uptake rate of 55 nℓ/hr at 84% relative humidity). Considering, on the other hand, usual ionic conductances of cell membranes or, more specifically, ion channels, it turns out that a very small fraction of the apical membrane might provide the conductivity for the required inward current. (The flow rate across known ion channels is 4 to 6 decades greater than across transport ATPases.) If the ion channels which link electric current to water flow are arranged at the apices of the folds, as illustrated in Figure 8, the narrow extracellular clefts between the folds might equilibrate with the cytoplasm. This would reduce the membrane area loaded with the osmotic gradient by about two orders of magnitude so that the whole uptake might work at reasonable efficiency.

We suppose that electro-osmsotic water transfer of the kind described is not confined to the recta of *Thysanura* or to cases where the gradients surmounted are extreme.[15] Former studies on water resorption by insect recta[35,36] already revealed that water uptake is independent of net solute transport. Attempts to explain this phenomenon by standing gradients were therefore, as a rule, to introduce recycling mechanisms. Site and kind of such solute recycling become obscure, however, if on the one hand tight seals of the intercellular clefts are assumed, and on the other hand the basal cytoplasm of the rectal cells is found to be hypo-osmotic to the lumen.[37]

Despite all anatomical diversity, there are some structural aspects which seem to be common to insect recta:[16] a more or less extended subcuticular space sealed by sheath cells, meandering septate junctions between the epithelial cells, and at least at some fraction of these cells, a folded apical membrane which is coated with portasomes and associated with mitochondria. This is a sufficient morphological substratum for electro-osmotic water transport. When water is absorbed against smaller gradients, the folds of the apical membrane may be shallow and inconspicuous compared to often-elaborated differentiations of the basolateral membrane and an extended system of intercellular spaces, as for instance in the rectum of *Calliphora*.[38] For this reason and by adhering to the search for standing gradients, the apical membrane of insect recta got very little attention with respect to water transport.

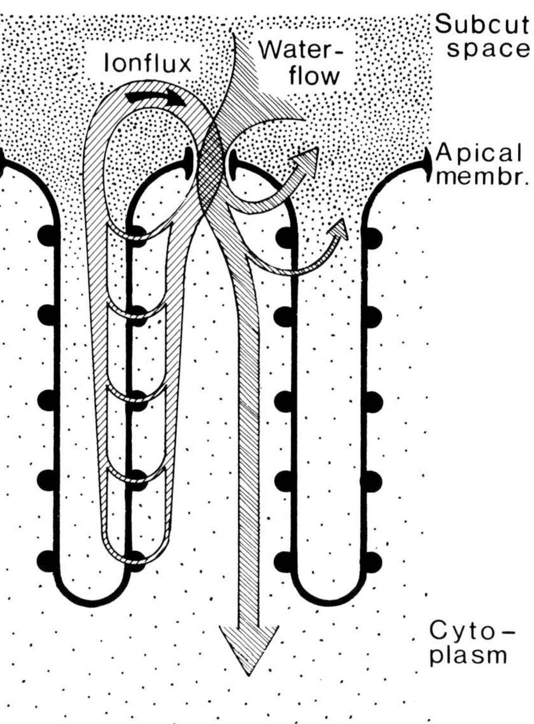

FIGURE 8. Suggested arrangement of active transport sites and passive ion channels and the pathways of current and water flow. High density of stippling symbolizes low water activity. (From Küppers, J., Plagemann, A., and Thurm, U., *J. Membr. Biol.*, 91, 107, 1986. With permission.)

Since the coupling ratio J_v/I expected for ion-selective membrane channels is small, electro-osmosis across cell membranes will use the energy of an electrogenic transport most efficiently when water is transferred against high and moderate gradients. For iso-osmotic water transport against negligible gradients, this form of electro-osmosis will be rather ineffective because of poor impedance matching. Hill[39] and McLaughlin and Mathias[40] propose the "classical" form of electro-osmosis for isotonic water transport. For renal proximal tubules, McLaughlin and Mathias "...postulate that electrogenic sodium pumps in the lateral membranes produce an electric potential within the lateral intercellular spaces, that the lateral membranes bear a net negative charge, and that fluid moves parallel to these membranes because of Helmholtz-type electro-osmosis, the field-induced movement of fluid adjacent to a charged surface."

ACKNOWLEDGMENTS

This work has been supported by the Deutsche Forschungsgemeinschaft. The cooperation of Mrs. A. Plagemann and Mrs. G. Peterseim-Neuhaus is gratefully acknowledged. Thanks are due to Mrs. I. Beständig and Dr. D.-Ch. Neugebauer-Gödde for their aid in preparing the manuscript.

REFERENCES

1. **Knülle, W. and Rudolph, D.,** Humidity relationships and water balance of ticks, in *Physiology of Ticks,* Obenchain, F. D., and Galun R., Eds., Pergamon Press, Oxford, 1983, 43.
2. **Edney, E. B.,** Water balance in land arthropods, in *Zoophysiology and Ecology,* Vol. 9, Springer-Verlag, Berlin, 1977, 196.
3. **Noble-Nesbitt, J.,** Active transport of water vapour, in *Transport of Ions and Water in Animals,* Gupta, B. L., Moreton, R. B., Oschman, J. L. and Wall, B. J., Eds., Academic Press, New York, 1977, 571.
4. **Machin, J.,** Atmospheric water absorption in arthropods, in *Advances in Insect Physiology,* Vol. 14, Treherne, J. E., Berridge, M. J., and Wigglesworth, V. B., Eds., Academic Press, New York, 1979, 1.
5. **Rudolph, D. and Knülle, W.,** Novel uptake systems for atmospheric water vapor among insects, *J. Exp. Biol.,* 222, 321, 1982.
6. **O'Donnell, M. J.,** Hydrophilic cuticle — the basis for water vapour absorption by the desert burrowing cockroach, *Arenivaga investigata, J. Exp. Biol.,* 99, 43, 1982.
7. **Ramsay, J. A.,** The rectal complex of the mealworm *Tenebrio molitor* L. (Coleoptera, Tenebrionidae), *Philos. Trans. R. Soc. London Ser. B,* 248, 297, 1964.
8. **Noble-Nesbitt, J.,** Reversible arrest of uptake of water from subsaturated atmospheres by the firebrat *Thermobia domestica* (Packard), *J. Exp. Biol.,* 62, 657, 1975.
9. **Machin, J., O'Donnell, M. J., and Coutchie, P. A.,** Mechanisms of water vapor absorption in insects, *J. Exp. Biol.,* 222, 309, 1982.
10. **Kirschner, L. B.,** The sodium chloride excreting cells in marine vertebrates, in *Transport of Ions and Water in Animals,* Gupta, B. L., Moreton, R. B., Oschman, J. L., and Wall, B. J., Eds., Academic Press, New York, 1977, 427.
11. **Noirot, C. and Noirot-Timothee, C.,** Ultrastructure du proctodeum chez le Thysanoure *Lepismodes inquilinus* Newman (= *Thermobia domestica* Packard). II. Le sac anal, *J. Ultrastruct. Res.,* 37, 335, 1971.
12. **Noble-Nesbitt, J.,** Water balance in the firebrat, *Thermobia domestica* (Packard). Exchanges of water with the atmosphere, *J. Exp. Biol.,* 50, 745, 1969.
13. **Beament, J. W. L., Noble-Nesbitt, J., and Watson J. A. L.,** The waterproofing mechanisms of arthropods. III. Cuticular permeability in the firebrat, *Thermobia domestica* (Packard), *J. Exp. Biol.,* 41, 323, 1964.
14. **Neuhaus, G., Siebler, H., and Thurm, U.,** Ein zur aktiven Aufnahme atmosphärischen Wassers befähigtes Organ (Analsack von *Lepisma*), *Verh. Dtsch. Zool. Ges.,* 71, 295, 1978.
15. **Küppers, J. and Thurm, U.,** Water transport by electroosmosis, in *Insect Biology in the Future,* Locke, M. and Smith D. S., Eds., Academic Press, New York, 1980, 125.
16. **Noirot, C., Smith D. S., Cayer, M. L., and Noirot-Timothee, C.,** The organization and isolating function of insect sheath cells: a freeze-fracture study, *Tissue Cell,* 11, 325, 1979.
17. **Keil, T. A. and Steinbrecht, R. A.,** Mechanosensitive and olfactory sensilla of insects, in *Insect Ultrastructure,* Vol. 2, King, R. C. and Akai, H., Eds., Plenum Press, New York, 1984, 477.
18. **Harvey, W. R., Cioffy, M., Dow, J. A. T., and Wolfersberger, M. G.,** Potassium ion transport ATPase in insect epithelia, *J. Exp. Biol.,* 106, 91, 1983.
19. **Küppers, J., Plagemann, A., and Thurm, U.,** Uphill transport of water by electroosmosis, *J. Membr. Biol.,* 91, 107, 1986.
20. **Blankemeyer, J. T. and Harvey, W. R.,** Identification of active cell in potassium transporting epithelium, *J. Exp. Biol.,* 77, 1, 1978.
21. **Thurm, U. and Küppers, J.,** Epithelial physiology of insect sensilla, in *Insect Biology in the Future,* Locke, M. and Smith, D. S., Eds., Academic Press, New York, 1980, 735.
22. **Rosenberg, J.,** Coxal organs in Scolopendromorpha (Chilopoda): topography, organization, fine structure and signification in Centipedes, *Zool. Jahrb. Abt. Anat. Ontog. Tiere,* 110, 383, 1983.
23. **Rosenberg, J. and Bajorat, K. H.,** Influence of the coxal organs of *Lithobius fortificatus* L. (Chilopoda) to the absorption of water vapour, *Zool. Jahrb. Allg. Zool. Physiol. Tiere,* 88, 337, 1984.
24. **Bernotat-Danielowski, S. and Knülle, W.,** Ultrastructure of the rectal sac, the site of water vapour uptake from the atmosphere in larvae of the oriental rat flea *Xenopsylla cheopsis, Tissue Cell,* 18, 437, 1986.
25. **Diamond, J. M.,** Coupling of water transport to active solute transport in epithelia, in *Water Transport across Epithelia,* Alfred Benzon Symposium 15, Ussing, H. H., Bindsley, N., Lassen, N. A., and Sten-Knudsen, O., Eds., Munksgaard, Copenhagen, 1981, 355.
26. **Miller, C.,** Coupling of water and ion fluxes in a K^+-selective channel of sarcoplasmic reticulum, *Biophys. J.,* 38, 227, 1982.
27. **Rosenberg, P. A. and Finkelstein, A.,** Interaction of ions and water in gramicidin A channels. Streaming potentials across lipid bilayer membranes, *J. Gen. Physiol.,* 72, 327, 1978.
28. **Levitt, D. G.,** Kinetics of movement in narrow channels, in *Current Topics in Membranes and Transport,* Vol. 21, Bronner, F., Ed., Academic Press, New York, 1984, 181.
29. **Barry, P. H. and Hope, A. B.,** Electro-osmosis in *Chara* and *Nitella* cells, *Biochim. Biophys. Acta,* 193, 124, 1969.

30. **Finkelstein, A.,** Water, nonelectrolyte and ion permeability of lipid bilayer membranes, in *Transport of Ions and Water in Animals,* Gupta, B. L., Moreton, R. B., Oschman, J. L., and Wall, B. J., Eds., Academic Press, New York, 1977, 169.
31. **Harvey, W. R.,** Water and ions in the gut, in *Insect Biology in the Future,* Locke, M., and Smith, D. S., Eds., Academic Press, New York, 1980, 105.
32. **Barry, P. H. and Hope, A. B.,** Electroosmosis in membranes: Effects of unstirred layers and transport numbers. I. Theory, *Biophys. J.,* 9, 700, 1969.
33. **Thurm, U.,** Basics of the generation of receptor potentials in epidermal mechanoreceptors of insects, in *Mechanoreception,* Schwartzkopff, J., Ed., Abh. Rhein.-Westf. Akademie der Wissenschaften, Opladen, 1974, 355.
34. **Dunham, P. B., Cass, A., Trinkaus, J. P., and Bennett, M. V. L.,** Water permeability of *Fundulus* eggs, *Biol. Bull. (Woods Hole, Mass.),* 130, 420, 1970.
35. **Phillips, J. E.,** Rectal absorption in the desert locust, *Schistocerca gregaria* Forskal. I. Water, *J. Exp. Biol.,* 41, 15, 1964.
36. **Grimstone, A. V., Mullinger, A. M., and Ramsay, J. A.,** Further studies on the rectal complex of the mealworm *Tenebrio molitor,* L. (Coleoptera, Tenebrionidae), *Philos. Trans. R. Soc. London Ser. B,* 253, 343, 1968.
37. **Gupta, B. L., Wall, B. J., Oschman, J. L., and Hall, T. A.,** Direct microprobe evidence of local concentration gradients and recycling of electrolytes during fluid absorption in the rectal papillae of *Calliphora, J. Exp. Biol.,* 88, 21, 1980.
38. **Gupta, B. L. and Berridge, M. J.,** Find structural organization of the rectum of the blowfly, *Calliphora erythrocephala* (Meig.) with special reference to connective tissue, trachea, and neurosecretory innervation in the rectal papillae, *J. Morphol.,* 120, 1683, 1966.
39. **Hill, A. E.,** Mechanisms of salt-water coupling in epithelia, in *Transport of Ions and Water in Animals,* Gupta, B. L., Moreton, R. B., Oschman, J. L., and Wall, B. J., Eds., Academic Press, New York, 1977, 183
40. **Mc Laughlin, St. and Mathias, R. T.,** Electro-osmosis and the reabsorption of fluid in renal proximal tubules, *J. Gen. Physiol.,* 85, 699, 1982.

INDEX

A

B

C

D

H

I

L

Errata

Water Transport in Biological Membranes, Volume II

Information on pages 274 to 278 is corrected as follows:

p. 274:	line 20	—	flow J to flow J_v
	line 26	—	$(a_{w1}a_{w2})$ to (a_{w1}/a_{w2})
p. 275:	line 21	—	4 μÅ to 4 μA; 9 μÅ to 9 μA
p. 276:	line 19	—	2.7 10^8 Pa to $2.7 \cdot 10^8$ Pa
	line 22	—	10 μÅ to 10 μA
	line 27	—	1 nℓ/hr μÅ to 1 nℓ/hr μA
	line 28	—	m^3/Å to m^3/As
	line 29	—	10^8 Å/m^3 to 10^8 As/m^3
p. 277:	line 5	—	10 μÅ to 10 μA
	line 7	—	20 μÅ to 20 μA
p. 278:	line 22	—	5000 ℓ/hr to 5000 nℓ/hr
	line 37	—	were to had